THE PROBLEM OF MENTAL DEFICIENCY

Eugenics, Democracy, and Social Policy in Britain c.*1870–1959*

MATHEW THOMSON

CLARENDON PRESS · OXFORD
1998

This book has been printed digitally and produced in a standard specification in order to ensure its continuing availability

OXFORD
UNIVERSITY PRESS

Great Clarendon Street, Oxford OX2 6DP

Oxford University Press is a department of the University of Oxford.
It furthers the University's objective of excellence in research, scholarship, and education by publishing worldwide in

Oxford New York

Auckland Bangkok Buenos Aires Cape Town Chennai
Dar es Salaam Delhi Hong Kong Istanbul Karachi Kolkata
Kuala Lumpur Madrid Melbourne Mexico City Mumbai Nairobi
São Paulo Shanghai Singapore Taipei Tokyo Toronto

with an associated company in Berlin

Oxford is a registered trade mark of Oxford University Press in the UK and in certain other countries

Published in the United States
by Oxford University Press Inc., New York

Reprinted 2001

ISBN 0-19-820692-5

ACKNOWLEDGEMENTS

WRITING this book has been an educational process; it is certainly not the book I would write today. Its origins reach back to a degree in Medieval and Modern History at University College London, when I was attracted by two courses at the Wellcome Institute for the History of Medicine, one on madness and society in the eighteenth century, the other on the idea of degeneration in nineteenth-century Europe. I am grateful to my two tutors, Roy Porter and Michael Neve, for stimulating an interest which I have been pursuing ever since. Funded by the Economic and Social Research Council, I completed a D.Phil. thesis at Oxford in 1992 on 'The Problem of Mental Deficiency in England and Wales, *c.*1913–1946'. During this time I found friendship and great hospitality at Linacre College. I was also fortunate to find an ideal academic base at the Wellcome Unit for the History of Medicine. Here, I benefited from long discussions with my supervisor Paul Weindling, who was a constant source of ideas and support. I would also like to thank those who commented on my work, including Richard Smith, Margaret Pelling, Charles Webster, Alan Caffrey, Joanna Innes, Phillip Waller, John Macnicol, and, in particular, Jose Harris who guided me in turning the thesis into a book. Valuable assistance was also provided by countless librarians, especially those at the Bodleian library, the History Faculty, and Social Science Faculty in Oxford, and by the archivists at the Greater London Record Office, the Public Record Office in Kew, the Contemporary Medical Archives Centre, the Royal College of Psychiatrists, MIND, Oxford City and Cambridge County Record Offices, the Fawcett Library, University College London, and the London School of Economics. I would like to thank all those who allowed me generous access to their archive collections. Funding to continue the project at Oxford came from the Wellcome Trust and a British Academy Postdoctoral Fellowship, while a grant from the Rockefeller Foundation enabled me to consult their archives in New York. The book was completed at the University of Sheffield, where I have been supported by a Wellcome University Award. I am grateful to all my colleagues in the Department of History for their encouragement. Finally, I am indebted to the Mackeith family, Michael, June, and Monica Stephen, Anna Guagnini, Humaira Ahmed, and my family, who all provided me with great support. Most of all, I thank Michelle O'Callaghan for her companionship and encouragement throughout.

Mathew Thomson

Sheffield
December 1996

CONTENTS

ABBREVIATIONS

ADD	Addison Papers
BJHS	British Journal for the History of Science
BL	Blacker Archive
BLPES	British Library of Political and Economic Sciences
BMA	British Medical Association
BOC	Board of Control
BOC	*Annual Reports of the Board of Control*
B. of Ed.	Board of Education
BUF	British Union of Fascists
CACMD	Central Association for the Care of Mental Defectives
CAMW	Central Association for Mental Welfare
CMO	Chief Medical Officer
CMOH	Chief Medical Officer of Health
COHSE	Confederation of Health Service Employees
COS	Charity Organization Society
DNB	*Dictionary of National Biography*
Ed.	Board of Education files (PRO)
EUG	Eugenics Society Archive
GLRO	Greater London Records Office
GNC	General Nursing Council
HO	Home Office
ILP	Independent Labour Party
JMS	*Journal of Mental Science*
LAMW	London Association for Mental Welfare
LCC	London County Council
LEA	Local Education Authority
LGB	Local Government Board
LSE	London School of Economics
MAB	Metropolitan Asylums Board
MACA	Mental After Care Association
MDP	Mental deficiency psychiatrist
MH	Ministry of Health
MHA	Mental Hospitals Association
MHIWJ	*Mental Hospital and Institution Workers' Journal*
MHIWU	Mental Hospital and Institution Workers' Union
MOH	Medical Officer of Health
MPA	Medico-Psychological Association
MRC	Medical Research Committee

NAMH	National Association for Mental Health
NAWU	National Asylum Workers' Union
NCCL	National Council for Civil Liberties
NCSS	National Council for Social Services
NHS	National Health Service
NSPCC	National Society for the Prevention of Cruelty to Children
Parl. Deb. (HC)	*Parliamentary Debates*, 5th series, House of Commons
Parl. Deb. (Lords)	*Parliamentary Debates*, 5th series, House of Lords
PEP	Political and Economic Planning
PRO	Public Records Office
RMPA	Royal Medico-Psychological Association
Roy. Com.	*Royal Commission*
Roy. Com. 1926	*Report of the Royal Commission on Lunacy and Mental Disorder* (Cmd. 2700, 1926)
Roy. Com. 1954–7	*Report of the Royal Commission on the Law Relating to Mental Illness and Mental Deficiency* (Cmnd. 169, 1957)
Roy. Com. Feeb.	*Royal Commission on the Care and Control of the Feeble-Minded* (cd. 4215–4221, 1908)
Roy. Com. Poor Law	*Report of the Royal Commision on the Poor Laws and Relief of Distress* (Cd. 4499, 1909)
SMOH	School Medical Officer of Health
T	Treasury
WWW	*Who Was Who*

Introduction

THEMES AND PARAMETERS..

The problem of mental deficiency, long ignored, is now becoming a common reference point in histories of twentieth-century British social policy. The proposals both for mass segregation of the 'feeble-minded' under the Mental Deficiency Act of 1913, and for the sexual sterilization of mental defectives in the 1930s, suggest an acceptance of eugenic ideas, an eagerness for state intervention, and an intolerance which fit uncomfortably with the prevailing view that British society during this period was characteristically 'peaceable' and 'liberal'.[1] The apparently exceptional nature of mental deficiency policy has not yet been adequately explained: historians have tended to dismiss the 1913 Act and the sterilization campaign as anomalies with little practical impact, but such an interpretation suffers from the fact there has been no thorough history of mental deficiency to date. The current literature leaves an important question unresolved. Why was it that segregation of mental defectives was accepted, whereas similar proposals for the inebriate, habitual criminals, and the unemployed all failed, and institutionalization more generally began to recede? If the answer really does lie in the impact of eugenics, then why was there such a significant impact in this one area, when, as is now generally agreed, eugenics had so little practical impact on other aspects of British social policy? Historians need either to reassess some of their broader ideas about social policy during the period, or to consider whether the problem lies in viewing the history of mental deficiency too narrowly, as simply an episode in the history of eugenics, and as merely an issue of coercion and control. As the first major study of the problem of mental deficiency, this book will explore these questions and provide answers which aim to be of relevance, not only to the history of psychiatry and the history of

[1] For instance: B. Harrison, *Peaceable Kingdom: Stability and Change in Modern Britain* (Oxford, 1982); F. M. L. Thompson, *The Rise of Respectable Society: A Social History of Victorian Britain, 1830–1900* (London, 1988); A. Reid, *Social Classes and Social Relations in Britain, 1850–1914* (London, 1992).

medicine, but also to our understanding of social policy and the nature and complexion of politics during the first half of the century.

Although there have been several recent studies of the passage of the 1913 Act and the sterilization campaign of the 1930s, little is known about how mental deficiency policy operated in practice. Historians of social policy have ignored the issue, even though contemporaries viewed it as a highly significant development, and even though the numbers under some form of care by the end of the interwar period—almost 100,000—make it one of the main areas of social welfare provision. The subject of mental deficiency has also been neglected in the history of psychiatry. The distinction between the mentally ill and the mentally disabled is at last beginning to be recognized in histories of mental health care during the early modern and nineteenth-century periods, but we still know very little about the major expansion of provision for mental defectives which took place in the twentieth century.[2] In part, this reflects a more general neglect of twentieth-century mental health policy, but it also reflects a bias towards the history of curing rather than caring, and a bias towards the history of psychiatrists and asylums, rather than the care provided in the community by voluntary groups, social workers, and families.[3] This study deliberately sets out to break such conventions and to reveal key areas which need further attention from historians of both social policy and psychiatry. Mental deficiency is an ideal subject for such a reorientation, for it was more obviously a problem of care than cure. It consequently brought into question the very role of the medical profession, stimulating a move towards non-institutional care, and creating opportunities for new or emergent professions. Because of the limitations of the current historiography this study has had to extend beyond the history of mental deficiency to sketch out histories of the psychiatric

[2] An important collection of essays, ranging from the medieval to the modern, was unfortunately published too late to integrate fully into this book: D. Wright and A. Digby (eds.), *From Idiocy to Mental Deficiency: Historical Perspectives on People with Learning Disabilities* (London, 1996).

[3] On neglected areas of the history of psychiatry see R. Porter, 'History of Psychiatry in Britain', *Journal of the History of Psychiatry*, 2 (1991), 279. Our understanding of the twentieth century still rests on the much criticized 'whiggish' work of K. Jones: *Mental Health and Social Policy, 1845–1959* (London, 1960); and *A History of the Mental Health Services* (London, 1972). The most important new work addresses the relationship of mental health to law, especially C. Unsworth, *The Politics of Mental Health Legislation* (Oxford, 1986). We also have a thorough study of special education policy: G. Sutherland, *Ability, Merit, and Measurement: Mental Testing and English Education, 1880–1940* (Oxford, 1984).

profession, the emergence of psychiatric social work, and central government's role throughout the mental health field. On the other hand, it has deliberately avoided close analysis of life within a single mental institution. Undoubtedly, there are rich records available for such microstudies. But the concentrated focus which is necessary for a study of life within the 'total institution' always runs the danger of being parochial, especially when, as in this case, there are such pressing broader issues at stake; moreover, such a focus would perpetuate the biases of psychiatric history which the study aims to avoid. Instead, the aim is to use institutional care as one avenue among many (including the neglected subject of care in the community) to explore the relationship between the problem of mental deficiency and the outside world.[4] As mental deficiency lay on the borders of the medical, penal, welfare, and educational divides, with mental deficiency defined under the 1913 Act according to social as well as medico-psychological criteria, it provides an ideal tool with which to explore the relationship between social norms and the changing boundaries of social pathology, and to examine how a sociomedical problem could be shaped by demographic, economic, and cultural factors.[5] In this sense, this is not simply a history of mental deficiency policy: it is also a history of the part played by the category of mental deficiency in constructing notions of intellectual, social, and moral normality.[6]

The problem of mental deficiency provides a fresh perspective on interwar social policy, and one which is narrow enough to encompass the policy-making process as a whole. Indeed, one of the key points of this book lies in the very structure of the analysis. Unlike most histories of social policy, which either concentrate on the creation of policy at the central level or undertake microstudies of local implementation, mental deficiency policy is here analysed as a complex, multilevel, and interactive process. The analysis descends down the strata of the policy-

[4] The limitations of an exclusive focus on life inside institutions are considered in: M. Ignatieff, 'Total Institutions and Working Classes', *History Workshop Journal*, 15 (1983), 167–73; M. Finnane, 'Asylums, Families, and the State', *History Workshop Journal*, 20 (1988), 134–48; B. Luckin, 'Towards a History of Institutionalization', *Social History*, 8 (1983), 87–94.

[5] P. Wright and A. Treacher, *The Problem of Medical Knowledge* (Edinburgh, 1982); L. Jordanova, 'The Social Construction of Medical Knowledge', *Social History of Medicine*, 8 (1995), 361–81.

[6] Mental deficiency is given an important place in Nikolas Rose's work on this theme: *The Psychological Complex: Psychology, Politics and Society in England 1869–1939* (London, 1985).

making process: from the 'high politics' of parliament and the civil service in Chapters 1 and 2; to the interest group politics of professions, voluntary groups, and pressure groups in Chapters 3–5; to local administration and party politics in Chapter 6; and finally, to the micropolitical interaction between social workers, families, and the defectives themselves in Chapter 7. Such a structure obviously poses problems for a narrative account. This is overcome by a second organizational strand, as the introduction and Chapters 1 and 2 provide an overview of the narrative, Chapters 3 and 5 each address different solutions to the problem, and Chapters 6–8 concentrate on the implementation of these solutions. Some repetition is inevitable when looking from so many perspectives, and is indeed to a certain extent desirable in that it reflects the integration of the strata, while at other times it has been avoided through cross reference. A second problem with taking such a broad approach and mapping out such a fresh territory is that there are inevitable limitations in the analysis of each single dimension.

The great advantage of the structure is that it enables the study to examine a social problem from all angles of the policy-making process, bringing together and interrelating a series of important issues which tend to remain disparate in histories of social policy. In the high-political spheres it examines the ability of the legislature to deal with the demands of complex new welfare systems in the twentieth century; it assesses the relationship between ideology and 'non-party' issues;[7] and it looks at the delegation of responsibility to the civil service, the independence and changing role of the expert, and the extent to which one can talk of 'official opinion'.[8] Chapters 3 and 5 use the issue of mental deficiency to explore the nature of what Harold Perkin has coined 'Professional Society'.[9] These chapters analyse the roles of professional, voluntary, and pressure groups, and assess their importance in shaping policy. In contrast to analyses which have pointed to common class and ideological interests among welfare providers,[10] this account will highlight the divisions between and within such groups

[7] On this important theme: M. Freeden, 'The Stranger at the Feast: Ideology and Public Policy in *Twentieth-Century Britain*', *Twentieth Century British History*, 1 (1990), 9–34.

[8] These issues are addressed in W. H. Greenleaf, *The British Political Tradition*, iii, *A Much Governed Nation* (London, 1988).

[9] H. Perkin, *The Rise of Professional Society: England Since 1880* (London, 1989).

[10] Perkin. *op. cit.*. For the view that a 'social hygiene' agenda united a wide range of interest groups, including those involved in mental deficiency: G. Jones, *Social Hygiene in Twentieth-Century Britain* (London, 1986).

caused by professional self-interests and insecurities, gender, and the changing relationship between professional groups and the political process. In Chapter 6, the neglected topic of the relationship of party to local provision of welfare is addressed, and an attempt is made to chart a geography of welfare provision. Chapter 7 uses detailed case histories to question theories of 'social control' and to consider the extent to which care and control are better seen as processes of negotiation.[11] Finally, a concluding chapter uses mental deficiency policy as a fresh perspective to address debates over the formation of the Welfare State. As a whole, by comparing interwar provision to that in the nineteenth century and to the Welfare State which followed, the study contributes to debates over whether the interwar period saw a transformation or continuation of the welfare system in Britain.[12] More generally, it is hoped that the study will rescue the problem of mental deficiency (and that of mental health care as a whole) from neglect by historians of social policy, and will thereby provide a new area which can be integrated into historical debates about the development of the modern Welfare State.[13]

The approach aims to demonstrate the value of integrating political and social history. Parliamentary and civil service politics are analysed

[11] This approach is explored in L. Gordon, *Heroes of Their Own Lives: The Politics and History of Family Violence* (London, 1989).

[12] For a Foucauldian account which emphasizes transformation of the vision of the social body: D. Armstrong, *The Political Anatomy of the Body* (Cambridge, 1983). Particular attention has been given to the way that social and penal problems were transformed into issues of medicine and welfare: D. Garland, *Punishment and Welfare: A History of Penal Strategies* (Aldershot, 1985). However, studies of institutions suggest that there were strong continuities reaching back into the nineteenth century: A. Crowther, *The Workhouse System, 1834–1929: The History of an English Social Institution* (London, 1981).

[13] Recent comparative work on the rise of welfare systems has explained international variation by seeing it as a reaction to the different timing of problems raised by industrialization (see Ch. 6); by contrasting ideological traditions: D. Ashford, *The Emergence of the Welfare States* (Oxford, 1986) (see Ch. 1); by the level of political pressure from beneath forcing welfare concessions from above (see Chs. 1 and 6); by the comparative strength of 'risk groups' struggling for the resources of the state: P. Baldwin, *The Politics of Social Solidarity: The Class Bases of the European Welfare State, 1875–1975* (Cambridge, 1990), 289–99 (see Ch. 7); by the impact of professional groups: Perkin, *The Rise of Professional Society* (see Chs. 3–5); or by comparing national models of 'state formation', 'institutional leverage' and 'policy feedbacks': M. Weir, A. S. Orloff, and T. Skocpol (eds.), 'Introduction', in *The Politics of Social Policy in the United States* (Princeton, 1988), 3–35 (see Chs. 2 and 8). Mental health care and eugenics are not generally integrated into such models. The main comparative study of eugenics (limited mainly to an Anglo-American perspective) concentrates on scientific development and pays little attention to the affect of different political structures: D. Kevles, *In the Name of Eugenics: Genetics and the Uses of Human Heredity* (London, 1985). As yet, there is virtually no comparative historical analysis of mental health systems.

through a prosopographical social history of individuals—their personal interests being key to understanding the ideological content behind the façade of consensus.[14] Conversely, the social spheres of profession, locality, community, and family are interpreted as having a political dimension. To demonstrate that 'high' and 'low' politics, and political and social history, are not mutually exclusive spheres, the analytical tools of gender, class, and professionalization are used as threads which run throughout the strata of the thesis. In order to emphasize further the potential for linking the political and social, high and low politics, and medical and 'mainstream' history, analysis of each policy-making stratum sets out to explore the impact of 'adjusting to democracy'—a theme which lies at the centre of the political history of the period, but which tends to be overlooked by social or medical historians. The study shows how our understanding of the problem of mental deficiency takes on new dimensions when we place it within the context of 'adjusting to democracy'—adjusting to the extension of the franchise, to the emergence to national power of a working-class party, to the increased demands on the legislative and administrative machinery brought about by the democracy, and to the redefinition of the very meaning of citizenship.[15]

Finally, the parameters of the study need further explanation. The main analysis will stretch from the passage of the Mental Deficiency Act in 1913 to the National Health Service Act of 1946 that removed independent power from the government department set up to administer the 1913 Act. Although this period is the main focus of the study, there will also be an introductory survey of the 'prehistory' of mental deficiency and an epilogue on the impact of the legislation of 1944–8 in the period leading up to the Mental Health Act of 1959.[16]

The second parameter is that geographically the study concentrates on England and Wales, and excludes Scotland and Ireland. With its separate legal system, Scotland passed its own Mental Deficiency Act in 1913 and administered this separately under its own Board of Control.[17] Ireland was excluded from the Mental Deficiency Act by the Home

[14] C. Webster, 'Conflict and Consensus: Explaining the British Health Services', *Twentieth Century British History*, 1 (1990), 115–51.

[15] R. Lowe, *Adjusting to Democracy: The Role of the Ministry of Labour in British Politics, 1916–1939* (Oxford, 1986).

[16] For a valuable analysis of the later period: Unsworth, *Politics of Mental Health Legislation*, 230–314.

[17] G. O. Thomson, 'Legislation and Provision for the Mentally Handicapped in Scotland since 1906', *Oxford Review of Education*, 9 (1983), 233–40; I. Levitt, *Poverty and Welfare in Scotland, 1890–1948* (Edinburgh, 1988).

Rule legislation, and it remained without any special legislation for mental defectives.[18] However, evidence from Ireland and Scotland will be included when it is useful as a supplement to that from England and Wales. This will be the case, for instance, with the analysis of the psychiatric profession which united all four countries in one body.[19]

The most problematic parameter is that of 'the problem of mental deficiency' itself. The first point which needs clarification is the use of the term 'mental deficiency'. Although this expression is now seen as socially and medically unacceptable, it is used nevertheless because it would be ahistorical to impose our own categories upon the past. The term 'mental deficiency' emerged in common medical use in Britain at the end of the nineteenth century.[20] It extended the medical differentiation between 'lunatics'—those who had acquired a mental illness, often a temporary and curable state—and those who had permanent mental damage or incapacity from birth, adding the 'higher-grade' category of feeble-mindedness to the 'lower grades' of imbecility and idiocy which were already legally differentiated from lunacy.

As a medical category, mental deficiency covered what we would now recognize as a wide range of mental disabilities and mental conditions. The classificatory schemes designed by contemporary doctors grouped together an often extraordinary range of 'types', running from microcephalics and hydrocephalics, to 'mongols' and 'wolf-children'.[21] In the early twentieth century, doctors continued to place considerable weight on physical stigmata as signs of mental deficiency.[22] In practice, more-

[18] Ireland had, however, come within the remit of the *Royal Commission on the Care and Control of the Feeble-Minded* (Cd. 4215–4221, i–vii; *Report*, Cd. 4202, viii, 1908) (hereafter, *Roy. Com. Feeb.*).

[19] T. Walmsley, 'Psychiatry in Scotland', in G. Berrios and H. Freeman (eds.), *150 Years of British Psychiatry, 1841–1991*, i. pp. 294–305; D. Healy, 'Irish Psychiatry', in Berrios and Freeman, *150 Years of Psychiatry*, i. 314–20.

[20] For a summary of its history: M. W. Barr, *Mental Defectives: Their History, Treatment, and Training* (Philadelphia, 1904). In its early usage the term 'mental deficiency' concentrated on children: E. Shuttleworth, *Mentally Deficient Children: Their Treatment and Training* (London, 1900). In the United States the expression 'feeble-minded' was used as a generic term for all grades of defectives, with the category of moron applied to those known as feeble-minded in Britain. The adoption of the term 'mental defective' by the 1913 Act was probably partly due to the influence of A. F. Tredgold whose *Mental Deficiency* (London, 1908) became the accepted textbook on the topic during the interwar period.

[21] For instance: W. Ireland, *The Mental Affections of Children*, 2nd edn. (London, 1900).

[22] M. Jackson, 'Images of Deviance: Visual Representations of Mental Defectives in Early Twentieth-Century Medical Texts', *British Journal for the History of Science*, 28 (1995), 319–37.

over, unrelated conditions such as partial sight, deafness, dyslexia, and even left-handedness could contribute towards a diagnosis of mental deficiency. Methods to test levels of intelligence were crude, and many high-grade defectives would not be seen as mentally handicapped today.[23] Mental deficiency is therefore highly problematic as a medical category. However, this study will not address the evolution of medical theories of mental deficiency, or the medical accuracy of the category, in any great detail.[24] There is obviously a place for such work, but the focus here is the interrelation between medical definitions and the legal, educational, and social definitions of mental deficiency in the construction of policy and the practice of administration.[25] Mental deficiency moved from being simply a medical term to being a medico-legal category with the passage of the 1913 Mental Deficiency Act. However, in practice mental defectives were only placed under care if they were also deemed to be socially at risk, to constitute a social risk themselves, or to be socially incompetent—which was usually judged in terms of their ability to support themselves, or to be supported, in the community. Moreover, the 1913 Act introduced the category of moral imbecility, which was designed to certify individuals on moral grounds alone.[26] Therefore the medico-legal definition of mental deficiency was effectively a social definition as well.[27] The issue

[23] The legislation did not define defectives according to 'intelligence quotients', which only came into practice gradually during the interwar period (for debates over timing and impact of testing see Sutherland, *Ability, Merit and Measurement*), but approximate IQ guidelines were soon laid down as follows: idiots under 20; imbeciles 20–50; feeble-minded 50–70. For the development and bias of mental testing techniques: S. J. Gould, *The Mismeasure of Man* (London, 1981), 154–71; B. Evans and B. Waites, *IQ and Mental Testing* (London, 1981). For a more positive view of the meritocratic and child-centred aims of educational specialization and the development of mental testing: A. Wooldridge, 'Child Study and Educational Psychology in England, c.1880–1950', D.Phil. thesis (Oxford, 1985), 320–5; A. Wooldridge, *Measuring the Mind: Education and Psychology in England, c.1860–1990* (Cambridge, 1994).

[24] For an overview of the history of this medical category: G. Berrios, 'Mental Retardation', in G. Berrios and R. Porter (eds.), *A History of Clinical Psychiatry: The Origin and History of Psychiatric Disorders* (London, 1995), 225–38; E. Miller, 'Mental Retardation', in Berrios and Porter, History of Clinical Psychiatry, 212–24; E. Doll, 'Historical Review of Mental Retardation, 1800–1965', *American Journal of Mental Deficiency*, 72 (1967), 165–89; L. Kanner, 'Medicine in the History of Mental Retardation, 1800–1965', *American Journal of Mental Deficiency*, 72 (1967), 170–5.

[25] On psychiatry as a site of medico-legal tensions: Unsworth, *Politics of Mental Health Legislation*, 4–35.

[26] In practice the new category of moral imbecility was not used on a large scale: S. Watson, 'The Moral Imbecile: A Study of the Relations Between Penal Practice and Psychiatric Knowledge of the Habitual Offender', Ph.D. thesis (Lancaster, 1988).

[27] For details of the Act see J. and S. Wormald, *A Guide to the Mental Deficiency Act 1913* (London, 1913).

was further complicated by the existence of a separate educational definition of mental deficiency under the 1914 Elementary Education Act, leading to considerable confusion over whether educationally defective children also qualified as mental defectives under the 1913 Act.[28] The tension between these various definitions of mental deficiency is a central concern of this study.

Finally, although the approach of this study is to analyse the 'construction' of the problem of mental deficiency, and although it is quite apparent that people were often certified as mental defectives for social as much as medical reasons, the study does not adopt a dogmatic social-constructionist position which denies the existence of individuals with disabilities altogether;[29] mental deficiency was not simply a fantasy of eugenic nightmares, even though these fears undoubtedly escalated and shaped perceptions of the extent of the problem. The history of psychiatry has too often been beset by a sterile battle over the reality of mental illness. By accepting that any social problem is necessarily socially, politically, ideologically, and linguistically constituted, yet also that there is a social reality of individuals within society who have special needs and different abilities, it is hoped to move beyond this epistemological impasse.[30] Through analysing the problem of mental deficiency in the context of its time and as far as possible in its own terms, we can at least develop a richer historical understanding of that era and hope to place our own attitudes and policies towards mental disability into comparative perspective.

[28] The 1870 Education Act had revealed the existence of many children who needed special schooling; already this had led to the Elementary Education (Defective and Epileptic Children) Act, 1899 (62 & 63 Vic., c. 32), which introduced permissive legislation for the creation of special schools for mental defectives. This legislation was made compulsory in 1914 (Elementary Education (Defective and Epileptic Children) Act, 4 & 5 Geo. V c. 45), while the Mental Deficiency Act, 1913 (3 & 4 Geo. V, c. 28), provided for the adult feeble-minded or those who were ineducable. For a summary: D. Pritchard, *Education and the Handicapped, 1760–1960* (London, 1963).

[29] Some anti-psychiatric histories have verged on such a denial. The problems of such an approach are addressed in the 'Introduction' in P. Wright and A. Treacher, *The Problem of Medical Knowledge: Examining the Social Construction of Medicine* (Edinburgh, 1982), 14–15.

[30] Of course such problems about the nature of knowledge are not confined to debates about mental illness, and are becoming increasingly important in social history: P. Joyce, 'The End of Social History?', *Social History*, 20 (1995), 73–91.

THE PREHISTORY OF MENTAL DEFICIENCY

Historians of the 1913 Mental Deficiency Act have almost all approached it as an event within the history of eugenics. The Act has been viewed as the one instance in which eugenic thinking was the primary force behind British social policy.[31] However, recent work casts some doubt on the view that eugenics was the determining force behind the legislation: Larson has shown that the scientific authority of eugenics received considerable criticism in the campaign for legislation, while the work of Barker and Simmons suggests that moral concern about the feeble-minded was probably more important than, and sometimes in conflict with, eugenics.[32] Nevertheless, a reinterpretation of the Act has been impeded, since these historians have continued to look at the issue from the perspective of the history of eugenics. In order to arrive at a more balanced view of the Act, and of the role of eugenics within it, we need to pay far more attention to the emergence and development of the problem of mental deficiency itself. By taking such a perspective it will become apparent that the demand for legislation arose out of a complex set of issues and attitudes, with eugenics an integral but interrelated part of the whole.[33]

Asylums for Idiots

Although idiocy was recognized in law as distinct from lunacy as early as the thirteenth century, there was no significant specialized provision in Britain until the nineteenth century. Idiots were cared for by family and community, within the poor law, or were certified under the lunacy

[31] J. Harris, *Private Lives, Public Spirit: A Social History of Britain, 1870–1914* (Oxford, 1993), 244–5; M. Freeden, *The New Liberalism: An Ideology of Social Reform* (Oxford, 1978), 190.

[32] E. Larson, 'The Rhetoric of Eugenics: Expert Authority and the Mental Deficiency Bill', *BJHS* 24 (1991), 45–60; D. Barker, 'How to Curb the Fertility of the Unfit: The Feeble-Minded in Edwardian Britain', *Oxford Review of Education*, 9 (1983), 197–211; H. Simmons, 'Explaining Social Policy: The English Mental Deficiency Act of 1913', *Journal of Social History*, 11 (1978), 387–403.

[33] Gillian Sutherland comes to a similar conclusion when looking at the question in terms of special education policy: *Ability, Merit, and Measurement*, 25–6.

legislation and placed in asylums alongside the mad.[34] Stimulated by the work of Itard and Séguin in France, several small private homes for idiots opened in the 1840s.[35] A number of these expanded into residential schools with a mission of educating and training individuals, who were described as 'idiots' but whose educability suggests that many of them were of a similar intelligence to those who would be labelled 'feeble-minded' in the second half of the century.[36] The emergence of this small group of idiot homes and schools can be largely attributed to active philanthropy and the expanding market for medical care and education, rather than a panic about idiocy as a social, moral, or eugenic problem.

Although the 'idiot asylums' attracted public attention to the problem of idiocy, the majority of those provided with residential care were almost certainly located in poor law institutions. On the other hand, it was clear that the less-eligibility principle of the workhouse and the stigma of pauperization were wholly inappropriate and inhumane when dealing with individuals who needed permanent care through no fault of their own. In response to such concerns, the Metropolitan Poor Act was passed in 1867, enabling the capital's poor law unions to join together to fund the Metropolitan Asylums Board (MAB) which provided specialized hospital care for the sick and disabled poor, including two asylums for seriously handicapped imbeciles.[37] Elsewhere, poor law authorities lacked the resources to fund, or the population to justify, such a specialization of care, and idiots were at best placed in a special wing or annexe of a workhouse, at worst placed in the general wards.[38]

[34] The early history is still largely uncharted. The most important recent work includes: P. Rushton, 'Lunatics and Idiots: Mental Disorder, the Community, and the Poor Law in North East England 1600–1800', *Medical History*, 32 (1988), 34–50; R. Neugebauer, 'Mental Illness and Government Policy in Sixteenth and Seventeenth Century England', Ph.D. thesis (Columbia University, 1976); and the essays by R. Neugebauer, P. Rushton, and J. Andrews in Wright and Digby, *From Idiocy to Mental Deficiency*, 22–117.

[35] For a discussion of these pioneers in educating the mentally handicapped: J. Trent, Jr., *Inventing the Feeble Mind: A History of Mental Retardation in the United States* (London, 1994), 40–59.

[36] Pritchard, *Education and the Handicapped*, 54–5.

[37] G. Ayers, *England's First State Hospitals and the Metropolitan Asylums Board, 1867–1930* (London, 1971).

[38] Although idiocy was recognized as a problem, it seems that 'higher-grade' defectives were viewed as useful workers in workhouses: Crowther, *The Workhouse System*, 138. This situation changed by the end of the century, in response to concern about feeble-minded women using the workhouse for support during pregnancy: J. Saunders, 'Quarantining the Weak-Minded: Psychiatric Definitions of Psychiatric Degeneracy and the Late-Victorian Asylum', in W. Bynum, R. Porter, and M. Shepherd, *The Anatomy of Madness* (London, 1985–8) iii. 282–4, 290–1.

The distinction between lunacy and idiocy was recognized in the passage of the 1886 Idiots Act, which made it possible to certify idiots separately from lunatics and without the approval of a magistrate, and in the 1891 Lunacy Act, which enabled local authorities to establish asylums which were for idiots alone.[39] However, these pieces of legislation had little affect on the way that idiots were provided for: local authorities took no action to build special asylums and continued to place idiots in lunatic asylums, workhouses, or charitable institutions. Nevertheless, demand for special care was increasing, in response both to the increased social awareness about idiocy, and to a humanitarian concern that idiots should not suffer the stigma of being placed alongside the mad or the indigent poor. The new legislation probably exacerbated the situation, since, in introducing an easier and less legalistic mechanism for certification, and in differentiating idiocy from stigmatized lunacy, it encouraged families to come forward to request care.[40]

By the last quarter of the century, there was an obvious shortage of care. Charitable institutions did expand, but not to the extent that they became the vast 'dumping' grounds which public lunatic asylums have been accused of being;[41] instead, entrance remained a privilege given out by subscribers to the respectable or gained through private payment.[42] Although poor law guardians sometimes funded the poor to enter these institutions (pauperizing whole families in the process), the majority either remained in the community or entered workhouses or asylums. In 1905 the largest of the charitable institutions, the Royal Albert Asylum in Lancaster, held 606 residents; the second largest, the Earlswood Asylum, in Redhill, Surrey, held 495; the Western Counties Asylum in Starcross, Devon, held 272; the Eastern Counties Asylum in Colchester, held 256; and the Midland Counties Asylum in Knowle, held 272; in all, a total of 1,800 residents. Already this number was

[39] Idiots Act, 1886 (49 & 50 Vic., c. 41); Lunacy Act, 1891 (54 & 55 Vic., c. 65).

[40] G. Shuttleworth, *Roy. Com. Feeb.*, i, qu. 9842, 9857. Hardest to estimate is the impact on demand of socio-economic factors, such as demographic change, employment opportunities, and urbanization.

[41] The theory that asylums were 'dumping institutions'—'a handy place to which to consign the awkward and unwanted, the useless and potentially troubling'—has been most vigorously argued by A. Scull, *Museums of Madness: The Social Organization of Insanity in Nineteenth Century England* (London, 1979), 240, 259. This view has been questioned by more detailed work on admissions: e.g. J. Walton, 'Lunacy in the Industrial Revolution: A Study of Asylum Admissions in Lancashire, 1848–50', *Journal of Social History*, 13 (1979), 1–22.

[42] D. Wright, 'The National Asylum for Idiots, Earlswood, 1847–1886', D.Phil. thesis (Oxford, 1993). For further discussion, see Ch 3.

significantly smaller than that under the MAB, which held 1,000 patients at Darenth Asylum and similar numbers at Caterham and Leavesden.[43] By the end of the century, the charitable idiot asylums faced the danger of gradually filling with chronic cases in need of long-term care, leaving less and less space to admit trainable cases.[44] They reacted by stressing that they only took educable cases, and a number of them changed their names to 'training institutions for the feeble-minded', in the hope that this would encourage parents to send their handicapped but educable children.[45] However, even in the case of the feeble-minded, there was now far less confidence in educating them to earn a living outside the institution than there had been earlier in the century.[46]

Although the emergence of charitable idiot asylums was undoubtedly important in creating a model of specialized residential care, their history cannot fully explain why mental deficiency came to be seen as such a significant problem by the end of the century. Their problems were not insurmountable; they could be solved by being more selective over the type of cases they accepted, and by leaving the chronic to be provided for elsewhere. Whereas the seemingly bottomless pit of public lunacy asylum provision created a vicious circle of escalating demand for further institutional care, a comparable role cannot be attributed to idiot asylums, whose growth was always restrained by their charitable status and funding. Instead, to understand why mental deficiency became a major social problem one needs to look at the way it became defined as a management problem which overlapped a wide range of institutional settings, including schools, workhouses, lunatic asylums, prisons, and inebriate reformatories.

Educating the Feeble-minded

In the long term, perhaps the most important trigger in the development of the problem of mental deficiency was the introduction of universal elementary education in 1870. Mass education revealed that there was a hitherto largely unrecognized section of the population,

[43] Fletcher Beach, *Roy. Com. Feeb.*, i, qu. 6936.

[44] M. A. Barrett, 'From Education to Segregation: An Inquiry into the Changing Character of Special Provision for the Retarded in England, 1846–1910', Ph.D. thesis (Lancaster, 1986).

[45] *Roy. Com. Feeb*, i, qu. 6936. Despite their earlier title, these institutions had always concentrated on higher-grade defectives.

[46] Charles Caldecott, Medical Superintendent of Earlswood Asylum, estimated that this was possible in only 1 per cent of cases, *Roy. Com. Feeb.*, i qu. 10321.

who were not handicapped to the extent that they could be diagnosed as idiots, but who were not normal either and who needed special training if they were to be educated.[47] Failure at school was easily detected, for payment by results had led to a system in which children advanced through a series of Standards based on annual examinations and were held back if they failed. Many children were unable to advance with their peers—in 1880 only 47 per cent of school-leavers had reached the Standard IV level expected of 10-year-olds. Because of payment by results, there was little incentive to keep on dull children and many left without even reaching Standard III.[48]

One of the earliest groups to become concerned about this problem was the Charity Organization Society. In defence of the less-eligibility principle of the poor law, and recognizing that some children were ineducable in normal schools and needed special training, the COS sympathized with the plight of poor families who were left with the choice of accepting financial assistance to send their mentally defective children to a charitable institution—and thereby dragging the whole family into pauper status through no fault of their own—or keeping the child at home, without training, with little hope of ever contributing to the family economy, and proving a burden which could undermine the whole family's ability to support itself. This concern over the feeble-minded was part of a wider awareness of the existence of 'defective' children within the nation's schools. In 1874 the COS undertook an investigation on the training of the blind, and in 1875 it set up an enquiry into 'The Education and Care of Idiots, Imbeciles, and Harmless Lunatics'.[49] The report recommended that a distinction be made between dangerous lunatics, on the one hand, and harmless idiots, imbeciles, and 'feeble-minded' (a term coined by the Secretary of the enquiry, Sir Charles Trevelyan), on the other, and that the harmless should be placed in special institutions with assistance from the state but without the stigma of pauperism.[50]

During the 1880s, special schools, rather than residential institutions, became the favoured solution to the problem of the feeble-minded child. Some large schools had already reacted to the problem by estab-

[47] Sutherland, *Ability, Merit, and Measurement.*

[48] Pritchard, *Education and the Handicapped*, 116–17.

[49] M. Rooff, *Voluntary Societies and Social Policy* (London, 1957), 100.

[50] H. Bosanquet, *Social Work in London, 1869–1912: A History of the Charity Organisation Society* (London, 1914), 195–6; C. L. Mowat, *The Charity Organisation Society, 1869–1913: Its Ideas and Work* (London, 1961), 58–9.

lishing a new Standard 0 for defective children. Backed by the COS and the British Association, Francis Warner undertook a survey of 50,000 school children in 1888 and found that 1 per cent had nervous and physical deficiencies and were failing in school.[51] The following year, the government published the report of the Royal Commission on the Blind, the Deaf, and the Dumb. Although the initial report included the feeble-minded in its proposals for compulsory special education, they were eventually excluded from the legislation of 1890.[52] Nevertheless, schools for the feeble-minded were being opened through local initiatives: the first in Leicester in 1892, followed by schools in London, Nottingham, Birmingham, Bradford, Brighton, and Bristol. By 1897, there were 1,300 feeble-minded children receiving education in thirty-one special schools.[53]

Pressure for the state to regulate and aid these special schools built up during the 1890s. Warner's investigation was extended to 100,000 school children and was published in 1893. The COS followed this up by mounting a campaign for state legislation, founding the National Association for Promoting the Welfare of the Feeble-Minded in 1896.[54] Special schools joined the campaign, lobbying for extra funding to cover the greater expense entailed in educating feeble-minded compared to normal children. The campaign was bolstered when the Report of the Poor Law Schools Committee recommended special provision for feeble-minded children.[55] The government responded by setting up the Departmental Committee on Defective and Epileptic Children, which reported in 1898 and led to the 1899 Elementary Education (Defective Children) Act.[56] However, recognizing that the expense of providing special schools was beyond the means of many local education authorities, legislation was restricted to an enabling act, with the result that provision continued to be very varied across the country.[57]

Ironically, the 1899 Act was passed just as confidence began to fade in the special school as a comprehensive solution to the problem of the

[51] Pritchard, *Education and the Handicapped*, 116–18.

[52] *Report of the Royal Commission on the Blind, the Deaf, and the Dumb* (C. 5781, 1889); Elementary Education (Blind and Deaf Children) Act, 1893.

[53] Pritchard, *Education and the Handicapped*, 120–31.

[54] P. Townsend, 'The Care of the Feeble-Minded', *Charity Organisation Review*, 44 (1900), 151–8.

[55] Pritchard, *Education and the Handicapped*, 132–5.

[56] *Report of the Departmental Committee on Defective and Epileptic Children* (C. 8746, 1898); Elementary Education (Defective and Epileptic Children) Act, 1899.

[57] Pritchard, *Education and the Handicapped*, 149–50, 186.

feeble-minded.[58] The opening of special schools had raised the question of what was to be done with the feeble-minded when they reached the leaving age. Despite specialized methods of training, it was difficult to see how the majority of the feeble-minded would be able to support themselves in an increasingly competitive job market. There was also considerable concern that feeble-minded women would be prey to sexual exploitation. In London, voluntary organizations such as the COS, the National Association for the Care of the Feeble-Minded, the National Vigilance Association, and the Metropolitan Association for Befriending Young Servants, opened refuge homes to protect feeble-minded women. The dual concerns of education and moral protection made the problem of the feeble-minded an ideal issue for women to carve out a new niche in voluntary organizations and local government. In large towns around the country they set up after-care services to monitor the feeble-minded after they had left school.[59] Led by women like Ellen Pinsent in Birmingham, these groups attempted to find work for feeble-minded school-leavers, visited them regularly in their homes, and gradually collected a mass of case records of the typical life of the feeble-minded. The results were so discouraging that women such as Pinsent began to advocate residential care for the adult feeble-minded.[60] In Manchester, Mary Dendy came to a similar conclusion. Finding that the School Board was not catering for the majority of the feeble-minded, and seeing the squalor experienced by the feeble-minded in the community, Dendy founded the Lancashire and Cheshire Society for the Permanent Care of the Feeble-Minded in 1898, and in 1902 opened a residential school at Sandlebridge which offered care beyond school age. By 1912, there were 150 children at Sandlebridge and a further 150 adults in an adjoining industrial college.[61] However, most of these voluntary homes and refuges were small and dependent on charitable funding. Once again, therefore, the voluntary sector began to lobby for state action, this time to provide permanent care for the feeble-minded.[62]

[58] Pritchard, *Education and the Handicapped*, 132–5.

[59] P. Hollis, *Ladies Elect: Women in English Local Government, 1865–1914* (Oxford, 1987), 145, 277–9.

[60] E. Pinsent, 'On the Permanent Care of the Feeble-Minded', *Lancet*, 1 (21 Feb. 1903), 513–15.

[61] H. McLachan, *Records of a Family, 1800–1933* (Manchester, 1935), 135–84.

[62] For further discussion of the role of women in this area see Ch 4.

The Weak-minded Criminal

The emergence of the problem of mental deficiency also needs to be understood within the context of a shift in the nature of penality, from punishment to welfare, which took place in the last third of the nineteenth century.[63] The Victorian prison system had been established on the premise that crime was a wilful act which could be deterred by a system of certain and just penal servitude, and that the character of criminals could be morally reformed through the disciplinary education of the prison regime.[64] Although crime rates did fall in the second half of the century, it was found that there were some criminals who were not deterred by punishment and who spent their lives entering and leaving prisons. Some were recognized as professional criminals, but there were others who seemed to have no wilful control over their actions and whose characters appeared unreformable. British writers tended to be cautious about accepting the view of continental positivists, such as Cesare Lombroso, that there was a criminal class with its own distinct physiognomy, but from the late 1860s psychiatrists and prison doctors began to claim that a large proportion of these criminals were predisposed to criminality because of 'weak-mindedness'.[65] Weak-mindedness helped to explain why some criminals, though not certifiably insane, could not restrain themselves from offending, were so easily caught, were so disruptive and difficult to manage within prisons, and why the penal regime was so unsuccessful in reforming their moral

[63] Despite disagreements about the nature and cause of this welfarist mode, liberal historians tend to agree with revisionists that such a shift took place. M. Wiener, *Reconstructing the Criminal: Culture, Law, and Policy in England, 1830–1914* (Cambridge, 1990), explains the shift as the result of a new view of human nature and recognizes the 'new view of the social underdog' (p. 381). D. Garland, *Punishment and Welfare: A History of Penal Strategies* (Aldershot, 1985), attributes the shift to an economic and political crisis after 1890 which led to the search for more acceptable but more effective and pervasive forms of control.

[64] Wiener, *Reconstructing the Criminal*, 92–156.

[65] J. B. Thomson (doctor at the Scottish State Prison at Perth), 'The Hereditary Nature of Crime', *Journal of Mental Science*, 15 (1869), 487–98; H. Maudsley (psychiatrist), *Responsibility in Mental Disease* (London, 1874). On the British reception of positivism see Wiener, *Reconstructing the Criminal*, 233–5; L. Radzinowicz and R. Hood, *The Emergence of Penal Policy in Victorian and Edwardian England* (Oxford, 1990), 3–33. 'Weak-mindedness' was the term which was generally used within the prison system. Sometimes it referred to the mentally ill as well as the mentally defective.

characters.[66] Although the classification of prisoners as weak-minded was encouraged by the entry of doctors into the prison service, the evidence suggests that the main reason for its successful adoption was that it provided such a useful tool of management and an ideal explanation for the failings of the penal system.[67]

As early as the 1870s it was being proposed that weak-minded criminals should be removed from prisons and placed in separate institutions. Although no such legislation was implemented, prisons did begin to separate off the weak-minded internally.[68] There were also proposals to establish systems of preventive detention for habitual and inebriate criminals. A few specialized institutions for criminal inebriates were established under the 1898 Inebriates Act, and indeterminate sentences for persistent offenders were experimented with, but neither had much success, in large part because the predominant view, despite the influence of positivist thinking, was still that those convicted of crime deserved a just sentence and were capable of responding to the certain promise of liberty in return for their moral reform.[69] By contrast, there seemed to be an obvious biological difference in weak-minded criminals that made permanent segregation appear to be both

[66] L. Zedner, *Women, Crime and Custody in Victorian England* (Oxford, 1991). Although Zedner points to the way that female prisoners came to be seen as a management problem because of their irrational natures, this applied to both male and female mental defectives. Indeed, Zedner's account of mental deficiency is misleading in that she views it as a label which was almost exclusively directed at women; in fact, men heavily outnumbered women among those detected within the penal system (the proportion of weak-mindedness among male and female prisoners was similar, but there were a far higher number of males within prisons). On the other hand, Zedner's material does suggest that it was more common to label female inebriates as being weak-minded (pp. 5–6).

[67] S. Watson has emphasized the importance of looking at how management problems at the local level of prisons (and other 'surfaces of emergence', such as schools, workhouses, and asylums) led to the construction of the problem of the weak-minded: 'Applying Foucault: Some Problems Encountered in the Application of Foucault's Methods to the History of Medicine in Prisons', in C. Jones and R. Porter (eds.), *Reassessing Foucault: Power, Medicine, and the Body* (London, 1994), 132–5; S. Watson, 'The Moral Imbecile: A Study of the Relations between Penal Practice and Psychiatric Knowledge of the Habitual Offender' (Ph.D., University of Lancaster, 1988), ch. 6.

[68] Saunders suggests that legislation was prevented by concerns about cost, the difficulty of proving the link between weak-mindedness and habitual criminality, and the fact that workhouses and asylums did not yet view the weak-minded as in need of special care: 'Quarantining the Weak-Minded', 273–96.

[69] On the limitations of these policies: Radzinowicz and Hood, *Emergence of Penal Policy*, pp. 231–315. Since reformatories for criminal inebriates were only established by Lancashire and London county councils, the two largest local authorities in the country, it suggests that local action was inhibited elsewhere by the cost and because specialization was only justified in areas with very large and concentrated populations.

justifiable and necessary. Moreover, if criminals were truly unreformable and undeterred by punishment—as the poor results from inebriate reformatories, for instance, suggested[70]—then there was a strong case for arguing that their root problem was an inherent weak-mindedness, rather than a rational disposition towards crime: they were mental defectives, rather than wilful criminals. Thus the Mental Deficiency Act can be seen as succeeding, where the inebriate reformatories and preventive detention failed, in providing (in theory) an administrative apparatus through which to provide the weak-minded, habitual criminal with permanent care, rather than temporary custody;[71] in doing so, the problem was shifted from the penal sphere to a borderline area, lying uneasily between penality and medicine.[72]

The Role of Eugenics

Although Francis Galton coined the term 'eugenics' in 1884, it was not until two decades later that a movement emerged, with the formation of the Eugenics Education Society in 1907 and the publication of its journal, the Eugenics Review, from 1909. Nevertheless, in the intervening two decades the core concerns of the eugenics debate were developed within the framework of Social Darwinism.[73] By the end of the century there was widespread anxiety that changes within modern society were reversing the progress of evolution and leading to the degeneration of the British people. The inexorable increase in recorded rates of lunacy—from 2,248 or 2.26/10,000 of the population in 1807, to 86,067 or 29.26/10,000 in 1890—was seen as one of the most important indicators of this process.[74] Since it was mankind's intelli-

[70] Zedner, *Women, Crime, and Custody*, 259–63.

[71] Significantly, legislation to compel local authorities to establish inebriate reformatories was delayed because of the prospect of the Mental Deficiency Act, and was seen as largely unnecessary once the 1913 Act had been passed: Radzinowicz and Hood, *Emergence of Penal Policy*, 307–21. The close relation between the inebriate and mental deficiency systems is evident in the conversion of Farmfield inebriate reformatory in London and Brockhall in Lancashire into mental deficiency institutions.

[72] It is too simplistic to describe this as a 'medicalization' or 'biologization' of the problem, since the definition of mental deficiency rested so heavily on assessment of character and manageability, rather than medical diagnosis. The fact that the category of 'moral imbecile' (a lack of moral but not necessarily intellectual sense) proved to be so controversial suggests that a wholesale 'biologization' of wilful morality was not generally accepted: for further discussion on this issue see Watson, 'The Moral Imbecile'.

[73] G. Jones, *Social Darwinism and English Thought* (Brighton, 1980).

[74] A. Scull, *The Most Solitary of Afflictions: Madness and Society in Britain, 1700–1900* (Avon, 1993), 337.

gence and wilful control over his more primitive instincts which made him the most highly evolved creature in the natural world, it was assumed that the mind was also the most recently evolved part of his nature and the most liable to succumb to the onset of degeneration. It was widely believed that these mental qualities were rooted in the physical structure of the brain and, like physical characteristics, were inherited between generations.[75] Degenerationists, such as psychiatrist Henry Maudsley, warned that an increasing number of people were inheriting minds which were in some way weakened, and that even if this did not manifest itself in lunacy it would lead to eccentric or anti-social behaviour, perhaps even criminality, among the offspring.[76] If an individual was aware of this predisposition, mental breakdown might be prevented by a strategic navigation through the biological and social stresses brought on by modern life.[77] There was less optimism about those who inherited congenital mental problems, whose mental weakness was incurable, and whose offspring were likely—some thought certain—to inherit a weakness of mind, leading to further insanity, social failure, and deviant behaviour. On the other hand, since mid-Victorian psychiatry tended to view idiocy and imbecility as terminal points on the path of degeneration, leading to a condition of sterility which extinguished the line of descent, they were not yet the subject of serious eugenic alarm.[78]

In the first decade of the twentieth century the rather diffuse debate over degeneration found a sharper focus, as mental defectives became defined as the central eugenic threat facing the nation. This shift is best explained, not by the pressure of an organized eugenics movement—none yet existed after all—but by the way the problem of mental deficiency provided a focus for a series of more general moral, social, and demographic anxieties. With the recognition that many of the social failures in schools, workhouses, and prisons were in fact feeble-minded, it became obvious that mental deficiency was not an isolated problem, but one which was integral to broader social issues and which extended

[75] M. Clark, 'The Rejection of Psychological Approaches to Mental Disorder', in A. Scull (ed.), *Madhouses, Mad-Doctors, and Madmen* (London, 1981), 271–312.

[76] On Maudsley, see T. Turner, 'Henry Maudsley—Psychiatrist, Philosopher and Entrepreneur', Psychological Medicine, 18 (1988), 551–74; D. Pick, *Faces of Degeneration: A European Disorder, c.1848–1918* (Cambridge, 1989), 203–16.

[77] J. Oppenheim, *'Shattered Nerves': Doctors, Patients, and Depression in Victorian England* (New York, 1991).

[78] Henry Maudsley argued that the endpoint of degeneration was 'sterile idiocy': *The Physiology and Pathology of Mind* (London, 1868), 246–7.

out to the increasingly murky borderline of the normal population; in such circumstances, it became far harder to view mental deficiency as a terminal point of degeneration, and easier to see how it could affect the fitness of the general population. These fears were fuelled, as the inclusion of the new category of the 'feeble-minded' in the 1901 census helped to increase the estimated population of mental defectives from 97,000 in 1891, to 133,000.[79] Even if this was the result of greater awareness, rather than a real increase, it demonstrated that the mental defectives housed in institutions were simply the tip of a much larger problem within the community.[80]

The heightened awareness of the extent of mental deficiency coincided with growing fears about the fitness of the British population. Anxiety about the character of the urban poor was far from new.[81] It escalated, however, in the late nineteenth century, in reaction to a combination of economic hardship and unemployment, changing relations between the classes, and changing ideas about poverty.[82] It has become something of a historical cliché to argue that this anxiety turned into panic over 'national efficiency' in reaction to the reported poor condition of British recruits for the Boer War.[83] However, the way in which the Interdepartmental Committee on Physical Deterioration came down so firmly against the view that the race was degenerating, and proposed environmental and public health reforms, suggests that the dominant view about national fitness was not so pessimistic.[84]

[79] *Roy. Com. Feeb.*, viii. 378. For a discussion of the measurement of the feeble-minded population, see Radzinowicz and Hood, *Emergence of Penal Policy*, 320–1.

[80] It was recognized that these figures were unreliable because of the public's confusion over the terminology: evidence of John Tatham, *Roy. Com. Feeb.*, i. 269.

[81] G. Himmelfarb, *The Idea of Poverty: England in the Early Industrial Age* (London, 1985).

[82] The impact of the 'great depression' and changing class relations are explored in: G. S. Jones, *Outcast London: A Study in the Relationship Between Classes in Victorian Society* (London, 1971), 281–315. This can be contrasted to G. Himmelfarb: *Poverty and Compassion: The Moral Imagination of the Late Victorians* (New York, 1991). Himmelfarb is sceptical about whether poverty became worse in real terms (pp. 21–39) and rejects Jones's interpretation of the London dock strike of the 1880s as creating a new 'outcast class' of casual labourers (pp. 51–3). Instead, she has argued that the period was marked by the emergence of a more compassionate and 'demoralized' view of poverty. However, integral to this 'demoralized' view there could be a paternalistic and authoritarian attitude towards those who could not be helped to help themselves (see her discussion of Charles Booth, pp. 164–6).

[83] G. Searle, *The Quest for National Efficiency* (Oxford, 1971).

[84] R. Soloway, 'Counting the Degenerates: The Statistics of Race Deterioration in Edwardian England', *Journal of Contemporary History*, 17 (1982), 137–64; Himmelfarb: *Poverty and Compassion*; Harris, *Private Lives, Public Spirit*, 243.

Probably more important in stimulating the eugenic debate was concern about the falling birth-rate among the middle classes and respectable working classes compared to the high rates among the poorest and least responsible sectors of society.[85] Mental defectives became such a potent subject for concern because they provided a symbol which linked these overlapping anxieties about moral, demographic, and racial decline: because of their mental defect they sank within society to join the residuum; lacking moral restraint they bred unchecked with similarly weak-minded individuals; and their offspring were brought up in such a socially and morally impoverished environment, and inherited such weak mental powers, that they perpetuated the vicious circle of decline, turning to crime or falling on the rates in order to survive. In an era in which the 'social question' was rising to the forefront of the political agenda, mental deficiency seemed to provide a biological explanation for social failure that had to be addressed by opponents and advocates of collectivist policies alike.[86]

Finally, to understand why eugenic concern began to focus on mental deficiency we need to look at developments within social purity work.[87] In the last quarter of the century an increasing number of the women in rescue homes were reported as being feeble-minded.[88] Mental defect provided both a convincing biological explanation for the plight of these women—their mental defect left them with no restraint over their sexual instincts and unprotected against abuse by men—and a powerful justification to place them under protection and control in refuges and homes. Gradually, mental defect and immorality became so closely associated that it was not uncommon for immorality to be viewed as evidence, rather than the consequence, of mental defect. It was because feeble-minded women were already such an important target for sexual

[85] R. Soloway, *Demography and Degeneration: Eugenics and the Declining Birthrate in Twentieth-Century Britain* (London, 1990).

[86] The relationship between attitudes towards eugenics and social expenditure was complex; support for eugenics did not necessarily correspond with anti-welfarism. For further discussion see Ch. 1.

[87] The campaigning of women on this issue can be seen as part of a wider movement to pressurize the state to introduce legislation to protect women and maintain social purity: F. Mort, 'Purity, Feminism, and the State: Sexuality and Moral Politics, 1880–1914', in M. Langan and B. Schwarz, *Crises in the British State, 1880–1930* (London, 1985), 209–25.

[88] In Birmingham, 37 per cent of those in rescue and Magdalen homes were reported as feeble-minded, while a nationwide survey revealed that 2,521 out of 14,725 in these homes were feeble-minded and that these women had given birth to over 1,000 illegitimate children: Zedner, *Women, Crime and Custody*, 279; H. Kirby, 'The Feeble-Minded and Voluntary Effort', *Eugenics Review* (1909), 87.

control that eugenic arguments, which were still highly controversial, were accepted so rapidly as an additional justification for control. Since the concern to regulate moral behaviour was so deeply rooted, it brought a much wider constituency of support to the eugenic campaign than it would ever have gained through scientific arguments alone.[89]

The Royal Commission on the Care and Control of the Feeble-Minded

The appointment of the Royal Commission on the Care and Control of the Feeble-Minded was the government's response to the growing feeling that many of the nation's social ills could be traced back to the root cause of mental deficiency, and that the present system of dealing with this was inadequate.[90] Despite the timing of the Royal Commission, following on so shortly after the Physical Deterioration Report of 1904, the two had less in common than is often assumed. The 1904 Report had paid little attention to 'mental deterioration' and made no recommendations for the setting up of a Royal Commission on the feeble-minded.[91] Pressure groups took advantage of the climate of opinion stimulated by the deterioration debate, but it was already obvious that some sort of investigation into provision would be necessary.[92] The absence of doctors on the original Committee suggests that it was not designed to produce a survey of mental deterioration to parallel the 1904 Report. The Commission was dominated by lay members: Ellen Pinsent, the pioneer in after-care; Charles Loch, of the COS; W. H. Dickinson, Liberal MP and Chairman of the National Association for the Promotion the Welfare of the Feeble-Minded; and

[89] Zedner, *Women, Crime and Custody*, 276–8; Simmons, 'Explaining Social Policy', 388. It is significant that contemporaries drew comparisons between the campaign over mental deficiency and the White Slave Trade—opponents suggested that both were based on false alarms and proposed excessively interventionist solutions, while supporters saw both as part of a broader struggle to protect defenceless women from immoral sexual exploitation by men: *H. of C.*, vol. 39, c. 571–627, 10 June 1912; Mort, 'Purity, Feminism, and the State', 219–25. Significantly, the Report of the Royal Commission recommended legislation to strengthen the legal impediments against sexual abuse of mental defectives: *Roy. Com. Feeb.*, viii. 154.

[90] There is no detailed account of the events leading up to the appointment of the Royal Commission. Pritchard claims that a petition calling for a Royal Commission signed by 140 influential persons was sent to the Home Secretary in April 1903: *Education and the Handicapped*, 183; Sutherland, *Ability, Merit, and Measurement*, 39. There is no evidence to suggest that eugenic pressure, *per se*, led to the Commission.

[91] *Report of the Interdepartmental Committee on Physical Deterioration* (Cd. 2175, 1904).

[92] Pritchard, *Education and the Handicapped*, 183; Sutherland, *Ability, Merit, and Measurement*, 39.

the Reverend H. N. Burden, owner of a number of homes for the feeble-minded as well as homes for inebriates. All of these members came to the Commission with a prior commitment to extending care beyond school age. There was also strong representation from lawyers and from the Home Office, suggesting that the problem of how to deal with the weak-minded in prisons was a vital issue.[93] In 1904 and 1905, two doctors, Horatio Donkin and James Dunlop, the latter also Superintendent of Statistics in the Office of the Registrar-General of Scotland and an inspector under the Inebriates Act, joined the Commission after it was recognized that issues of provision could not be divorced from the medical questions of defining and calibrating feeble-mindedness.

The Commission has been described as a puppet in the hands of the eugenicists: a 'packed jury' which,

> too readily and too uncritically accepted the hereditary thesis and too perfunctorily dismissed the evidence on the influence of environment [whose] proposals were an amalgam of social Darwinism, eugenics, and Lombrosianism [and whose] outcome was a foregone conclusion.[94]

Undoubtedly, this critical view of the Commissioners' objectivity has some truth in it; however, whether it was a 'packed jury', or simply reflected the predominant views within the establishment, is a more debatable point. A close examination of the Minutes of Evidence suggests that the Commission's conclusions were not out of step with the views of the vast majority of witnesses, many of whom had considerable practical experience of dealing with mental defectives. Moreover, to read the Commission simply in terms of eugenics and Social Darwinism is to neglect a series of other crucial concerns which informed the investigation.

The Commission's warrant had set out its aims as:

> to consider the existing methods of dealing with idiots and epileptics, and with imbecile, feeble-minded, or defective persons not certified under the Lunacy Laws; and in view of the hardship or danger resulting to such persons and the community from insufficient provision for their care, training, and control, to report as to the amendment in the laws or other measures which should be adopted in the matter, due regard being had to the expense involved in any proposals and to the best means of serving economy therein.[95]

[93] The Committte included the Home Office civil servant, W. P. Byrne, the Commissioner in Lunacy, Frederick Needham, and three lawyers.

[94] Radzinowicz and Hood, *Emergence of Penal Policy*, 329.

[95] *Roy. Com. Feeb.*, viii. 1.

This warrant was extended in November 1906:

> to enquire into the constitution, jurisdiction, and working of the Commission in Lunacy and other Lunacy Authorities in England and Wales, and into the expediency of amending the same or adopting some other systems of supporting the care of lunatics and mental defectives; and to report as to any amendment of the law which should, in their opinion, be adopted.[96]

Neither warrant made any mention of the eugenic idea. Instead, attention was drawn to 'the hardship or danger resulting to such persons and the community from insufficient provision'.[97] Thus the main problem set for the Commission was the administrative one of creating a more rational and efficient system for those mental defectives currently under care and those for whom no care was available.

The Commission took evidence from 248 witnesses on sixty-eight days, with the hearings open to the press and public. For the first two years the majority of witnesses came from a multitude of institutional agencies which had experience in dealing with mental defectives. It rapidly became apparent that the feeble-minded were to be found in a huge range of institutional settings, and that the care provided was usually wholly inadequate. In the later stages of the hearings, most of the witnesses dealt with the reorganization of the central administration. Although outside its official remit, the Commission also solicited evidence on the scale, definition, and aetiology of mental deficiency, justifying these lines of enquiry on the grounds that they were integral to the issue of provision. Since there were such marked differences of opinion on these issues, the Commission set up its own medical survey of the problem, directed by two doctors who already had experience in surveying rates of mental deficiency—W. Potts, Medical Officer of Birmingham, and A. F. Tredgold, one of the country's leading medical experts on mental deficiency. On the basis of the survey it was estimated that the number of mental defectives in England and Wales was 149,628, or 0.46 per cent of the population, with 66,509 deemed to be in need of some form of care.[98]

[96] *Roy. Com. Feeb.*, viii. 1. [97] Ibid.

[98] This estimated number in Scotland was 11,627 or 0.26 per cent of the population, while in Ireland it was 25,415 or 0.57 per cent. Separate estimates were produced because Scotland and Ireland were dealt with by different legal systems. It was acknowledged that there was little prospect that much could be done for Ireland, since there was little existing provision and since Irish rate-payers would be unable to bear the expense alone. By contrast, in Scotland, the low rate of mental deficiency and existing forms of provision made the prospects for a Scottish Act much more likely. Thus a distinct but similar Mental Deficiency Act covering Scotland followed that for England and Wales in 1913, but no legislation was introduced to cover Ireland.

The scale of the problem and the fact that high levels of mental defect were even reported in rural areas supported the view that mental deficiency was a truly national problem.

The Commission published its Report in 1908, accompanied by seven further volumes of evidence. It concluded that mental deficiency was a major problem, that existing systems of care were wholly inadequate, and that the introduction of a specialized system of care was urgent. Such a measure was necessary, first, because lack of training and control led to injury to the defectives themselves, secondly, because it resulted in injury to others, and thirdly, because it led to 'much continuous expenditure wasteful to the community and to individual families'.[99] The Report stressed that 'the mental condition of these persons, and neither their poverty nor their crime, was the real ground of their claim for help from the state';[100] not only was it inhumane to place those who could not be held responsible for their actions in workhouses and prisons, it was also inappropriate and unjust, for punishment and less-eligibility could have little reformative influence over the feeble-minded. Again, the Commissioners presented themselves as being sensitive towards the defectives and their families, as they recommended that the feeble-minded should be ascertained 'without any undue invasion of the privacy of the family', and care should be provided 'so long as it is necessary for his [the mental defective's] own good'.[101]

The Report concluded that the current system of special schools was not enough on its own:

> We find a local and permissive system of public education which is available, here and there, for a limited section of mentally defective children, and which, even if it is to be useful during the years of training, is supplemented by no subsequent supervision or control, and is in consequence often misdirected and unserviceable.[102]

With an eye to economy, the Report also criticized placing mental defectives in lunatic asylums: 'We find lunatic asylums crowded with patients who do not require the careful hospital treatment that well-equipped asylums now afford, and who might be treated in many other ways more economically, and as efficiently'.[103] Instead, the Report came down heavily in favour of specialized residential homes, providing economical yet humane conditions suited to long-term, or even lifetime,

[99] *Roy. Com. Feeb.*, viii. p. 3. [100] Ibid. 7. [101] Ibid. 8. [102] Ibid. 3.
[103] Ibid.

care. Particular praise was given to the colony model, in which residents were to be provided with a protected environment in which they could live, work, and socialize as if they were in the outside world.[104] However, the Report also recommended systems of care and control within the community, including 'family supervision', 'friendly guardianship', and 'wardship'.[105]

The Report's condemnation of current forms of institutional provision was well supported by the evidence of witnesses. The main area of contention was whether the special school was an adequate alternative to residential care. But the majority of witnesses, including some who worked in this area, dismissed them as inadequate: they did not provide care beyond school; they were expensive (with a recommended ratio of one teacher to eighteen students, compared to one to fifty in elementary schools); and they could only be of benefit to academically educable (and therefore by definition non-defective) children.[106] Witnesses who worked in prisons, reformatories, lunatic asylums, and workhouses were virtually unanimous in calling for the removal of mental defectives to a new system of care.[107] Even those running the existing charitable and MAB mental deficiency institutions tended to look favourably on a system of care which would relieve them from having to cater for chronic cases.[108] Managers of prisons, reformatories, and workhouses frequently viewed the feeble-minded as a disruptive influence within their ordered, disciplinary regimes. As G. B. Griffiths, Governor of Holloway Prison put it: 'It seems unscientific to introduce a person of deficient self control into such an establishment and then to punish him because he is unable to keep its necessary regulations.'[109] Likewise, those working within the poor law tended to regard the workhouse as the wrong place for the feeble-minded. Mary Clifford, a poor-law guardian in Bristol for over twenty years, argued that as feeble-minded women were

[104] This is of course a highly idealized view of these institutions. For further discussion see Ch. 3.

[105] *Roy. Com. Feeb.*, viii. 9. For further discussion on 'community care' see Ch. 4.

[106] On cost: M. Lawrence, *Roy. Com. Feeb.*, i, qu. 7738–43; on inadequacy of day schools: Miss E. F. Coward, Head Mistress Burnley Coal Clough Special School, *Roy. Com. Feeb.*, i, qu. 12473.

[107] It is ironic that the desire to remove mental defectives from systems of institutional care was such a common theme behind a policy which is now viewed solely in terms of the expansion of institutional segregation.

[108] *Roy. Com. Feeb.*, i, qu. 1784–95, 6936, 11011, 10321.

[109] Ibid., i, qu. 3549.

not responsible for their infirmity they deserve better treatment in diet, in quiet accommodation, in opportunities to go out under supervision, and in recreation. They are good and religious-minded, and ought not to sink to the level of the low people who come in and out of the usual able-bodied wards.[110]

Clifford's vision of special care epitomizes the paternalistic and humanitarian ethos which paralleled eugenic concern about the feeble-minded and which was equally important in shaping proposals for a specialized system of care.

Although the tone of the Report's conclusions was certainly alarmist, with terms such as 'moral danger' and 'wayward and irresponsible lives' used freely, eugenic danger was not mentioned by name (though it may, in one instance, have been obliquely alluded to in the phrase 'the source of lasting injury to the community').[111] A large number of the witnesses had expressed considerable concern about the inability to control the sexual activity of feeble-minded women under existing circumstances; but this was as often a moral as a eugenic concern. In fact, the evidence of the witnesses suggests that recent historiography has gone too far in representing the problem of mental deficiency as an issue which was, almost exclusively, one of controlling the reproduction of feeble-minded women. Even Mary Dendy, who is often cited as one of the leaders of the campaign to control the sexuality of feeble-minded women, argued that there were three boys to every two girls, that boys were 'quite as dangerous', that mental defect was just as likely to be passed down the paternal line of the family, and that the only reason more adult women were reported was because men were more easily lost sight of.[112] Evidence on prison populations showed that a far higher number of men were reported as weak-minded: of those admitted in the six months ending May 1905, 2,612 or 2.7 per cent of men were feeble-minded, compared to 623 or 2.5 per cent of women.[113] Likewise, in local prisons the ratio of weak-minded prisoners in 1904 was 1.79/1000 for men and only 0.88/1000 for women.[114] And in the MAB asylums, the largest single organization providing specialized institutional care for the mentally defective, there were 3,234 males and 3,407 females under treatment in 1907—hardly evidence of an overwhelming bias against women. Undoubtedly, the problem of the sexuality of feeble-minded women in the community did bear heavily on the Commission, and it attracted considerable attention in

[110] Ibid., ii, qu. 16267. [111] Ibid., viii. 3. [112] Ibid., i, qu. 814–15.
[113] Ibid., i. 683. [114] Ibid. 175.

the reception of the Report, but this should not hide the fact that both women and men were the subject of concern, and that ideas about gender constructed male, just as they did female, mental deficiency.

Direct reference to eugenic problems, though again not to eugenics by name, only arose after 180 pages of the Report—and then only for two pages. The reticence suggests that the Commissioners remained cautious about eugenics, realizing that such issues were well outside the scope of the investigation. Nevertheless, the Commissioners claimed that it was impossible to avoid commenting on the evidence which had been supplied by many of the witnesses, especially since the eugenic question had a direct bearing on the relative importance of providing residential care. It was acknowledged that there had been considerable confusion and inconsistency in the terms used when discussing the aetiology of mental deficiency, especially in discussion of 'heredity', and that the statistical evidence on the issue had been inadequate.[115] However, these reservations were virtually ignored, as the Report went on to give virtually full backing to those who argued that the causation of mental deficiency was primarily hereditary.[116]

The Report's support for the theory that mental deficiency was in large part hereditary was defended on the basis that this was the majority view of the witnesses—of thirty-five who dealt with the issue, twenty-five attached supreme importance to heredity as a factor in causation, and of the remaining ten only three were strongly hostile.[117] Support of such medical luminaries as the Cambridge Regius Professor of Medicine, Clifford Allbutt, gave credibility and authority to the hereditarian position. The Commissioners were impressed by the fact that many of those who supported this viewpoint had experience of a huge number of cases through their work in institutions. By contrast, the evidence of the critics of heredity was dismissed on the grounds that they lacked experience and that their views were based largely 'on the theories which they hold'.[118]

The Commissioners argued that proponents of the environmentalist position, such as the Board of Education's chief medical inspector of special schools, Alfred Eicholz, overestimated the efficacy of special education and environmental reform because they based their views on experience with backward rather than mentally defective children.[119]

[115] Ibid., viii. 180–1. [116] Ibid. 184. [117] Ibid. 181. [118] Ibid. 181–2.

[119] Ibid. 182. On Eicholz: Pritchard, *Education and the Handicapped*, 141–3.

In examination, Eicholz's evidence was treated very sceptically.[120] After his contribution to the Interdepartmental Committee on Physical Deterioration, it was probably well known that he would take a hostile line to the notion that mental defect was inherited. On the other hand, before the Deterioration Committee, he had been open to the idea of a system of segregation for feeble-minded school-leavers, arguing that 'most of them are only fit for custodial treatment'.[121] Yet, in his evidence to the Royal Commission, his position shifted firmly against colonies for the feeble-minded, except for those who were morally dangerous.[122] In part, this reflected the educationalist viewpoint that it was impossible to tell whether a child was ineducable until it had been proven that schooling and special training could not overcome physical defects and mental backwardness. However, the objections also stemmed from the Board of Education's suspicions that the Home Office was attempting to highjack their special school policy, replacing it with a system of permanent confinement instead.[123]

Advocates of residential care were obviously afraid that if mental deficiency was attributed to environmental causes the wind would be taken out of their campaign: the notion that mental defect was a permanent condition would be put in doubt, and the emphasis of policy would be forced to shift from providing institutional care to improving the living conditions of the poor. It was this threat which probably encouraged so many witnesses to stress that mental defect occurred equally in rural and urban areas and among all classes, and therefore could not be directly linked to the conditions of the environment. Typical was the influential Mary Dendy, who emphasized that: 'Everything leads one to suppose that there are as many children to be found

[120] *Roy. Com. Feeb.*, i. 205–11.

[121] *Report of the Royal Commission on Physical Deterioration*, qu. 456, also qu. 572.

[122] *Roy. Com. Feeb.*, i, qu. 3614; ibid., viii, qu. 3614. See also, A. Eicholz, 'The Treatment of the Feeble-Minded', *British Medical Journal*, 2 (1902), 683–7.

[123] Sutherland, *Ability, Merit, and Measurement*, 25–56. Suspicion of the Commission was probably not helped by the fact that it included no representatives from the Board of Education but several from the Home Office. This may have been because the Board of Education had assumed that the Royal Commission would only be dealing with the issue of provision for adults: Sutherland, *Ability, Merit and Measurement*, 39.

amongst the upper classes of society who are feeble in intellect as amongst the poorest.'[124]

The Commissioners attempted to point out inconsistencies within Eicholz's evidence. When pushed, he was forced to admit that heredity did play a part in some cases and that 70 per cent of mental defectives had a congenital condition. Moreover, his argument that feeble-minded families were not a danger to the race—since they were on a 'racial down grade', heading towards extinction as they bred weak stock and as their fertility faltered over generations—still implied that the condition was inherited.[125] Bearing in mind these contradictions, the weight of evidence in the opposite direction, and the flimsy statistical basis of Eicholz's own evidence, it is hardly surprising that the Commission was not swayed towards an exclusively environmentalist position.[126] In fact, the Commissioners were willing to accept that environmental factors influenced the problem, and were far less hostile to witnesses who argued that environmental factors were important, though secondary to heredity. What the Commissioners objected to was Eicholz's suggestion that the problem of mental deficiency could be solved by improving the physical environment alone, for this implied that there was little need for a new system of institutional care—a suggestion which was counter to virtually all the evidence gathered by the Commission, to all the pressure which had been building up over the past few decades, and seemed to be of little immediate or practical use to those feeble-minded who were already in prisons, inebriate institutions, and workhouses.

Undoubtedly, the Report's support of the views that heredity was far more important than environment and that the fertility of mental defectives was abnormally high went on both counts far beyond what could reasonably be asserted from the evidence. However, to a large degree this can be seen as tactically driven: by adopting this stance, the

[124] *Roy. Com. Feeb.*, i, qu. 843, 961–4. There seems to have been particular interest in whether mental deficiency legislation might be used to help control spendthrift members of the wealthy classes who threatened to squander their inheritance: *Roy. Com. Feeb.*, i, qu. 17243. In their final report, the Commissioners used the idea that the wealthy produced mentally defective offspring and could obtain care for them to justify specialized care for those who could not afford it: 'the division between pauper and non-pauper is quite unreal in the case of the mentally defective. The son of respectable parents, who is permanently supported wholly or in part by relatives and friends, requires, as mentally defective, the same treatment' (*Roy. Com. Feeb.*, viii. 53). Ironically, it was those, such as Eicholz, who supported an environmentalist view of causation, who came closest to an underclass view of the problem, arguing that mental deficiency was far more prevalent in slums and among paupers: *Roy. Com. Feeb.*, i. 205.

[125] Ibid. i. 206.

[126] Harris, *Private Lives, Public Spirits*, 244–5.

conclusions were given an urgency which they would otherwise have lacked. Legislation was presented as vital, not only for the welfare of the defectives and the social order of the community, but also for the racial health of the nation: 'especially in view of the evidence concerning fertility, the prevention of mentally defective persons becoming parents would tend largely to diminish the number of such persons in the population.'[127] In this brief passage the Commission at last clearly articulated its eugenic concerns about mental deficiency. On the other hand, it should be borne in mind that these comments were given a low profile, delayed until 180 pages into the report, and were followed by a statement reiterating that the racial issue provided an additional reason for measures which were already necessary. It is also worth noting that the Commissioners took a hostile line towards witnesses who favoured sterilization of mental defectives.[128]

In summary, although the eugenic thinking informing the Commissioners' views of the feeble-minded is undeniable, it was not central. This can be seen in the way that the proposed mental deficiency policy fitted into a broader vision of reorganization of mental health care which had little to do with eugenics. In its conclusion, the Report recommended that the proposed specialization of care for the feeble-minded, imbeciles, and idiots should serve as a model for all types of mental defect.[129] The full system of classification was to include: 'persons of unsound mind' (which was to replace the term lunatic, since the latter had become a term of abuse and could be confusing as it was often used to include idiots), 'persons mentally infirm' (which was to be applied to the senile, who were recognized as the chief cause of overcrowding in asylums), idiots, imbeciles, feeble-minded, moral imbeciles, and then a series of groups if also mentally defective, including epileptics, inebriates, the deaf and dumb, and the blind. This specialized but integrated system was designed to make the whole problem of mental deficiency manageable—not only to place new categories under care, but to remove idiots, imbeciles, the senile, and other groups from lunatic

[127] *Roy. Com. Feeb.*, viii. 185.

[128] As in their disapproval of D. Harrison, Medical Officer of Special Schools in Liverpool, who sterilized a feeble-minded girl: *Roy. Com. Feeb.*, i, qu. 10673, 10729–31. The Report condemned 'surgical or other artificial measures' to prevent hereditary transmission, on the grounds that 'the general feeling of the people would at present rightly condemn any legislation': *Roy. Com. Feeb.*, viii. 185.

[129] In its conclusion the Report suggested that mental defect should serve as a generic term embracing congenital as well as acquired mental disorders.

asylums so that they could be rescued from being sites for permanent care and returned to their intended role as curative hospitals.[130]

Towards the 1913 Mental Deficiency Act

Already an issue which was attracting considerable interest among the educated classes, the 1908 Report of the Royal Commission on the Care and Control of the Feeble-Minded promoted mental deficiency to the forefront of national debate. The Report was a stimulus to, as much as a direct product of, the eugenics movement. It is no coincidence that the Eugenics Education Society was founded in 1907; the threat of the feeble-minded provided the Society with what was to be the most successful campaigning issue in its history, helping to thrust the Society into national prominence in a very short space of time. However, the Report also attracted support because, in calling for further specialization of institutional care, it was in sympathy with broader recommendations for reform of the welfare system. Despite their disagreements over other issues, both the Majority and Minority Reports of the Royal Commission on the Poor Law in 1909 firmly supported specialized institutions for the feeble-minded. The 1908 Report also found favour at the Home Office, where Home Secretary Winston Churchill supported institutions for the feeble-minded as part of a broader vision of specialization of penology.[131] However, the idea of residential institutions alarmed the Board of Education, which saw the expansion of the special school system as one of the cornerstones of the school medical service. Interdepartmental tension over this issue helped to delay action on the 1908 Report—already made difficult because of very full legislative programmes—until 1913.[132] The delay provoked the development of a high profile lobbying campaign in which the Eugenics Society played a prominent role. Therefore, when the government introduced legislation in 1912 and 1913, it was unsurprisingly viewed by many

[130] *Roy. Com. Feeb.*, viii. 186–8.

[131] P. Addison, *Churchill on the Home Front, 1900–1955* (London, 1992), 116–26. By 1910, Churchill was also convinced that action was necessary because the feeble-minded presented a danger to the race. He even flirted with the idea of sterilization, considering it more humane than confining the feeble-minded in what amounted to prisons. It should be noted that, in contrast to his support of segregation for mental defectives, Churchill vehemently opposed indeterminate sentences for habitual criminals on the grounds that they were unjust; once again, this emphasizes the importance of biological difference in excluding mental defectives from the possession of rights in liberal thinking.

[132] Sutherland, *Ability, Merit, and Measurement*, 40–7.

people as first and foremost a piece of eugenic legislation. Through looking at the prehistory of the Act, it should be apparent that this is to accept the eugenicists' own propaganda at face value; in fact, the roots of the Act were far more complex.[133]

There were two key contexts for mental deficiency becoming a 'problem' in the nineteenth century: first, civil society was becoming permeated by new voluntary and statutory welfare agencies which greatly increased the visibility of mental deficiency; and secondly, this society invested considerable resources in establishing systems of institutional care. Mental defectives were recognized and defined in terms of both their failure to support themselves or act responsibly within the community, and their failure to respond to regimes of management and reform in schools, the penal system, asylums, and workhouses. The concept of mental deficiency attracted so much attention because it seemed to provide a scientific explanation for individual irresponsibility and social and moral failure. The mental defective provided a symbol for a shift in the view of human nature, from seeing people as wilful masters of their own destinies, to acknowledging that human action was limited by the influence of environment, society, and, in particular, by heredity.[134] In fact, the shift from wilfulness to the notion of a socially and biologically determined character can be exaggerated. Early twentieth-century social thought acknowledged that it was the state's duty to help citizens overcome external impediments, but only so as to enable and facilitate wilful improvement of moral character.[135] However, because of the very nature of their defect, mental defectives were the one group who were widely accepted as lacking will or the potential for self-improvement. This fundamental biological division led to the exclusion of mental defectives from the contract of citizenship rights and responsibilities which lay at the heart of the liberal British polity. As a consequence, a policy such as permanent segregation became a possibility; equally, it became society's responsibility to provide care and

[133] As Sutherland has argued, the mental deficiency campaign shows that 'the labels "Social Darwinist" and "eugenist" are not in themselves very informative about either configurations of attitudes or policy committments; they need to be both elaborated and qualified before they can be used to "place" people or campaigns with any precision': *Ability, Merit and Measurement*, 39.

[134] Wiener, *Reconstructing the Criminal.*

[135] J. Harris, 'Political Thought and the Welfare State, 1870–1940: An Intellectual Framework for British Social Policy', *Past and Present*, 135 (1992), 116–41; S. Collini, 'The Idea of "Character" in Victorian Political Thought', *Transactions of the Royal Historical Association*, 35 (1985), 29–50.

protection without the stigma of pauperization or punishment in return. In summary, to understand why the problem of mental deficiency had become acute by the early twentieth century we need to go beyond an explanation which rests on the eugenic threat posed by the feeble-minded, to consider how this fear interacted with anxieties about regulating the boundaries of responsible citizenship and managing an increasingly sophisticated network of welfare institutions.

I
The Parliamentary Politics of Mental Deficiency

INTRODUCTION

The Mental Deficiency Act was passed by an overwhelming majority in 1913. Thereafter, the British legislature avoided seriously addressing the problem again until the Mental Health Act of 1959. At first sight, this legislative void would seem to suggest that there was little parliamentary or party-political debate about mental deficiency during the interwar period, and that mental deficiency was a largely non-political subject, part of a political consensus over the expansion of specialized medical services.[1] However, a number of historians have already demonstrated that mental deficiency was an area of medical care where such a consensus did not exist. In particular, they have stressed the conflict which emerged along lines of class and party when mental deficiency policies were justified on eugenic grounds.[2] The main limitation of this work is that it tends to view debate as simply revolving around the issue of eugenics. In fact, the politics of mental deficiency was shaped by a more complex terrain of ideological, ethical, and social issues. Moreover, to understand the real nature of conflict and consensus we need to go beyond class and party to integrate the influence of the gender, religion, and professional and social backgrounds of politicians.

New insights are gained if the problem of mental deficiency is situated within the context of the central political issue of the per-

[1] For criticism and a survey of this position: C. Webster, 'Conflict and Consensus: Explaining the British Health Service', *Twentieth Century British History*, 1 (1990), 115–51.

[2] G. Searle, 'Eugenics and Class', in C. Webster (ed.), *Biology, Medicine, and Society, 1840–1940* (Cambridge, 1981), 217–42; G. Searle, 'Eugenics and Politics in the 1930s', *Annals of Science*, 36 (1979), 159–69; J. Macnicol, 'Eugenics and the Campaign for Voluntary Sterilization in Britain between the Wars', *Social History of Medicine*, 2 (1989), 147–70; G. Jones, 'Eugenics and Social Policy between the Wars', *Historical Journal*, 25 (1982), 717–28.

iod—the process of adjusting to democracy.[3] For the legislature, adjusting to democracy meant adjusting to a complex set of interrelated changes in the political environment: a widening of the political franchise to include the working class and women; the emergence of a national working-class party; the increased volume and technicality of demands placed on the parliamentary machinery by the move towards collectivism; the need to design increasingly interventionist policies which were still acceptable within the framework of rights and duties which underpinned the core ideology of this democracy; and the redefinition of the democractic subject which took place with the extension of the political franchise.[4] Finally, adjusting to democracy meant resisting as well as reacting to these new conditions.

THE MENTAL DEFICIENCY ACT OF 1913

In 1909, the Eugenics Society began to lobby the government to act on the Report of the Royal Commission on the Feeble-Minded. Also the Reverend Inge and C. S. Loch held a meeting with Home Secretary Herbert Gladstone who promised action within a year.[5] In 1910, the government held a second meeting with a group of over thirty MPs, led by Charles Masterman and Charles Trevelyan.[6] On 16 July a deputation led by the National Association for Promoting the Welfare of the Feeble-Minded met members of the Cabinet, including the Prime Minister, the Lord Chancellor, the Home Secretary, and the President of the Local Government Board. Although Winston Churchill and John Burns were sympathetic, the government refused to be hurried.[7]

Frustrated by the government's inaction, the Eugenics Society and the National Association drafted their own 'Feeble-Minded Control Bill', which cautiously avoided any recommendations involving new expenditure or administrative reorganization, and which simply proposed to legalize the certification and detention of the feeble-minded. After a non-party meeting in the House of Commons on 5 December

[3] The phrase is borrowed from Rodney Lowe, *Adjusting to Democracy: The Role of the Ministry of Labour in British Politics, 1916–1939* (Oxford, 1986).

[4] For an exploration of the relationship between democracy and subjectivity: P. Joyce, *Democratic Subjects: The Self and the Social in Nineteenth-Century England* (Cambridge, 1994).

[5] *Eugenics Review*, 1 (1909), 142.

[6] Ibid. 2 (1910), 89–90.

[7] Ibid. 2 (1910), 163–4.

1911, the bill gained the support of MPs George Barnes, Charles Bathurst, Noel Buxton, Arthur Henderson, J. W. Hills, Sir Alfred Mond, and Dr Walter Rea.[8] A second draft bill—the Mental Defect Bill—produced by the Charity Organization Society, was dropped because its recommendations entailing expenditure were likely to ruin its chances as a private member's bill.[9] Instead, the COS added its support to the Feeble-minded Control Bill, as did a cross-party group of politicians, including Lord Claud Hamilton, Sir George Younger, Pike Pease, Sir Charles Nicholson, Dr Walter Rea, William Pearce, Fred Jowett, Will Crookes, Ernest Pollock, Dr Christopher Addison, and Willoughby Dickinson.[10] The Bill was presented to the House on 19 February and had its second reading on 17 May 1912. Conservative back-bencher Gershom Stewart introduced the bill, with the names of six Liberals, four Conservatives, and two Labour members to back him. Despite the fact that its supporters downplayed its eugenic content, the bill provoked heated debate, with Liberal MP Josiah Wedgwood leading the opposition.[11] However, such private initiative was now unnecessary, for the Government was in the midst of introducing its own bill, which had its first reading on 16 May, and was vigorously debated on 10 June and 19 July. The Bill was then sent to committee for consideration of some of the objections which had been raised, but the Government could not find enough space to reintroduce it to the House. It was not until 28 May 1913 that a second Mental Deficiency Bill was introduced by Home Secretary Reginald McKenna. The bill was debated over the summer and passed by an overwhelming majority of 358 to 15 just before the summer recess. With less controversy, the Elementary Education (Defective and Epileptic Children) Bill, compelling all local authorities to provide special schools, was also introduced in the summer of 1913 and became law in 1914, with a majority of 341 to 37.

Historians have tended to view the legislation which eventually emerged in 1913 as a considerable dilution of the 1912 proposals; indeed, it has been assumed that the Mental Defective Act which resulted was largely ineffectual.[12] This overestimates the extent of the

[8] Ibid. 3 (1911), 355.

[9] This is perhaps the 'Mental Defect Bill' which was proposed by Acland Allen, Willoughby Dickinson, Godfrey Locker Lampson, Sir Alfred Mond, and Mr Pollock on 15 April 1912: *Parl. Deb.* (HC), v. 37, 38.

[10] *Eugenics Review*, 4 (1912), 108–9.

[11] *Parl. Deb.* (HC), v. 38, 1443; E. J. Larson, 'The Rhetoric of Eugenics: Expert Authority and the Mental Deficiency Bill', *BJHS* 24 (1991), 52.

[12] Larson, 'Rhetoric of Eugenics', 59–60; H. G. Simmons, 'Explaining Social Policy: The English Mental Deficiency Act of 1913', *Journal of Social History*, 11 (1978), 400–1.

modifications which were made. The original Mental Deficiency Bill contained some very clumsy and contentious phrasing which was bound to raise opposition: for instance, the feeble-minded were defined as those who were incapable 'of competing on equal terms with their normal fellows', and it was proposed to segregate all those 'in whose case it is desirable in the interests of the community that they should be deprived of the opportunity of procreating children'.[13] Taking note of the objections, the second bill dropped all reference to the 'Eugenic Idea' and strengthened the safeguards protecting individual liberties.[14] But these changes were largely superficial; the final Act still provided the apparatus for the compulsory and permanent segregation of the feeble-minded. The most serious limitation resulted, not from modifications to the Act itself, but from the passage of the Education Act, which provided a separate system of care for all educable mental defectives under the age of 17.

Support

The Mental Deficiency Bill gained such overwhelming support because it seemed to provide a rational adminstrative solution to a series of social problems, and it had the backing of the Royal Commissions on the Feeble-Minded and the Poor Law and the vast majority of expert opinion. Home Secretary McKenna introduced the first bill as a necessary reform leading to the specialization of care for those whose liberty was already restricted in prisons and workhouses, drawing heavily on the evidence of the Royal Commission on the Feeble-Minded for support.[15] Likewise, he introduced the second bill as a humanitarian measure which was in the best interests of defectives and the community alike: 'at present they are left, uncared for and unprotected, to live miserable existences, and, quite apart from their own habits, become a source of degredation, misery, and poverty to others.'[16] The dominant view among those who spoke was that it would be inhumane not to act. As Conservative MP Dr William Chapple recounted:

> I have gone into a home and heard the screeching, half-barking sound of an idiot tethered in the backyard like a dog, with the pathetic, grief stricken mother doing all she could to hide it. Do you mean to say it would not be a kindness to

[13] Larson, 'Rhetoric of Eugenics', 54. [14] *Parl. Deb.* (HC), 53 (1913): 219–21.
[15] *Parl. Deb.* (HC), 39 (1912): 627–9.
[16] McKenna, *Parl. Deb.* (HC), 53 (1913): 219–20.

go to the mother and say, 'we have a magnificent institution; we will take the child and protect it'?[17]

It was argued that the legislation would simply make available the type of care which was already provided by the specialized charitable and poor law idiot homes, and which had been developed successfully in the United States.[18] It was constantly stressed that mental deficiency institutions were places for protection and humane care, and should not be seen as a type of prison. On the contrary, the Act aimed to reform the current situation in which the feeble-minded were inhumanely and inefficiently sent to workhouses and prisons. As Charles Roberts put it: 'We have been guilty of neglecting these unfortunate people in the past. We have far too often branded them with the stigma of pauperism and criminality when we should have given them sympathetic treatment.'[19]

The Liberal Government's support needs to be understood in the context of its broader programme of welfare reform between 1906 and 1914 and an ideological shift towards 'collectivism' which was epitomized by the 'New Liberalism'.[20] It was gradually being accepted within the Liberal Party that the liberty of the individual sometimes had to take second place to the good of the community. Liberal MP C. A. McCurdy was attracted to the Mental Deficiency Bill as one of the first real attempts to deal with social evil through a scientific approach, and argued that this type of subject should be more important to the Party than the old Liberal campaigns over Home Rule and Welsh Disestablishment:

It is not by building up temples of liberty and by polishing and repolishing political constitutions that the people can grow up to proper manhood and healthy national life. . . . The new century has brought new social problems, and it is with one of these problems this Bill makes a courageous effort to deal.[21]

To McCurdy, and an increasing proportion of those in his party, it was wrong to be unduly deterred by worries about protecting the liberty of the subject:

there was a danger of making a fetish of the word 'liberty', and of attempting to deprive these children in the name of a useless liberty, of what I believe under

17 *Parl. Deb.* (HC), 53 (1913): 820.

18 Ibid. 38 (1912): 1482.

19 Ibid. 1509.

20 M. Freeden, *The New Liberalism: An Ideology of Social Reform* (Oxford, 1978).

21 *Parl. Deb.* (HC), 53 (1913): 834.

the Act will be the care and guardianship of the State, exercised for their interests, and for their interests alone.[22]

As the Liberal Westminster Gazette put it: 'It would, indeed, be the last crime in the name of liberty if we were prevented, by some fanciful or superstitious regard for our own personal freedom, from applying a rational treatment to these unfortunate beings.'[23] Because New Liberals placed so much emphasis on the importance of character, they could adopt surprisingly draconian policies towards mental defectives, who, by definition, had a biological lack of character, and who threatened to disrupt social policies which were designed for those citizens who did.[24]

It is an exaggeration to regard eugenics as either the most important or effective argument for the legislation of 1913. As Larson has pointed out, during the parliamentary debate on the Feeble-Minded Control Bill eugenic rhetoric and evidence based on the science of heredity were notably absent from the arguments of the bill's proponents (though not from those of its opponents, who exposed this link to undermine the bill).[25] Considerable concern was expressed about the danger of feeble-minded women in the community, but this was more often a 'common sense' argument, about the need to protect these women and their unsuitability as mothers, than a eugenic one.[26] Similarly, when the problem of inheritance was discussed it was done so in a practical, rather than scientific, manner, recounting examples of mental defectives who had given birth to defective offspring and detailing the terrible social and moral, as much as eugenic, results. As Leslie Scott, a Conservative MP who was active in the campaign for legislation, emphasized, the Government's bill was the result of experience, rather than eugenics.[27] There were, of course, exceptions to this; however, whether the main advocate of eugenics, Dr William Chapple, a surgeon by background and author of a book advocating sterilization of the unfit, can be taken as representative is rather debatable.[28]

The main concern among supporters of the Act was to counter the criticism that it was coercive and a dangerous infringement of individual liberty. They pointed out that the legislation was accompanied by an elaborate system of legal safeguards, that in the majority of cases familial

[22] Ibid. 38 (1913): 835.
[23] *Westminster Gazette* (4 June 1913), quoted in Freeden, *New Liberalism*, 192.
[24] Freeden, *New Liberalism*, 185. [25] Larson, 'Rhetoric of Eugenics'.
[26] Ibid. 51. [27] *Parl. Deb.* (HC), 53 (1913): 235.
[28] Ibid. 815–21. Chapple, a New Zealander by birth, was Liberal MP for Stirlingshire. His book advocating sterilization was *The Fertility of the Unfit* (London, 1905).

consent would be obtained before a mental defective was placed in an institution, and that institutional care would only be recommended if the care at home was inadequate. But more fundamentally, there was considerable doubt over whether a 'liberty of the subject' argument could be applied to individuals who lacked the mental ability to exercise liberty in a meaningful sense; supporters argued that the only liberty that defectives would gain if the bill was defeated was a liberty to be exploited in the community.[29] As Tory MP Leslie Scott argued: 'The liberty of the subject, for which Mr Wedgwood contends, is, in the last resort, merely the liberty to die in the workhouse, the brothel, or the gaol.'[30]

The perceived moral, social, and eugenic threat posed by mental defectives was sufficient to overcome most Conservatives' reservations towards the erosion of individual liberty under an advancing tide of state collectivism and rising taxation. The comments of Tory MP Gershom Stewart, introducing the Feeble-Minded Control Bill, suggest a different set of concerns to those of the Liberals:

> The whole social body now obsequiously and willingly submits itself to great changes for the aged, infirm, the weak, the blind, and the unfortunate in every possible direction, because these have been or may be, effective citizens; but there is no need for an army on the march, as we are, to run grave risks and seriously increase its responsibilities by consenting to being escorted by a helpless army of camp followers who can never be efficient people. Any army which has to march through difficult country must do something to embarasss itself of such a weight as that, and we, with industrial competition very severe and with the grave possibility of national danger always in front of us, are, I think, within our rights in taking steps to try and improve the present condition of things.'[31]

Stewart's anxieties about national and martial efficiency were part of a broader tendency within a riven early twentieth-century Conservative Party.[32] What is also significant, however, is the way that Stewart used mental deficiency as a boundary marker to draw a line beyond which the onward march of welfare for the weak could no longer be justified as a tool for producing 'effective citizens'.

[29] For instance, Leslie Scott, *Parl. Deb.* (HC), 53 (1913): 236. Mary Dendy was influential in developing this argument: ibid. 809; *Nation* (5 July 1913), 531.

[30] *Nation* (6 Nov. 1912), 310–11.

[31] *Parl. Deb.* (HC), 34 (1912): 1448–9.

[32] E. Green, 'The Strange Death of Tory England', *Twentieth Century British History*, 2 (1991), 67–88.

The silence or support of the majority of Labour MPs suggests that they did not regard the Mental Deficiency Act as anti-working class.[33] They were not even swayed by Josiah Wedgwood's jibe that the proposed definition of mental defect would be used to put away people like Keir Hardie.[34] The most prominent Labour supporter was Will Crooks, who had experience of the subject of institutional care in both a professional sense, through his work on the Metropolitan Asylums Board, and in a personal sense, having been placed in a workhouse as a boy of 8. Nevertheless, Crooks scorned the 'liberty of the subject' argument, arguing: ' There is talk of liberty of the subject. I sometimes feel inclined to prosecute the fathers and mothers of those children for not allowing them to be cared for.'[35] As far as Crooks was concerned, the bill was a humane and Christian measure which was greeted by the working class themselves.[36] Greta Jones has suggested that those Labour MPs who actively supported the Act, apart from Fred Jowett of the ILP, were from the right of the party, representing a conservative strain within the early Labour Party.[37] Certainly, Crooks fits this description, described in one recent account as 'a right wing Fabian'.[38] And it may well be the case that Labour MPs, who were usually drawn from the 'respectable working class', supported the Act as a way to ensure control over members of the 'profligate and unrespectable poor'. Crooks, for instance, though keen for a humane reform of the poor law system, had a harsh attitude towards the 'unemployable', and it seems that this influenced his attitude towards the mental defectives: with considerable distaste, he explained that many of the unemployable,

> were formerly mentally defective children who had been allowed to drift about the world, and to become absolutely useless. There is only one fitting description; they are almost like human vermin. They crawl about, doing absolutely nothing, except polluting and corrupting everything they touch.[39]

[33] Analysis of Labour's attitude towards the Act is absent from accounts of Labour's position on social policy during this period: H. Pelling, 'The Working Class and the Welfare State', in id., *Popular Politics and Society in Late Victorian Britain* (London, 1968); P. Thane, 'The Labour Party and State Welfare', in K. D. Brown (ed.), *The First Labour Party* (Beckenham, 1985), 183–216.

[34] *Parl. Deb.* (HC), 53 (1913): 248. Hardie, himself, neither spoke nor voted on the issue.

[35] Ibid. 38 (1912): 1499.

[36] Ibid. 53 (1913): 826–31. G. Haw, *From Workhouse to Westminster: The Life Story of Will Crooks, MP* (London, 1907).

[37] G. Jones, *Social Hygiene in Twentieth-Century Britain* (London, 1986), 58–60.

[38] A. M. McBriar, *Fabian Socialism and English Politics, 1884–1918* (New York, 1962), 120.

[39] *Parl. Deb.* (HC), 38 (1912) 1500.

Crooks supported institutional care as a way to prevent defectives from producing more children and to provide an environment where they could be set to work to contribute towards their own support.[40] The remarks of Mrs Nevitt, of the Manchester District Co-operative Guild, quoted in the parliamentary debate, again reveal the often harsh attitudes of representatives of the working class towards mental defectives:

> Our women also feel very keenly at the present time that there is no restraint on this class of people perpetuating their race. We find that the working-class people at the present time are restricting their families—that I am absolutely certain of—because they feel they cannot bring their children up in the manner they would like to do. Yet at the same time they have to keep, or help to do so, this class of person, and there is no restraint whatever put on their bringing families into the world. We feel very strongly that we are populating the country with the wrong class, and the people who are worthy citizens have to restrict their own families.[41]

As far as Mrs Nevitt was concerned, these mental defectives had no real liberty if left at home unattended while their parents went out to work.[42] Mr McGlasson of the Trades Council, also quoted in the debate, was equally unsentimental about the liberty of mental defectives or the working-class family, pointing out that unless compulsory care was introduced, parents would continue to exploit their mentally defective children by setting them to work.[43]

Opposition

Although the case for legislation did not rest on scientific arguments alone, its opponents strategically attacked this, its weakest, basis. It was argued that the hereditary basis of mental deficiency remained unproven and that the definition of mental deficiency was wholly unscientific. However, underlying these objections was a more fundamental concern about liberty and democratic accountability which ran through much of the opposition to the Act, particularly that of its leader, self-styled 'last of the Radicals', Josiah Wedgwood.[44]

Wedgwood opposed the science behind the legislation, not just

[40] *Parl. Deb.* (HC), 38 (1912) 1500. [41] Ibid. 53 (1913): 235.
[42] Ibid. 236. [43] Ibid.
[44] C. V. Wedgwood, *The Last of the Radicals: Josiah Wedgwood MP* (London, 1951). *Dictionary of National Biography, 1941–50*, pp. 941–3.

because he thought it was misconceived, but because he felt that legislation which was led by scientific conjecture and placed power into the hands of experts was a dangerous precedent for democracy. To Wedgwood, who, it should be stressed, had no prior experience or knowledge of the issue, mental deficiency seemed to have been inflated far beyond its real proportions, foisted on the government by outside experts and pressure groups, such as the Eugenics Society.[45]

Wedgwood turned his virtually single-handed opposition into something of a crusade in defence of liberty. In two late-night sittings, sustained by barley water and milk chocolate, Wedgwood threw 120 amendments and 150 speeches at the bill.[46] He castigated the Government, not only over the content of its bill, but for the way that parliamentary procedure was, in Wedgwood's opinion, becoming increasingly undemocratic. He objected to measures being forced on the legislature through the recommendations of Royal Commissions, the Government cutting off debate before the details of proposed legislation had been properly discussed by the House, the practice of sending bills into committee stage for redrafting by government appointees, the increasing pressure to vote according to party lines rather than independent judgement, and the passage of legislation which placed power into the hands of non-elected experts who were outside democratic control: all of these trends removed power to scrutinize or veto legislation from the elected Members of Parliament.[47] Thus Wedgwood's concerns reflected some of the more fundamental problems of democratic accountability faced by the British legislature as it adjusted to the increased volume of parliamentary business brought about by the onset of collectivism and the complexity of new measures, such as the Mental Deficiency Act. Wedgwood's concerns about the erosion of legislative power were echoed in the House of Lords by the Marquess of Salisbury, who objected to the shortening of parliamentary debate, the secretive nature of the committee stage, and the waning influence of the Lords as a chamber for revision.[48] Like Wedgwood in the Commons, Salisbury proposed an endless series of amendments to the details of the Act. In doing so, the two of them demonstrated that, though an individual Member could not defeat a measure single-handedly,

[45] *Parl. Deb.* (HC), 38, (1912): 1474.

[46] Wedgwood, *Last of the Radicals*, 94–5; J. Wedgwood, *Memoirs of a Fighting Life* (Plymouth, 1940).

[47] *Parl. Deb.* (HC), 38 (1912): 1470, 1476; ibid. 53 (1913): 243–53; ibid. 56 (1913) 429, 435, 442, 497.

[48] *Parl. Deb.* (Lords), 14, (1913): 1705–15, 1757–826.

parliamentary procedure did still enable them to make its passage far more difficult and controversial.

That the Mental Deficiency Act symbolized a point of transition and tension in Britain's adjustment to democracy is again suggested by the remarks of the jurist A. V. Dicey, who highlighted the Act as a sign of the problems of balancing the interests of the individual and the community:

> Most of its provisions for the protection of defectives from themselves and from their neighbours, recommend themselves to common sense. They would probably have been welcomed by a humanitarian and a jurist, such as Bentham. Yet the Act would hardly have been passed by the Parliament, say of 1860. The interference which it involves with the dangerous liberty of defectives would at least have raised suspicion in the minds of men who had hailed the individualism of Mill's *Liberty* with indiscriminating applause. They would have felt that the measure was open to one serious objection. The Mental Deficiency Act is the first stage along a path on which no man can decline to enter, but which if too far pursued, will bring statesmen across difficulties hard to meet without considerable interference with individual liberty.[49]

To Dicey, the Mental Deficiency Act, like compulsory medical inspection and vaccination, highlighted the tension between democracy and collectivism:[50]

> The ideal of democracy, in short, is government for the good of the people by the people, and in accordance with the wish of the people; the ideal of collectivism is government for the good of the people by experts, or officials who know, or think they know, what is good for the people better than any non-official person or than the mass of people themselves.'[51]

Wedgwood recognized that, in practice, the Act would be directed against the working class, and therefore opposed it on the grounds that it was a 'class measure'.[52] He argued that, rather than punishing the victims, the social causes of the problem needed to be addressed.[53] Although he acknowledged that some people might really need institutional care, he saw no reason why this could not be offered on a voluntary

[49] A. V. Dicey, *Law and Opinion in England*, 2nd edn. (London, 1914), pp. l–li.

[50] Ibid., pp. lxxiv–v. [51] Ibid., p. lxxii.

[52] On radical Liberal opposition to 'class legislation' in the 19th cent., such as the Contagious Diseases Act and labour legislation: E. Biagini, *Liberty, Retrenchment and Reform: Popular Liberalism in the Age of Gladstone, 1860–1880* (Cambridge, 1992), 159. For the view that the legislation of the period was pervaded with bias against the working class: P. Johnson, 'Class Law in Victorian England', *Past and Present*, 141 (1993), 147–69.

[53] *Parl. Deb.* (HC), 38 (1912): 1473; *Nation* (2 Nov. 1912), 'Dangers of the Mental Deficiency Bill', 214–16.

basis.[54] He regarded the legislation as part of a dangerously coercive tendency to place the interests of the state above those of individual conscience, and drew parallels with the incarceration of those who had objected to vaccination and vivisection in Britain and of civil rights agitators in India.[55] Wedgwood reasoned that if the Act was really about protection, rather than control, voluntary homes would be sufficient, and that, instead of punishing the innocent, the laws protecting minors might be extended to protect the feeble-minded in the community.[56]

Wedgwood was supported by only a small group of fellow Liberals. Analysis of this group suggests that, like Wedgwood, they tended to represent an older radical liberalism which still placed individual liberty above the interests of the community. His Liberal allies included: the land reformer R. L. Outhwaite; W. M. R. Pringle, who later became Chairman of the Liberal and Radical Candidates Association; Francis Neilson, author of pamphlets on land tax, socialism, and free trade, and editor of *Democratic Monthly*; and J. M. Hogge, journalist, social investigator, and radical.[57] However, the libertarian concerns of this group were dismissed as 'sloppy sentimentalism' by another Party member, and seem to have held little sway within the parliamentary party.[58]

Wedgwood found more support outside Parliament, particularly in sections of the Liberal press. The *Nation*, for instance, took strong exception to the eugenic and anti-working class nature of the 1912 bill, describing it as: 'the first essay in legislation of the Eugenist School. . . . It looks like a harmless proposal for segregating idiots. It really is a first essay in the scientific breeding of the poor.'[59] Although this view was criticized in letters from psychiatrists, doctors, and those working in the field, lay correspondents tended to back the paper's line. However, the *Nation*'s opposition modified after amendments had been made to the legislation, and perhaps also as it became more informed. By October 1912 the *Nation* acknowledged that the feeble-minded were a distinct class who 'may claim of a humane society care and supervision'.[60] And, although it was still argued that the evidence on the hereditary basis of mental defect was far from conclusive, it was admitted that:

[54] *Parl. Deb.* (HC), 38 (1912): 1473. [55] Ibid. 1468–9.
[56] *Nation* (2 Nov. 1912), 'Mental Deficiency Bill', 214–16.
[57] Unless otherwise stated, biographical material on MPs is from M. Stenton and S. Lees, *Who's Who of British Members of Parliament*, ii–iii (Brighton, 1978, 1979).
[58] Harry Lawson, *Parl. Deb.* (HC), 53 (1913): 836.
[59] *Nation* (15 June 1912), 'The Crime of Being Inefficient', 390–1.
[60] *Nation* (26 Oct. 1912), 'Mental Deficiency Bill', 169.

it is at least so far hereditary as to make the inter-marriage of two feeble-minded persons undesirable, and the marriage of a feeble-minded person with a normal one undesirable only in a lesser degree. For these reasons segregation of the sexes is the natural accompaniment of care and supervision. It is the return which the feeble-minded person has to make for the guardianship of society.[61]

Where the *Nation* differed from supporters of the Bill, was over whether institutional care should be provided on a voluntary or compulsory basis—not over whether such care was needed.[62]

Liberal suspicions about the mental deficiency legislation were also provoked by a feeling that it was a Tory measure which had been foisted on the Liberal Government. There was a feeling that the Liberals had been pressurized into action by Tory jibes about their poor record of social reform during the session,[63] and that the measure was part of a dangerous, new, collectivist tendency within Conservatism, which threatened liberty and democracy and could lead towards a militarist state. As the *Nation* warned:

The Tory Party is plainly leaning to the most coercive side of eugenist theory. Shut up the poor, conscript them, tax their bread—all for the sake of the Empire, and the elder and younger sons who run it. . . . Tory members rallied with zeal to the coercive side of the Mental Deficiency Bill. Tory newspapers, with military rather than moral and industrial efficiency in view, favoured the elimination of the unfit.[64]

Liberal MP, W. Pringle, suspected that the mental deficiency issue was a ploy to distract the Liberal Government's attention from the more important issues of Home Rule and Welsh Disestablishment, and that once an Act was passed the Tories would exploit its unpopularity to undermine the Liberal Party.[65]

Tory opposition to the legislation was minimal. Only Frederick Banbury raised objections on grounds of individual liberty. Otherwise, objections were largely directed to the increased expenditure which the Act would demand.[66] However, only in the case of Banbury did this lead to voting against the final bill. In fact, the more common complaint was that not enough money had been set aside to support

[61] *Nation* (26 Oct. 1912), 'Mental Deficiency Bill', 169.

[62] *Nation* (6 April 1913), 'Second Thoughts on the Mental Deficiency Bill', 8–9.

[63] Astor, *Parl. Deb.* (HC), 39 (1912): 165–6; Bonar Law, ibid. 50 (1913): 28.

[64] *Nation* (23 Nov. 1912), 339; see also, ibid. 340–1, 'The New Tory Programme'. Tory MP Leslie Scott criticized this attempt to bring party into the debate: *Nation* (30 Nov. 1912), 392–3.

[65] *Parl. Deb.* (HC), 53 (1913): 821–5.

[66] Ibid. (HC), 38 (1912): 1491.

local authorities: as Montagu Barlow warned, 'it will produce amongst the rate payers and amongst those who have to adminster local government such a disgust'.[67] However, such rate-payer objections were largely mitigated in the second bill, as the Government proposed to match local expenditure with a central grant.[68] Other Tories, such as Mr Ward, objected to the extension of the 'host of paid locusts' which would be created to administer the Act, and pointed out that, 'it is providing soft jobs for professional people, the very classes who support this type of legislation'.[69]

Tory objections were stronger when it came to voting on making special schools compulsory, with all thirty-seven opponents to the 1914 bill coming from the Conservative and Unionist Party. In part, this was again based on opposition to increased public expenditure: education in existing special schools had already shown itself to be a costly business, and this now threatened to be imposed on all local authorities. There was also a feeling that it was inefficient, inequitable, and nonsensical to spend more in the vain hope of educating those who were virtually ineducable than one devoted to normal children who would grow up to make a real contribution to society. Further Tory resistance reflected efforts to protect parental rights and responsibilities. Frederick Banbury opposed compulsion on the grounds that: 'What we are really doing is preventing the fathers and mothers of such persons discharging their proper responsibilities towards their children.'[70] Banbury presented himself and other 'reactionary' Tories as sympathetic to the poor, in contrast to the Liberals who had no qualms about intruding on their liberty.[71] In the Lords, the Marquess of Salisbury reiterated these points, arguing:

> The people who really look after the liberty of the poor are to be found on the despised Tory benches. The Liberal and Labour Members do not care for the liberty of the poor at all, and it is to be remembered that this bill only affects the poor.[72]

Such concern about the rights and duties of the parents of defectives, again suggests that many MPs remained unconvinced by the

[67] Ibid. 53 (1913): 815. On the way that such financial concerns were forcing social policy into national politics: J. Harris, 'The Transition to High Politics in English Social Policy, 1880–1914', in M. Bentley and J. Stevenson (eds.), *High and Low Politics in Modern Britain* (Oxford, 1983), pp. 58–75.

[68] *Parl. Deb.* (HC), 53 (1913): 222.

[69] Ibid. 56 (1913): 423, 431.

[70] Ibid. 64 (1914): 2383.

[71] Ibid. 59 (1914): 477.

[72] *Parl. Deb.* (Lords), 17 (1914): 111.

argument that mental deficiency was a hereditary condition—why concern oneself about the rights of parents if they too were assumed to be mentally defective? We should not underestimate the strength of Tory misgivings over special education; Wedgwood, in fact, expected the Tory objections to defeat the bill. However, the opposition subsided once parents were given the right to refuse to send their children to a residential school.[73]

Although the majorities behind the legislation of 1913–14 had been resounding, the opponents had made its passage stormy and provoked considerable public controversy. Moreover, the issue had exposed the problems which the early twentieth-century legislature faced in adjusting to the increased volume and technicality of welfare measures. In the House of Lords, Lord Ribblesdale argued that the delay in acting on the recommendations of the Royal Commission on the Feeble-Minded highlighted major problems with the legislative system: not only was action hampered by the divisiveness of the party system, but,

> you have got the whole parliamentary machine into such a condition of over pressure and jar that what has happened to this unfortunate Bill is not likely to prove an isolated instance, and I believe it is perfectly possible at any time that agreed social legislation in the true sense of the word 'social' may have to be dropped under the rigid and standardised machinery which the Parliament Act has brought into operation.[74]

Because of such difficulties, the legislature was reluctant to readdress the question of mental deficiency during the interwar period, avoiding substantial reform of the 1913 legislation until the 1959 Mental Health Act. Apart from a government modification of the Mental Deficiency Act in 1927, the only times that the issue was raised were in private member's questions. In so doing, as Wedgwood had feared, considerable power to modify the system fell into the hands of those 'experts' who administered the system. Despite the absence of major legislation, the reluctance of interwar governments to address the issue, and the lack of a clear party divisions, the parliamentary politics of mental deficiency did continue after 1913/14. It emerged out of divisions of class, gender, religion, and professional background, and was fought over the balance between individual and community interests, morality, the power of experts, and the role of eugenics as a tool of social engineering. As

[73] Including that of Wedgwood: *Parl. Deb.* (HC), 56 (1913): 1717.
[74] *Parl. Deb.* (Lords), 13 (1912): 62–5.

such, the history of mental deficiency during the interwar years indicates that apparent consensus on 'non-party' issues, such as this, may hide a rich vein of debate and even conflict.

THE FRANCHISE FACTOR AND THE PROBLEM OF MENTAL DEFICIENCY

The 1918 Representation of the People Act tripled the size of the British electorate.[75] The shift in the class-basis was less dramatic, already set in place by the Reform Acts of 1867 and 1884–5. However, the tone of politics was modified by the extension of the vote to women (until 1928, only those over 30) and to many young, single, and mobile men, who had been unable to register under the old system. Labour's emergence as a party of national significance in 1918 was as much a coincidence as a direct result of these changes.[76] However, as a result, the interests of the working class, as well as those of women, had greater force in post-1918 legislative debate.

By replacing the householder and property qualifications, which were so fundamental in determining who was 'responsible' to vote, with a qualification based on residency, the 1918 franchise heightened anxiety about whether the 'irresponsible' could still be excluded from the electorate. Lord Hugh Cecil was one of those MPs who was worried about universal suffrage extending the vote to 'wastrels and vagabonds and a real class in the community who do not belong to what can be called the working classes'.[77] Cecil pointed out that a man who had resided in prison for six months might now have the right to vote, and that the removal of the property qualification might even give vagrants the vote.[78] Such concerns extended to those who resided in mental institutions—an issue which was given an added dimension by sympathy for shell-shocked soldiers.[79] The Secretary of State for the Home Department, Sir George Cave, responded that the mental competence of voters to exercise the franchise was to be determined by the returning officer.[80] This suggested that people who had recovered from mental illness, such as shell-shocked

[75] C. Matthew, J. Kay, and R. McKibbin, 'The Franchise Factor in the Rise of the Labour Party', *English Historical Review*, 91 (1976), 723–52.

[76] D. Tanner, *Political Change and the Labour Party, 1900–18* (Cambridge, 1990).

[77] *Parl. Deb.* (HC), 93 (1917): 2190.

[78] Ibid. 94 (1917): 221.

[79] Ibid. 95 (1917): 1718.

[80] Ibid.

troops, would be free to vote. It also assumed that the returning officer would be able to determine who had mental competence—a task which was surely made far harder by the extension of the definition of mental defect into the hazy borderline area of feeble-mindedness. Some MPs remained unsatisfied and called for the legislation to specifically exclude certain groups from the franchise. Colonel Gretton, for instance, proposed that:

> A person shall not be entitled to be registered or to vote as a Parliamentary or local government elector if within the qualifying period such person has been an inmate after conviction, of any prison, or an inmate of any lunatic asylum or Poor Law institution, other than for medical relief, except as a member of the staff of such asylum or institution.[81]

There was alarm about the possibility of the inmates of huge institutions, particularly those sited in sparsely populated areas, voting *en masse* and thereby deciding the outcome of elections.[82] The issue became more acute after it became clear that those who obtained poor law relief for themselves and their dependents would no longer be disqualifed from the franchise. Cave proposed that disqualification should be reinstated for anyone who resided for thirty days in a prison, asylum, or workhouse.[83] However, it was pointed out that this would unfairly disenfranchise many shell-shocked soldiers and people who had been placed in poor law institutions to receive medical care because of lack of hospital space elsewhere.[84] Cave's ideas, which would have introduced a new type of disqualification based on a person's past record of character and mental stability, were given serious consideration but were not accepted.[85] Mr Dickinson argued that it was unfair to add a political penalty to those who had already served a legal penalty, and that, since many people had been imprisoned simply because of their inability to pay fines, to exclude them from the franchise would be a class measure.[86] Similarly, Mr Harvey pointed out that to disenfranchise those who had been 'maintained' as lunatics would discriminate against the poor and not affect those in private asylums[87] (one speaker claimed that it was not unknown for members of the aristocracy to be brought out of private asylums for crucial parliamentary votes).[88] In the end a compromise was reached whereby current residency as an inmate

81 *Parl. Deb.* (HC) 306. 82 Ibid. 297–8.
83 Ibid. 99 (1917): 1017, 1115. 84 Ibid. 85 Ibid. 1117–24.
86 Ibid. 1124. 87 Ibid. 1126. 88 Ibid. 1131.

in an asylum or prison disqualified a person from voting, but left such persons free to vote once they were released.[89]

As the debate demonstrates, it was no longer considered acceptable for the franchise to discriminate according to economic or class status. A logical, though more contentious, implication of this was that a person should not permanently lose this right simply because of a past record as a maintained lunatic, criminal, or pauper. A striking omission in the debate was the position of mental defectives, which scarcely attracted any discussion. This silence is difficult to explain, considering that the 1913 Act was still fresh in the memory of this cohort of MPs, and that the majority of the feeble-minded still resided in the community. A number of possible conclusions can be drawn. The first is that the majority of MPs were relatively unconcerned about the problem; though this does seem to be inconsistent with the alarmist tenor of the debate just five years earlier. The second, and more convincing, possibility is that MPs assumed that mental defectives were already automatically excluded from the franchise because of their permanent mental incompetence. If this is the case, then apart from disqualification through age, mental defect was now the main reason for excluding individuals who resided in the community from the right to vote. As such, the Act did potentially provide a way to exclude some of those whom Cecil described as not even belonging to the working class and who would otherwise have been enfranchised by the 1918 Act.[90]

The fact that there was so little debate about this suggests that the exclusion of such individuals was not yet seen as a class issue. In fact, it is very difficult to determine whether Returning Officers really did exclude mental defectives from voting, and it seems unlikely that they would have found it easy to distinguish high-grade defectives. Probably more important in preventing this vote was the deeply ingrained view among the population as a whole that mental defectives could not and should not vote. This view was already well established in the nineteenth century,[91] and it seems fair to assume that the extension of the definition of mental deficiency in 1913 would have confirmed or even strengthened such prejudices. In 1912, at least one contemporary politician was explicit about the relationship between the mental deficiency and franchise issues:

89 *Parl. Deb.* (Lords), 27 (1917): 94.

90 *Parl. Deb.* (HC), 93 (1917).

91 This is evident in the way that 19th-cent. radical proposals for franchise reforms invariably marked out idiots and imbeciles as groups of people who did not deserve the vote: Biagini, *Liberty Retrenchment and Reform*, 272–4, 287, 311.

as to the Franchise Bill, I confess that I view it as not an altogether unmixed benefit if the extended franchise which the Government propose is to be shared by a very large and increasing number of mentally deficient persons.[92]

It would be wrong to present the 1913 Act as a crudely deliberate attempt to establish a new biological definition of citizenship; however, the ongoing concern over redefining the boundaries of responsible political citizenship should be seen as an important part of the context in which mental deficiency came to be regarded as a serious social problem.

THE NATURE OF THE CONSENSUS

The huge majority behind the 1913 Mental Deficiency Act had been based on two fundamental assumptions. First, it was agreed that some sort of specialized care needed to be provided for the welfare of mental defectives. Secondly, mental defectives were marked out as lying outside the boundaries of responsible citizenship: in need of care and control, with the right to be given humane care, but too irresponsible to exert their own civil rights or to have the right to liberty. These assumptions remained in place, restricting the parameters of debate, and to that extent there was a consensus over policy during the interwar years. However, the assumptions contained enough ambiguities for there to be areas of division and conflict within this consensus.

Apart from occasional announcements by the Minister of Health, the main time that the issue of mental deficiency was raised in Parliament during the 1920s was in private members' questions. A few MPs raised questions because they had a special interest in the problem, as was most notably the case with Leslie Scott, Conservative MP and President of the Central Association for Mental Welfare (CAMW). However, most of the MPs who addressed the issue did so because of cases involving their constituents. They were likely to ask the Minister of Health about the conditions within institutions, whether the rights of patients and their families were being respected, and whether their cases were fairly reviewed, their property protected, and their private correspondence left uncensored. However, the basic lack of freedom for patients was left largely unquestioned; indeed, criticism was far more

[92] Lord Ribblesdale, *Parl. Deb.* (Lords), 13 (1912): 62.

likely to be directed at the lack, rather than existence, of segregative provision.[93] With mental health care being placed under the Ministry of Health in 1919, and with reform of the mental hospital system in the 1920s, there was a growing acceptance of the desirability of expanding mental welfare, and a feeling that an issue such as mental deficiency was a medical rather than a political problem.[94]

The main discussion of mental deficiency in the 1920s was over the 1927 Mental Deficiency Amendment Act. The amendment, passed with far less controversy than the 1913 Act, represented a confirmation of the legislature's faith in dealing with mental deficency through a specialized system of medical care. The Act was, in part, provoked by the encephalitis lethargica epidemic which had swept Britain after the First World War. The disease resulted in symptoms of unruly and backward behaviour in adolescents, symptoms which were very similar to mental deficiency, but could not be dealt with as such, since defectives had to have had their condition from 'birth or early age'. The experience of the epidemic strengthened the tendency to view mental deficiency as a medical, rather than social or moral, problem.[95] The waning power of moral arguments is evident in the failure of campaigns to expand the use of the moral imbecile category. Despite claims that a large number of sexual offenders were moral imbeciles, and the backing of several women MPs, the Departmental Committee on Sexual Offences dismissed such a link, and jurists remained reluctant to certify sex offenders to a life in a mental institution instead of a temporary penal sentence.[96] The replacement of the label 'moral imbecile' by that of 'moral defective' in the 1927 Amendment Act reflected the view that a person should only be certified as a mental defective if he or she had a mental (intellectual) defect in the first place; the moral defect was to be only a secondary factor.

The acceptance of mental deficiency services as medically and socially beneficial is demonstrated by the scarcity of opposition in 1927 to a series of amendments which potentially amounted to a substantial extension of the Act: the requirement that a person should be

[93] See the analysis of mental illness (though not mental deficiency) in Parliament between 1927 and 1932 in: S. Ramon, *Psychiatry in Britain: Meaning and Policy* (London, 1985), 97–126.

[94] For the debate on whether to place mental health under the Ministry of Health: Leslie Scott, *Parl. Deb.* (HC), 112 (1919): 1868–73.

[95] N. Chamberlain, *Parl. Deb.* (HC), 203 (1927): 2344–5.

[96] S. Jeffries, *The Spinster and Her Enemies: Feminism and Sexuality, 1880–1930* (London, 1985), 69–70; *Parl. Deb.* (HC), (1923): 1651–8.

proved to be mentally defective from 'birth or early age' was liberally reinterpreted to include those who had not been detected until the age of 18; education authorities were given greater power to notify school leavers directly to the mental deficiency authorities; it was made easier to obtain care for defectives who came from homes where the care was already adequate; and local authorities were compelled to provide training facilities in day centres for those defectives who remained in the community.[97]

When the amendments had first been discussed, Josiah Wedgwood, now on the Labour benches, had put up his usual staunch libertarian objections.[98] However, in the final debate, though still highly suspicious that the legislation unnecessarily infringed individual liberties and was based on dubious eugenic premises, Wedgwood was forced to admit that there were some beneficial aspects to the legislation, such as its encouragement of community care, and he therefore reluctantly acceded, acknowledging: 'It is merely continuing and accentuating proposals of previous measures.'[99] Wedgwood's libertarian concerns carried even less force than they had in 1913. Minister of Health, Neville Chamberlain, dismissed him as an 'eccentric' and 'fanatic', who had not visited a mental deficiency institution for the last twenty years and was therefore unaware of the great improvements which had taken place.[100]

The 1927 Amendment Act coincided with the Report of the Royal Commission on Lunacy.[101] Although this commission studiously avoided the issue of mental deficiency, its recommendations on making mental treatment more easily available, on a voluntary and temporary basis, contributed to the new optimism in the medical benefits of psychiatric care which was evident in the debates on the 1927 Act. Henry Snell, Labour MP, and a member of the Royal Commission on Lunacy, epitomized this optimism, overcoming his concerns about liberty of the subject on the basis that expansion of services was necessary for defectives' 'higher liberty' and to 'enable them to enter into their rightful heritage of the kingdom of the mind'.[102] In a telling shift, it was now argued that reform was necessary not so much to protect the civil liberty of the defectives, but to make doctors more

97 *Parl. Deb.* (HC), 202 (1927): 3219–24. 98 Ibid. 2319. 99 Ibid. 2337.

100 Ibid. 2343–5.

101 *Report of the Royal Commission on Lunacy and Mental Disorder* (Cmd. 2700, xiii. 373, 1926).

102 *Parl. Deb.* (HC), 202 (1927): 2355. Snell had been a COS agent in the 1890s, a lecturer to the Fabian Society, and Chairman of the British Ethical Union.

legally secure when certifying defectives.[103] There was far greater respect given to the views of experts than there had been in 1913. Many of the speakers demonstrated some personal experience or understanding of the mental deficiency services, and several medic MPs made influential contributions. Most of those speaking admitted that many aspects of the system were currently undesirable—there were too few places in institutions, there were doubts about the standard of care, and there was a desire for more flexibility and a greater flow between institutions and the community—but they supported the amendments as a way to improve and expand the system of care. The degree to which the issue was now being driven by a medical and administrative dynamic was perceived by R. Morrison, who pointed out that the real question was not whether Parliament passed the reform or not, but whether, and in what way, the law was interpreted and implemented.[104]

Beneath this apparent consensus over policy there were significant differences between Labour and Conservative MPs. Although, the Conservative Government expressed its confidence that the Labour Party would see the expansion of mental welfare as a positive step, there was evident concern that Labour MPs would object if they suspected that the Act unduly infringed the liberties of their working-class constituents.[105] This was evident in T. P. Beamish's provocative argument that there was a basic contradiction in Labour's attitude to many welfare issues:

> it is a very curious thing that the Labour Party, who in certain respects are tyrannical in their endeavours to control people, the very instant there is a suggestion of anything in the nature of vaccination or purification of the race or its improvement, are inclined at once to get up and express the most extreme fears that the liberty of the individual is being interfered with.[106]

In fact, most of the Labour members who spoke, though evidently sensitive and sympathetic towards protecting working-class liberties, did not feel that the problems were sufficient to justify opposing a measure which offered to improve the delivery of services.[107] Labour

[103] Sir Douglas Newton, ibid., 2317; Dr Vernon Davies, ibid., 2381–4; C. Fremantle, ibid. 2387.

[104] Ibid. 2389.

[105] Crompton Wood, ibid. 2355.

[106] Ibid. 2355.

[107] Similarly, during the debates over the 1930 Mental Treatment Act (under a Labour Government) the main concerns about extension of the power of experts came from the Labour side, but this was not strong enough to result in a significant opposition: C. Unsworth, *The Politics of Mental Health Legislation* (Oxford, 1986), 224–9.

member Harry Charleton, himself the father of a mentally defective child, questioned the assumption that working-class parents were eager to place their defective children in institutions, and expressed concern over the standard of some of the care in these institutions, but he preferred to support the reform of the system rather than oppose institutions altogether.[108] Another Labour MP suggested that the weakness in the 'liberty of the subject' position had been exposed by the complaints from families of mental defectives after Wedgwood's objections had delayed the bill the previous year.[109] The Party therefore supported extension of mental deficiency services as part of its broader programme for increased hospital and medical provision.[110]

Only rarely did Labour MPs show signs of following Wedgwood in arguing that the solution was to remove poverty, rather than provide a system of specialized care. Ironically, considering that he had signed the 1909 Minority Report on the Poor Law, one of these critics was George Lansbury. Commenting on reform of the poor law in 1929, Lansbury argued that the way to prevent disease was not to provide more and more experts, but to improve standards of living. He went on to extend this argument to the problem of mental deficiency:

> I have never heard such rubbish and nonsense talked and written as I have heard on this particular subject of the mentally deficient. I am not a medical man, and, I thank God, I am not an expert either, but I know perfectly well that where I live most of the children come from the same kind of homes that the phthisical cases come from and the others who suffer from poverty diseases. I believe that often it is the worry, the struggle to live, the being pressed down because they have not the means to get their daily bread, that creates in a woman, when she is carrying a baby, the sort of things that bring about mental deficiency in the child when it is born.[111]

The problem with Lansbury's position was that, in stressing the link between mental deficiency and poverty, it was in danger of stigmatizing the poor and lending support to those who wanted to portray mental

[108] H. Charleton, *Parl. Deb.* (HC), 203 (1927): 2357–8. [109] Ibid. 2365.

[110] In 1922 the Labour National Executive complained to the Government about the lack of places in mental deficiency colonies: PRO Treasury Files (T), 161, 149/5, 13551/01, 11 Jan. 1922. On Labour's support for expansion of general hospital provision: The Labour Movement, *The Labour Movement and the Hospital Crisis. A Statement of Policy* (London, 1922); The Labour Party, *The Hospital Problem* (London, 1924); C. Webster, 'Labour and the Origins of the National Health Service', in N. A. Rupke (ed.), *Science, Politics, and the Public Good: Essays in Honour of Margaret Gowing* (London, 1988), 184–202.

[111] *Parl. Deb.* (HC), 227 (1929): 1609.

deficiency as something which was confined to the poorer classes. Moreover, in explaining mental deficiency as a problem of poverty, the argument was in danger of weakening the case for increased provision of mental deficiency services which were seemingly in desperate demand. For these reasons, the majority of Labour members, though recognizing that poverty contributed to the problems of the mentally defective, kept such arguments in the background and continued to support expansion of services.[112]

Conservative MPs tended to have rather different reasons for supporting the extension of mental deficiency provision. In the 1927 debate it is apparent that concerns over the health of the race and an interest in eugenics were far more important among the Tory speakers, and that such concerns had not disappeared since 1913, though they had been modified. There now seems to have been a greater confidence in basing arguments on the expertise of eugenics. By contrast, the moral dimension of the argument for segregation of defectives, which had been so strong in 1913, was now less marked, or perhaps less openly acknowledged. But Tory concerns were not simply biological and eugenic: there was also a continuing anxiety about the fitness of mental defectives as political citizens. Debates about citizenship provided a framework which could embrace both biological and political arguments for segregation. This is made clear in the comments of Conservative MP, Mr Smithers:

> Upon whose character does happiness of our country depend to-day? It depends upon the character of every man and woman who has a vote, and that is why it is so important to lead public opinion to tighten-up the possibility of people who are not fit for it bearing the responsibilities of citizenship through their votes, and by measures such as this, restricting the possibility of mentally-deficient people being produced by mentally-deficient parents.[113]

There was, necessarily, a tension in the Conservative position. The evidence presented in the 1927 debate made it clear that if the system of care was to work effectively it demanded far more resources from central and local government. The 'interests of the community' argument was strong enough to convince most of the Party that expenditure on providing care for defectives was justifiable, but there were some signs of this consensus fracturing at the margins. For instance, in 1926,

[112] A similar argument has been made for Labour attitudes on policy for the mentally ill: Ramon, *Psychiatry in Britain*, 118.

[113] *Parl. Deb.* (HC), 202 (1927): 2364–5.

James Erskine of the Anti-Waste League enquired whether sterilization of mental defectives, as practised in the certain areas of the United States, might provide a cheaper and eugenically effective solution to the problem.[114] Similar questions were raised by other Tory MPs in the late 1920s.[115]

It seems that the tenor of the debate was becoming more conflictual by 1930. The publication of the Wood Report on Mental Deficiency, revealing a substantial increase in the estimated scale of the problem and associating mental deficiency with 'the social problem group', surely contributed to this.[116] So did the atmosphere of class politics surrounding the General Strike and the escalating levels of unemployment. Pressure to restrict levels of public expenditure made the option of a sterilization policy harder to ignore. And this, in turn, was to provoke the type of response made by Lansbury in 1929—that the real issue was to address the causes of economic inequality and poverty, not to throw more money into provision of expert services.[117] The Labour Government, faced by the awkward revelation that the expense of supplying new institutions for mental defectives was more per person than building new working-class homes, was forced to consider ways to reduce the cost of care.[118] Just how serious the fracturing of consensus really was would be tested during the 1930s, as the issue of sterilization came to be openly debated in Parliament.

THE STERILIZATION DEBATE

On 21 July 1931, a short bill proposing voluntary sterilization of mental defectives was introduced by Labour MP, A. G. Church. Fellow Labour MP, Dr Hyacinth Morgan spoke against the motion, and the bill was defeated by 169 votes to 89.[119] A breakdown of the vote provides one of

[114] Ibid. 191 (1926): 302.

[115] Dalrymple White, *Parl. Deb.* (HC), 214 (1928); Sir Berkely Sheffield, ibid. 1204.

[116] *Report of the Interdepartmental Committee on Mental Deficiency, 1925–29 (Wood Report)*, 3 vols. (London, 1929). On the social problem group: J. Macnicol, 'In Pursuit of the Underclass', *Journal of Social Policy*, 16 (1987), 293–318.

[117] *Parl. Deb.* (HC), 227 (1929): 1609. Labour suspicion of experts helps to account for Labour MPs providing the main opposition to the Mental Treatment Act of 1930 (while it was still a Labour Party Bill): Ramon, *Psychiatry in Britain*, 97–126.

[118] PRO MH 58/97, Robinson to Minister of Health, 22 June 1931.

[119] *Parl. Deb.* (HC), 255 (1931): 1249–56.

the few insights we have into the attitudes of politicians on this issue. Support for sterilization crossed party lines: it included 43 Conservatives, 3 Independents, 13 Liberals, and 31 Labour MPs. On the other hand, the main opposition to the bill came from the Labour Party.[120]

Support

To understand the support for the bill, it is worth looking in some detail at the case presented by Church.[121] The first point which he stressed was that sterilization was to be a voluntary option: the operation would only be performed after gaining the consent of defectives and their parents or families and the approval of the government and the judiciary. Secondly, Church presented the operation as a breakthrough in medical progress: it had been thoroughly tested in California, would not endanger the patient physically or emotionally, and was in no sense an act of mutilation. Moreover, the safety and the need for the operation had expert support from the Eugenics Society, local authorities, the medical profession, and social workers. Thirdly, again co-opting the authority of 'expert' opinion, Church argued that the Wood Report had demonstrated that the defective population was far larger than hitherto realized and was inexorably increasing. In such circumstances, segregation of all defectives was no longer economically feasible. The majority would inevitably remain in the community, free to breed, and a dysgenic threat to the nation, unless sterilization could be legalized.

The presence of forty-three Tory MPs among the supporters of the bill is not surprising. There had been growing signs of interest in sterilization in the late 1920s, and the economic rationale was stronger than ever in the midst of the financial crisis of 1931. The support of some Tories reflected an active involvement in the social hygiene movement. For instance, Basil Peto was former Honorary Secretary of the Society for the Protection against Venereal Diseases, and F. E. Fremantle was a former Medical Officer of Health and an advocate of preventive medicine.[122] However, the majority supported the measure for a

[120] Labour's role in defeating sterilization is emphasized in the secondary literature on the debate, but inadequate attention is given to the all-party nature of the support: G. Jones, 'Eugenics and Social Policy between the Wars', *Historical Journal*, 25 (1982), 717–28; Macnicol, 'Eugenics and the Campaign for Voluntary Sterilization', 147–70.

[121] *Parl. Deb.* (HC), 255 (1931): 1249–52.

[122] Jones, *Social Hygiene*, 43–52. Fremantle was MOH for Hertfordshire, 1902–16, and President of the Epidemiological Section of the Royal Society of Medicine, 1928–30.

much more confused mix of economic, moral, and eugenic rationales, each of which was given greater urgency by the findings of the Wood Report. The Report's claim that mental deficiency was concentrated in an expanding social problem group suggested that sterilization might be the key to preventing a whole host of problems among those at the bottom of the social scale, and would help to offset the supposed dysgenic effect of falling middle-class fertility.

The presence of the thirty-one Labour supporters is more surprising. Their support suggests that they, at least, did not view sterilization as an anti-working-class measure. It seems likely that those who supported the bill accepted Church's portrayal of sterilization as a purely voluntary measure, based on firm scientific and medical expertise, and with safeguards to protect against abuse. As such, their support differed little from that given by the majority of Labour MPs for the Mental Deficiency Act. However, Church's motion was also based on two other rationales—saving money and preventing the reproduction of future generations of defectives. This makes the support of these Labour MPs more difficult to explain.

The Labour support can be explained, in part, as a reflection of the class attitudes of middle-class, and to a lesser degree 'labour-aristocrat', members within the Party. This constituency was more likely to be concerned about preventing the development of a 'social problem group'—a social residuum which lay outside of the 'respectable' working class, and was a financial drain and a moral and eugenic threat to the rest of the working class. Such anxieties help to account for the strong support for sterilization which was reported to exist among 'the weekly wage earners, who were most directly affected, both socially and economically, by the subnormal people among them'.[123] From white-collar, middle-class, and intellectual backgrounds, the Labour MPs who supported the sterilization bill were socially distant from the realities of working-class life, and were likely to be less sympathetic or sensitive than other Labour MPs to the interests of those at the bottom of the social scale. These MPs were also drawn to sterilization because in some respects it appeared to be a 'progressive' policy: it showed how the latest medical technology could be directly applied to solving a social problem, efficiently saving the expense of institutional care, and humanely enabling mental defectives to remain in the community; the

[123] Blacker Archive, Wellcome Unit for the History of Medicine, Oxford (BL), 2.2.13, *Report of the Voluntary Eugenics Sterilization Conference* (23 May 1932), 20, 32.

prominence of religious moralists in opposition added to the impression that this was an issue of progress against taboo.

Church provides a good example of the middle-class and technocratic orientation of those Labour MPs who supported sterilization. He had been Private Secretary to the Fabian Sidney Webb in Labour's 1924 Government. He was General Secretary of the National Union of Scientific Workers, a group which subsequently deregistered from being a union to become a professional organization—the Association of Scientific Workers. And he was one of the founders of the monthly, *The Realist*, where he attacked the Labour Party and the unions for their lack of support for science. Church's impatience with the Party was again evident when he was one of the few Labour MPs to follow Ramsay MacDonald into the National Government in 1931.[124] G. Lathan, voting alongside Church on the Labour benches, had a similar type of background, serving as President of the National Federation of Professional Workers from 1921 to 1937. Sterilization also attracted the support of planners, intellectuals, and perhaps a younger generation within the Party. Such supporters included: Somerville Hastings, President of the Socialist Medical Association; George Benson, who became Chairman of the Howard League for Penal Reform from 1938 to 1961; James Horabin, the journalist and Fabian; and William Henderson, journalist and son of Arthur Henderson (who, by contrast, voted against sterilization).

The fact that the highest proportion of supporters from any one party came from the Liberal Party again suggests that support for sterilization reflected a middle-class progressivism. This progressivism crossed party lines, embracing such significant figures as the former Liberal and now Labour President of the Board of Education, C. P. Trevelyan, the Liberal E. D. Simon, and the Independent Eleanor Rathbone.[125] As the progressive journal, the New Statesman, explained:

> The legitimate claims of eugenics are not inherently incompatible with the outlook of the collectivist movement. On the contrary, they would be expected to find their most intransigent opponents amongst those who cling to individualistic views of parenthood and family economics.[126]

Sterilization also seems to have been one of those issues which attracted people who were increasingly dissatisfied with the strife of a class-based

[124] G. Werskey, *The Visible College* (London, 1978), 41; Jones, *Social Hygiene*, p. 106.

[125] M. Freeden, 'Eugenics and Progressive Thought: A Study in Ideological Affinity', *Historical Journal*, 22 (1979), 645–71.

[126] 'Sterilisation of Defective', *New Statesman and Nation* (25 July 1931), 102–3.

party politics and looked instead to a new politics of the centre based around scientific planning.[127] In this respect, it is interesting to see that two MPs associated with Oswald Mosley's New Party voted in favour of the 1931 Bill.[128]

The final significant trend revealed by the vote was the strong support from women. Although only three women MPs voted on the issue, they all supported sterilization. As women, and holders of independent and strong views, all three—Ellen Wilkinson of the Labour Party, the Independent Eleanor Rathbone, and the Duchess of Atholl of the Conservative Party—can again be seen as occupying a position on the margins of the party system.[129] Their support cannot be dismissed as merely coincidental, for it was matched by strong female support for the issue throughout the decade: of twenty-eight major organizations supporting sterilization, nine were women's groups; and of 335 smaller groups, over half were women's organizations, while many of the remainder had high female membership.[130] Neither was their support insignificant, for, although women were still underrepresented in Parliament, they made up 52.8 per cent of the electorate by 1931, making it increasingly important to address women's concerns.[131] Their support gains further significance when it is borne in mind that women rather than men were likely to be the main target, despite the greater danger involved.

[127] A. Marwick, 'Middle Opinion in the Thirties: Planning, Progress, and Political Agreement', *English Historical Review*, 79 (1964), 285–98.

[128] Major C. R. Dudgeon was a Liberal at the time of the vote. Evelyn John St Loe Strachey was a Labour MP until he became a member of the New Party from February to June 1931, and then an Independent, rejoining Labour in 1943. His support for sterilization is given added significance by the fact that in the 1900s he had been editor of the *Spectator*, a great defender of libertarianism.

[129] They each had a strong personal interest in medical and social issues. The Duchess of Atholl served as a member of the Departmental Committee on the Highlands and Islands Medical Services, 1920–4, and was the first Conservative woman to hold a government office as Parliamentary Secretary of the Board of Education, 1924–9. Her independence is shown by the withdrawal of the Tory Whip in 1935, and her resignation from the party to stand as an independent in 1938. Ellen Wilkinson had been a member of the National Union of Women's Suffrage Societies, 1913–14, and was Parliamentary Private Secretary to Susan Lawrence in the Ministry of Health, 1929–31. Eleanor Rathbone is best known for her campaigning on the issue of family allowances during the interwar period, and was President of the National Union of Societies for Equal Citizenship, 1919–29.

[130] C. P. Blacker, 'Voluntary Sterilization: The Last Sixty Years', *Eugenics Review*, 54 (1962), 9–23.

[131] B. Harrison, 'Women in a Men's House: The Women MPs, 1919–1945', *Historical Journal*, 29 (1986), 623–54; M. Pugh, 'Popular Conservatism in Britain: Continuity and Change, 1880–1937', *Journal of British Studies*, 27 (1988), 254–84.

To understand the strength of female support we need to appreciate that many women saw positive advantages in the operation.[132] Women tended to be strong supporters of the extension of medical and social welfare services, and saw sterilization as part of this broader welfare platform. Many women (and some men) supported sterilization as part of their campaign to make birth control freely available to all women; for instance, the birth control activists Eleanor Rathbone, Ernest Thurtle, and E. D. Simon all supported Church's bill.[133] As Labour's Dr Edith Summerskill argued, as late as 1936, sterilization could be defended on purely medical grounds, 'as a form of preventive medicine'.[134] Women tended to be particularly concerned about the strain and suffering which could be placed on mothers who bore mentally defective children, and felt that it was sensible and humane to try to prevent this. They also tended to feel strongly that it was socially irresponsible to allow feeble-minded women to become mothers—not necessarily on eugenic grounds, but because they would be bad mothers, breeding large families who would never receive proper care. Sterilization would provide a much more effective form of birth control than other types of contraception, which depended on the responsibility of the mother, and would improve the lives of mentally defective women, enabling many of them to remain in the community. Such ideas crossed class and political boundaries. Strong support for sterilization even came from the members of the National Conference of Labour Women and the Women's Cooperative Guild, who passed resolutions in favour of making voluntary sterilization freely available to all, not just to the rich who could obtain the operation already through private practice.[135] Eleanor Barton, Secretary of the Women's Cooperative Guild, was wary about

[132] The level of support from women for Nazi sterilization has been vigorously debated between C. Koontz and G. Bock, as summarized in A. Grossmann, 'Feminist Debates About Women and National Socialism', *Gender and History*, 3 (1991), 350–8. Bock argues that sterilization gained little support from women and was primarily directed against them: 'Antinatalism, Maternity and Paternity in National Socialist Racism' in G. Bock and P. Thane (eds.), *Maternity and Gender Politics: Women and the Rise of the European Welfare States, 1880–1950* (London, 1991), 233–55. Whatever the case in Germany, the British case shows that women could be vigorous supporters of sterilization.

[133] Thurtle's support is given added interest by his marriage to the daughter of George Lansbury (who had spoken against sterilization in 1929) and his position as Private Secretary to Lansbury in 1929.

[134] 'Legalisation of Voluntary Sterilisation', *Manchester Guardian* (22 June 1936).

[135] *Voluntary Sterilization Conference Report, 1932*, p. 30; for support of women's Labour groups: Blacker, 'Voluntary Sterilization', 18–19; Jones, *Social Hygiene*, 103–4; 'Labour Women's Support', *Manchester Guardian* (22 June 1936).

sterilization being applied to non-defectives, but was adamant that: 'The biggest problem was the case of the mentally defective women who kept on having children. That obviously was a case where sterilization was necessary.'[136]

Opposition

Although the Labour support for sterilization is interesting, far more important was the fact that the majority of Labour members voting opposed the bill. It is not difficult to explain this opposition. Labour MPs might accept that the liberty of mental defectives was not sacrosanct and therefore support a policy of institutional care, but most of them drew the line at sterilization. Morgan, speaking against the bill, set out Labour's main objections. Like Wedgwood, he had considerable doubts about whether the measure adequately safeguarded the liberty of the subject; in particular, he questioned whether sterilization could really be 'voluntary' when the subjects of the operation were, by definition, people who lacked the mental ability to make responsible decisions about their own lives. In his view, sterilization was an anti-working-class measure. By this, he meant that it was the type of policy which would, in reality, only be applied to mental defectives in the working class; he did not go as far as arguing that it was a eugenic attack on the working class as a whole. He was sceptical about the scale of the problem, but he did not deny that mental deficiency existed. Indeed, one of his main objections to sterilization was that it might provide a cheap alternative which diverted resources from institutional care. Moreover, releasing sterilized defectives into the community would leave them to live 'degraded lives, spreading disease, mainly venereal, still impulsive sexually and a dangerous menace to innocent women and children, spreading bad habits of hygiene and citizenship, drifting to the lowest and possible criminal haunts of society like the damned in Dante's "Inferno"'. As this notably unsympathetic view reveals, Morgan himself believed in a close relationship between mental deficiency and the 'social problem group'. However, like Lansbury in 1929 and Eicholz in 1908, he argued that the real way to prevent mental deficiency was not through controlling breeding, but through removing the

[136] *Voluntary Sterilization Conference Report, 1932*, p. 30. On Mrs Barton: J. M. Bellamy and J. Saville (eds.), *Dictionary of Labour Biography* 9 vols. (London, 1972–93), i. 38–40.

conditions of poverty which produced a defective 'germplasm'. Morgan concluded his colourful speech by condemning the proposal as a 'pagan, anti-democratic, anti-Christian, unethical Bill'.[137]

In the tense political atmosphere of the economic crisis of the 1931, the majority of Labour members were suspicious of sterilization as a cost-cutting measure.[138] Such feelings were shared by the Labour Government of 1929–31. Arthur Greenwood, the Minister of Health, dragged his feet in reacting to the Wood Report.[139] And the Labour Party Research Department raised a whole series of objections. It regarded sterilization as an economy measure, which would only be used on the poor, and which would divert attention away from an environmental approach to the problem. Like many others, it also pointed out that it would be highly imprudent to act as long as scientific research on the subject remained so primitive and contentious.[140]

Labour opposition gained an important extra dimension from the 'Christian' objections, mentioned in passing by Morgan—himself a Catholic. The strong non-conformist lobby within the Party was generally opposed to sterilization: in total, thirteen of the Labour opponents of the bill had a history of serving as lay preachers or ministers, while only one such member supported the proposal. But the 'non-conformist conscience' within the Labour Party was more important at an individual rather than institutional level.[141] International comparison

[137] *Parl. Deb.* (HC), 235 (1931): 1254. Morgan's objections carried both political and medical weight as he was medical adviser and consultant specialist on industrial diseases to the TUC General Council, and a member of the Council of the BMA.

[138] The run on gold started on 15 July, reaching a peak in the last week of July: R. Skidlesky, *Politicians and the Slump: The Labour Government of 1929–31* (London, 1967), 339. On sterilization as an economy measure: Searle, 'Eugenics and Politics', 159–69.

[139] PRO MH 58/103, Greenwood to Scott, 9 Jan. 1930, Brock 12 Nov. 1929; MH 79/292, Brock to Robinson 16 Mar. 1932. Greenwood himself had researched into the questions of juvenile labour and deterioration between 1908 and 1911, paying particular attention to environmental factors though not discounting hereditary work: Greenwood Papers, Bodleian Library, Shelf 1.

[140] PRO MH 58/103, n. 168a, Labour Party Research and Information Department, Public Health Advisory Committee, 'The Causes and Prevention of Mental Deficiency' (July 1929). On the other hand, it should be noted that the Labour Party never made an explicit policy statement against sterilization. Labour's response was tame and mute compared to that of G. K. Chesterton's Distributist League: PRO MH 58/103, Birmingham Branch of the Distributist League to A. Greenwood, Sept. 1929.

[141] P. Caterall, 'Morality and Politics: The Free Churches and the Labour Party between the Wars', *Historical Journal*, 36 (1993), 67–85. The influence of this individual conscience is indicated by the estimate that seventy-four Labour MPs in 1935 were teetotallers: S. G. Jones, 'Labour, Society, and the Drink Question in Britain, 1918–1939', *Historical Journal*, 30 (1987), 105–22.

would suggest that Protestantism was less likely to prove a barrier to sterilization than Catholicism. By 1931, Catholics constituted about two million of the population of Great Britain, and the majority of them, being of Irish origin and working class, were usually Labour Party supporters.[142] Although this constituency was probably too small and too poorly organized to be crucial in deciding elections, its importance was accepted within the Labour Party.[143] Moreover, during the 1920s Labour's attention to Catholic interests on welfare issues had been fostered by the Catholic, John Wheatley, who was Labour's first Minister of Health in 1924.

The Catholic Church made its hostility towards sterilization known very early on through announcements in the Catholic press, and placed considerable pressure on some Labour MPs.[144] In 1934, the *Catholic News* warned: 'The Labour Party has suffered more disadvantage than it is probably aware of because of its association in the public mind with the promotion of sterilisation.'[145] The *Universe* felt the call to vigilance was having some success: 'All over the country Catholic citizens must be on the alert and act promptly, as has been done in London and Northumberland, and also elsewhere.'[146] The influence of the Catholic vote is most apparent in areas with a concentrated Catholic constituency. In Glasgow, for instance, all five Labour MPs voted against the 1931 motion, and a rally backed by 200,000 Catholics against sterilization in 1934 gained the support of three of the city's MPs.[147] Both Morgan and his seconder, David Logan, were Catholics. Logan came from the Scotland Division of Liverpool where there was a large Irish population. In total, six Catholic MPs opposed the 1931 motion, with none voting in favour. It was this Catholic opposition, rather than that of Labour, which the Eugenics Society felt would be hardest to overcome. As C. P. Blacker, the Secretary of the Eugenics Society, argued as late as 1937: 'it is very

[142] C. Cook and J. Stevenson, *Longman Handbook of Modern British History, 1714–1987* (London, 1988), 186.

[143] R. Swift and S. Gilley (eds.), *The Irish in Britain, 1815–1939* (London, 1985). Labour's defeat at the Stockport by-election of 1920, after W. O'Brien had split the Irish vote in protest against Labour for ignoring the Home Rule issue, had alerted Labour's leaders to the importance of the Irish vote: C. Howard, 'Expectations Borne to Death: Local Labour Party Expansion in the 1920s', in J. Winter (ed.), *The Working Class in Modern British History* (Cambridge, 1983), 67.

[144] J. Dwyer 'The Catholic Press, 1850–1950', in G. A. Beck (ed.), *The English Catholics* (London, 1950), 473–514. PRO, Ministry of Health files (MH) 58/103, 26 Feb. 1929, letter from Salford Diocesan Catholic Federation.

[145] PRO MH 51/541, *Catholic News* (14 July 1934).

[146] PRO MH 51/557, *The Universe* (3 Aug. 1934).

[147] PRO MH 51/557, *Catholic Times* (9 Mar. 1934); *Catholic News* (7 Apr. 1934).

difficult to get through any eugenic legislation if the Labour Party is strongly opposed. I do not myself believe that their opposition is implacable like that of the Catholics.'[148] Labour's ambivalent attitude towards sterilization has parallels with its position on birth control during the 1920s. There was strong support for the extension of birth control among women in the Party;[149] however, the Party hierarchy was very cautious, both because of its Catholic constituency, and because of the generally morally conservative and patriarchal views of working-class men.[150] Thus when a bill to enable local authorities to support birth control clinics was voted on in 1926, the main support came from the Conservatives rather than Labour.[151]

Perhaps foolishly, the Secretary of the Eugenics Society, C. P. Blacker, himself a Catholic by background, had thought that the Catholic opposition to sterilization might be overcome if the operation was presented as being in the medical interests of women, rather than as a birth control measure[152] and as a voluntary measure, with Catholics free to decline the operation. However, Catholic opinion was definitively turned against sterilization on doctrinal grounds after the papal pronouncement of 'Casti Connubi', which condemned all artificial contraceptive practice on the theological grounds that 'private individuals have no other power over the members of their bodies than that which pertains to their natural ends', and that it was a sin to interfere with the sacrosanct nature of marriage. 'Casti Connubi' can be seen as a reflection of the Catholic Church's anxieties in an age of secularization: by making such a definitive statement on this most intimate aspect of its followers' lives, it reasserted the Church's control against the encroaching power of the secular state, and it provided a bulwark to defend its own constituency against fertility decline. Since most British Catholics were working class, the Catholic lobby combined its theological and moral arguments with an attack on the class bias of the proposed legislation, and thereby provided a glue which united the otherwise odd bedfellows of the Catholic Church and and the Labour Party.[153]

[148] BL 1.3.74. Blacker to L. Darwin, 24 Mar. 1937.

[149] P. Thane, 'The Women of the British Labour Party and Feminism, 1906–1945', in H. Smith (ed.), *British Feminism in the Twentieth Century* (Aldershot, 1990), 136–8.

[150] R. Soloway, *Birth Control and the Population Question in England, 1877–1930* (London, 1986), 280–333.

[151] Ibid. 290–2.

[152] BL 1.3.70, Blacker to Darwin, 22 Mar. 1937; BL 1.1.53, Darwin to Blacker, 20 Nov. 1930.

[153] L. Fairfield, *The Case Against Sterilisation* (Catholic Truth Society: London, 1934), 4–5.

The final key to the defeat of sterilization in 1931 was the low number of Tories who voted: the lobby fell from 387 on the previous vote, to just 156. The majority of those not voting were Conservatives: although 44 Conservatives voted for sterilization, just 24 voted against it; by contrast, 31 Labour MPs voted for the Bill and 130 opposed it. It is not clear how this should be interpreted. It may be the case that Tories were simply uninterested and absented themselves from the chamber, or they may have been present but abstained because they found the vote awkward for one reason or another. Whatever the case, the majority of Tories chose not to vote for the bill. Therefore it seems that most Conservative MPs agreed with the majority of the opposition that sterilization was either unnecessary, did not replace the need for institutional care, was premised on some still unproven scientific arguments, or raised too many libertarian problems; as such, the consensus over mental deficiency policy, though shown to be seriously fractured in 1931, still held.

THE NATIONAL AND CONSERVATIVE GOVERNMENTS OF THE 1930s

After the collapse of the Labour Government in the financial crisis of 1931, the party political environment took on a new complexion. The Labour Party was seriously weakened, as it split between those who followed its leader Ramsay MacDonald into the National Government and those who withdrew. Labour performed abysmally in the subsequent general election of 1931, leaving the Conservative Party to dominate the National Government. In the general election of 1935, Labour recovered somewhat, but the Conservatives still won by a landslide and formed their own government. Thus, from the summer of 1931 onwards, Parliament was dominated by the Conservatives. Under these changed conditions, and as a government strategy document pointed out in 1932, the failure of Church's 1931 bill 'could not be taken as an indication of what might happen to a similar bill in the present House of Commons if properly handled'.[154] Despite this, and despite the positive recommendations of the Interdepartmental (Brock) Committee on Sterilization, which reported in 1934, neither the National nor the Conservative Governments made any attempt to introduce a sterilization bill.

[154] PRO MH 58/103, Sterilization strategy document (*c.*1932), app. A, p. 8.

Although its political position had been severely weakened, the Labour Party remained an important bulwark against the introduction of a sterilization bill. The Eugenics Society attempted to draw sympathetic Labour Party members into its sterilization campaign through funding the National Workers' Committee for Sterilization, but it was generally only able to attract those, such as MP Ernest Thurtle and Labour women's groups, who supported sterilization as part of a broader campaign for extension of birth control technology. Similarly, of the ten members of the Parliamentary Committee of the Joint Committee on Voluntary Sterilization, only one member—Holford Knight—was from the Labour Party.[155] Meanwhile, opposition to sterilization emerged in the trades union movement, when George Gibson of the National Asylum Workers' Union (NAWU) brought forward a resolution at the Trades Union Congress of 1934 to delay any legislation on sterilization until the social aspects of the question had been fully considered.[156] Although Gibson's was a lone voice in the trades union movement, his statement, like the ongoing opposition from Catholics, made it clear that powerful groups within civil society would make sure that the Labour Party resisted any attempt to introduce sterilization. In such circumstances, and weakened by the split of 1931 and the dominance of the unions within the Party, Labour's intellectuals and planners were unwilling to repeat their support of 1931.[157]

Also relevant in explaining the lack of success of the sterilization campaign was the relative stability of the British party system during the 1930s, which continued to be dominated by the existing parties rather than new extremist parties that might have been less inhibited in supporting radical policies such as sterilization. A few of the dissatisfied intellectuals of 'middle opinion' drifted out of the mainstream of party politics and into the fascist New Party established by former Labour MP Oswald Mosley, but even the New Party was forced to act within the confines of the British ideology of liberty of the individual and was therefore reticent to give full support to sterilization. With his populist antipathy to the state's 'hag ridden' interference with privacy,[158] Mosley

[155] Jones, *Social Hygiene*, 103–4; Macnicol, 'Eugenics and Voluntary Sterilization', 162–3.

[156] *Report of the 66th Trades Union Congress (1934)*, 399–400; Jones, *Social Hygiene*, 102.

[157] B. Pimlott, *Labour and the Left in the 1930s* (Cambridge, 1977); G. D. H. Cole, *Fabian Socialism* (London, 1943), 80, 84–5.

[158] O. Mosley, *The Greater Britain* (London, 1934), 49–55.

restrained himself from supporting compulsory sterilization or birth control; rather, he argued that the greater social awareness of citizens within a fascist state would ensure that they voluntarily directed themselves towards eugenic ends.[159] Although the scientific aura of eugenics should have been in sympathy with Mosley's modernizing tendencies, there is little evidence to show direct links between the New Party and the Eugenics Society.[160] Collaboration was hampered by the British Union of Fascist's anti-intellectualism, its violence, which placed it outside the political mainstream, and its crude anti-Semitism, which was not generally shared by the British eugenics movement.[161]

Most crucial in explaining the political failure of sterilization was the lack of official support from the Conservative Party. In fact, support for sterilization among Conservatives showed no obvious sign of dying away after 1931; if anything, it became stronger in the middle of the decade. Conservative MPs Hugh Molson and Wing Commander A. W. H. James both attempted to introduce private member's bills in 1934.[162] And the Parliamentary Committee of the Joint Committee on Voluntary Sterilization was also dominated by Conservatives, including James, Sir Basil Peto, the Duchess of Atholl, Vyvyan Adams, and C. T. Culverwell.[163] A number of indicators suggest that support for sterilization was strengthening by 1934. A Eugenics Society survey of the views of

[159] O. Mosley, *Fascism: 100 Questions Asked and Answered* (London, 1936), Q. 76.

[160] On Mosley as modernizer see R. Skidelsky, *Oswald Mosley* (London, 1975). P. Hayes, 'The Contribution of British Intellectuals to Fascism', in K. Lunn and R. Thurlow (eds.), *British Fascism: Essays on the Radical Right in Inter-War Britain* (London, 1980), 168–86, argues that the Galtonian tradition should be seen as an important part of the fascist tradition. However, this must be distinguished from fascism in the institutional sense of the BUF or other parties where the link is tendentious and where the anti-Semitism was not based upon eugenics. R. Griffiths, *Fellow Travellers of the Right* (Oxford, 1983) shows little direct linkage between fascists and eugenics in Britain, his only mention being Pitt Rivers (p. 323). There is some evidence of a dialogue between Blacker and William Joyce: BL 5.1.1, Joyce to Blacker 7 Nov. 1939.

[161] On the anti-intellectual style of BUF: D. S. Lewis, *Illusions of Grandeur: Mosley, Fascism and British Society, 1931–81* (Manchester, 1987), 78. Despite the occasional anti-Semitic comments from Eugenics Society members in private, the fact that the Society aided Jews emigrating from Nazi persecution and the presence of Jews in the British Society warn against equating cultural prejudice with racism: see BL 1.4.50, Blacker to L. Darwin.

[162] *Parl. Deb.* (HC), 286 (1934): 79–86; James, ibid. 291 (1934). James was a keen eugenicist, taking a second wife named Eugenia. Molson, became Chairman of the Society for the Protection of Rural England, 1968–71. This link between interest in protection of the land and of the race is also evident in Leslie Scott, Tory MP and CAMW President, who was also a member of the Council for the Preservation of Rural England. For discussion of this link see Jones, *Social Hygiene*, 45.

[163] Blacker, 'Voluntary Sterilization: The Last Sixty Years', 14.

prospective candidates for the 1935 general election found that 202 of the 259 respondents were in favour of legislation.[164] By 1935, the Eugenics Society claimed that all except four Church of England bishops were privately in favour of sterilization, that the Methodists and Congregationalists were also sympathetic, and that it could even boast the support of three leading Roman Catholics.[165] One Anglican cleric—Bishop Barnes of Birmingham—was forthright in his support for sterilization as a progressive and ethical measure.[166] Finally, there was also a generally positive response in the national press to the Brock Report, and an opinion poll in the Morning Post reported that 78.7 per cent of the national population were in favour of voluntary sterilization.[167] Of course, all of this evidence needs to be interpreted cautiously: the Eugenics Society had a vested interest in claiming widespread support, and national opinion polls were unreliable. However, there is enough evidence to suggest that support was becoming stronger and more acceptable, particularly among the middle-class constitutency of the Conservative Party, whose views the press, the Anglican Church, and prospective MPs most closely reflected.

If it is accepted that potential Conservative Party support for sterilization was stronger by the mid-1930s, then the Governmental inaction on the issue becomes crucial in explaining the failure of the legislative campaign. The evidence suggests that this inaction did not stem from a lack of sympathy: Hilton Young, Conservative Minister of Health, made his personal support clear in a private departmental memorandum of 1932:

> On the merits of the issue alone there would be but one question in doubt—what should be the method of authorising sterilisation? There may be doubt to what extent the children of mental defectives are themselves mentally defective. There can be none that no reasonably governed community would under modern conditions of population, permit the multiplication of descendants from such stock.[168]

With pressure to keep public expenditure down, and resources strained by the high level of unemployment, sterilization had potential appeal to the Conservatives as an alternative to costly institutional care. The

164 *Eugenics Review*, 28 (1936), 123.

165 PRO MH 51/560, W. A. Robinson, 20 Mar. 1935, 4 June 1935.

166 J. Barnes, *Ahead of his Age: Bishop Barnes of Birmingham* (London, 1979); Obituary, *Eugenics Review*, 46 (1954), 12.

167 For national press opinion: PRO MH 51/556.

168 PRO MH 58/103, Young to Sec. 15 Jan. 1932.

policy also fitted in with the Conservative Party's growing interest and advocacy of preventive approaches to mental and physical health care, and its increasing concern over population questions in the second half of the decade.[169] Arguing that the number of workdays lost because of neurosis was a primary cause of national inefficiency, Conservative MP F. E. Fremantle urged the Government to include mental health targets within its 1937 campaign to improve national physical fitness.[170] Of course, support for mental hygiene was not limited to those on the right, but it is notable that the Conservative Party of the 1930s was the home for many of those politicians who were most actively interested in this area and that these ideas gained at least some sympathy from the Party leaders; as such, the decision to avoid the sterilization issue is even more noteworthy.[171]

There is no simple or single explanation for this governmental inaction on sterilization. In part, it was a question of timing: the financial situation was improving by the mid-1930s, thereby weakening the economic case for sterilization, the panic provoked by the Wood Report of 1929 had now largely subsided, and criticism of the Nazi sterilization programme was beginning to taint the policy. The second reason was caution: despite large majorities, the National and Conservative Governments did not want to become entangled in an issue which would be contentious and would give potent political capital to

[169] On this interest in the population question: Conservative Party Research Department Records, Bodleian Library, Oxford (CRD), 1/63/1, Wentworth Shields to Viscountess Davidson, 20 Apr. 1937. On interest in mental hygiene: CRD 1/15/4, Captain Blakely (National Institute of Industrial Psychology) to Sir Joseph Ball, 13 Sept. 1938. On preventive health care: CRD 1/60/3, 'Health and Physique of the Nation'; also see the report published by *The Times* on the government's preventive health campaign of 1937, *The Nation's Health* (London, 1937). The Research Department's interest in these areas was probably far more adventurous than that of the Party leaders: J. Ramsden, *The Making of Conservative Party Policy: The Conservative Party Research Department since 1929* (London, 1980). The links between the Conservative establishment and social hygiene are explored in Jones, *Social Hygiene*, 43–52.

[170] *Parl. Deb.* (HC), 338 (1937): 3303. That Fremantle had at least some influence within the Party is suggested by the fact that the introduction to his book of 1927 was written by Minister of Health Neville Chamberlain (the book was dedicated to the Prime Minister Stanley Baldwin): *The Health of the Nation* (London, 1927).

[171] The prominence of future Conservative Chancellor R. A. Butler (who voted for Church's sterilization bill), and of Montagu Norman, the Governor of the Bank of England and a close ally of the Conservative Party, and his wife Priscilla within the post-Second World War mental hygiene movement suggests that this tradition continued: P. Norman, *In the Way of Understanding* (Godalming, 1982); B. Schreiber, *The Men Behind Hitler* (London, *c.*1975). Interestingly, Butler's Parliamentary Private Secretary between 1936 and 1938 was Wing Commander James.

the Labour Party. The National Government, in particular, attempted to place itself 'above party'.[172] Political caution also resulted from the fact that the policy depended on a scientific understanding of the genetics of mental deficiency—a subject which lay politicians could never be certain about, and on which there was still no clear consensus among the experts. The governments of the 1930s wanted to avoid the type of criticism which one national newspaper directed at Minister of Health, Hilton Young, after the publication of the Brock Report, calling him: 'A bad Minister, because he is only the echo of his experts on diet, lunacy, and slums'.[173] As Hilton Young observed:

> the authority is a doubtful question. To leave it to the assent of the mental defective is absurd. He is disqualified *ex hypothesi*. I would not trust the parent or guardian, in the interest either of the mental defective or the community. Unfortunately the democracy will not allow such a question to be decided on its merits.[174]

Thus, despite a significant body of sympathy within the Party, the Tories were still very aware of the restraints set by the British democratic tradition; to be effective, sterilization would, in practice, have to involve coercion of individuals, who may have lacked positive rights but, in return, were to be protected in their own best interests by the rest of society, and for this reason the policy remained beyond the bounds of acceptability.

CONCLUSION

The parliamentary politics of mental deficiency between 1913 and 1939 provides an example of conflict within consensus over the expansion of medical services. Divisions running across lines of class and party did play a part, and an increasingly significant one, but they were matched in importance by divisions which emerged out of the gender, religion, and social and professional backgrounds of politicians. Therefore the politics of mental deficiency cannot be understood simply in terms of being a class or party issue. Apart from a few maverick exceptions, there was an overwhelming consensus about the desirability of creating and

[172] J. Ramsden, *The Age of Balfour and Baldwin, 1902–1940* (London, 1978) 325–7.
[173] PRO MH 51/566, *Sunday Express* (21 Jan. 1934).
[174] PRO MH 58/103, Young to Sec., 15 Jan. 1932.

then expanding specialized institutional provision for mental defectives. There was a tension within this consensus, about how much power could be placed into the hands of experts, both in defining the problem of mental deficiency and then in controlling its administration. However, the consensus held because it was rooted in a view of mental defectives as being fundamentally different from normal responsible citizens: they lacked the normal citizens' rights to positive liberties, but in return it was society's responsibility to provide them with care and control in institutions.

Such attitudes crossed political, class, and gender lines; since they were often held most firmly by members of the 'respectable' working class, the segregation of mental defectives cannot be dismissed as a simple 'class issue'. Because defectives were unable to exercise voluntary choice, and since it was difficult to view the policy as being in the best interests of the recipients, most politicians opposed sterilization. The most significant break with this consensus was among those who believed that it was more important for society to accept expert and scientific knowledge than to be restrained by ideas about individual rights. Since such politicians were scattered across the political spectrum, and in the absence of a realignment of the party system during the period, this position never gained acceptance within any of the major parties. However, just as Wedgwood had predicted, the power to define mental deficiency and interpret policy had been delegated to the specialists and experts who implemented it at the local and central government levels. It was here that the relationship between democratic government and the expert would really be fought out.

2
The Board of Control

INTRODUCTION

The 1913 Mental Deficiency Act was important not simply because it set up a new system of care for mental defectives, but also because it introduced a new government department—the Board of Control—to administer all mental health care; the two developments were integrally linked as part of a broader vision of the specialization of welfare, and the one cannot be understood apart from the other. As the legislature neglected the issue of mental deficiency after 1913–14, control and responsibility were effectively devolved to the Board of Control. This chapter will contain a profile of the Board, to show how its character affected the interdepartmental and intradepartmental politics of mental deficiency and shaped both the medical and administrative dimensions of the problem.

The civil service has been blamed by some historians for contributing to what is perceived as the negativism of interwar social policy.[1] The period has been contrasted to the innovative era of late-Victorian and Edwardian civil servants, such as Robert Morant and William Beveridge.[2] And the assertion of Treasury control after the First World War has been viewed as smothering innovation under a blanket of fiscal conservatism.[3] Recently, however, the call to reassess the role of the civil service has led to a new genre of departmental histories which modify the thesis of negativism.[4] It is now recognized that the Treasury

[1] R. Skidelsky, *Politicians and the Slump: The Labour Government of 1929–1931* (London, 1970), 430; P. Abrams, 'The Failure of Social Reform', *Past and Present*, 24 (1963), 53–64.

[2] R. Davidson and R. Lowe, 'Bureaucracy and Innovation in British Welfare Policy, 1870–1945', in W. J. Mommsen (ed.), *The Emergence of the Welfare State in Britain and Germany* (London, 1981), 263–95.

[3] R. Lowe, 'The Erosion of State Intervention in Britain, 1917–1924', *Economic History Review*, 2nd ser., 31 (1978), 270–86.

[4] M. Beloff, 'The Whitehall Factor: The Role of the Higher Civil Service', in G. Peele and C. Cook (eds.), *The Politics of Reappraisal* (London, 1975), 209–31; R. Lowe, 'Bureaucracy Triumphant or Denied? The Expansion of the British Civil Service, 1919–1939', *Public Administration*, 62 (1984), 291–310.

had to negotiate with spending departments;[5] that each government department, though influenced by the dominant Treasury-imposed culture of the civil service, had its own traditions and culture, and its own links to a 'special public' of pressure groups;[6] and that criticisms of lack of innovation or inefficiency need to take into account the political and financial constraints which existed at the time.[7]

In this chapter, I shall begin with an analysis of the 'character' of the Board of Control. This is of particular relevance to debates about the nature of the interwar civil service, since it highlights the difficulty of drawing an absolute division between public and private, or governmental and non-governmental, spheres; the Board occupied an intermediate position, linking government with civil society. By exploring the clash of bureaucratic cultures and the political intrigue, both within the Board itself and between the Board and other government departments, I shall demonstrate how heterogeneous the interwar civil service still was. Secondly, I shall explore the ways in which the character and image of the Board influenced its political relations with other government departments, and how this affected the funding and staffing of the Board and its legal and medical roles. Thirdly, I shall examine the extent to which the Board was evolving during the interwar period, away from its rather unique character, towards a bureaucratic ideal type—one which was closed off from the outside world, with a culture of disinterested objectivity, and which was organized into a rational hierarchy based on bureaucratic, rather than professional, expertise.[8] And finally, I shall use the Board of Control as a case study to explore the relationship between bureaucratic modernization and the adjustment to democracy. Did bureaucrats go beyond their roles as servants of the democracy to become independent political actors and the source of effective power, or did democratic constraints ensure that there was little danger of the bureaucracy providing a 'new despotism'?[9]

[5] A. McDonald, 'The Geddes Committee and the Formulation of Public Expenditure Policy, 1921–1922', *Historical Journal*, 32 (1989), 643–74.

[6] S. R. Stacey, 'The Ministry of Health, 1919–1929: Ideas and Practice in a Government Department', unpub. D.Phil. thesis (Oxford, 1984); H. E. Dale, *The Higher Civil Service of Great Britain* (Oxford, 1941).

[7] R. Lowe, *Adjusting to Democracy: The Role of the Ministry of Labour in British Politics, 1916–1939* (Oxford, 1986); C. Bellamy, *Administering Central–Local Relations, 1891–1919* (Manchester, 1988).

[8] W. Mommsen, '"Toward the Iron Cage of Future Serfdom"? On the Methodological Status of Max Weber's Ideal-Typical Concept of Bureaucratization', *Transactions of the Royal Historical Society*, 5th ser., 30 (1980), 157–81.

[9] Lord Hewart, *The New Despotism* (London, 1929).

THE CHARACTER OF THE BOARD OF CONTROL

Developing from the Commission in Lunacy, the Board of Control was firmly rooted to a nineteenth-century bureaucratic tradition, yet it was also a model for the specialization of government in the early twentieth century. The former was characterized by its legal, inspectorial role, the latter by its ability to provide administrative and medical expertise. The tension between these identities is fundamental to the problems encountered by the Board.

In the nineteenth century, central governmental regulation of the insane had been performed by the Commissioners in Lunacy, a body of legal, medical, and lay inspectors. From a rather whiggish perspective, historians have tended to criticize the Commissioners for their emphasis on legalism and their neglect of a reforming medical role.[10] Such criticism probably overestimates the value of any greater psychiatric role they might have played and undervalues the importance of the legal role during a period of involuntary confinement.[11] Criticism of inspection under the Commissioners needs to take into account both their lack of effective power to enforce standards, and the limitations of having a staff of just fifteen Commissioners during a period in which the number of institutionalized insane rose from about 30,000 in the 1850s to 80,000 by the 1890s;[12] this was certainly one area where the dynamic of bureaucratic expansion did not precede and drive the motor of government growth.[13] To overcome lack of real power the Commissioners had to depend on cultivation of an image of authority. However, this did little to enhance their popularity among the psychiatric profession or the general public. Neither did the recurrent public scandals about wrongful confinement, abuse of patients, and inhumane conditions of

[10] K. Jones, *A History of the Mental Health Services*, (London, 1972), 153; N. Hervey, 'A Slavish Bowing Down: The Lunacy Commission and the Psychiatric Profession, 1845–60', in W. Bynum, R. Porter, and M. Shepherd (eds.), *The Anatomy of Madness* (London, 1985–8), ii. 118–19.

[11] D. J. Mellett, 'Bureaucracy and Mental Illness: The Commissioners in Lunacy, 1845–90', *Medical History*, 25 (1981), 246–7.

[12] D. J. Mellett, *The Prerogative of Asylumdom* (London, 1982), 51–2.

[13] O. MacDonagh, *A Pattern of Government Growth, 1800–1860: The Passenger Acts and their Enforcement* (London, 1961), 337–50; id., 'The Nineteenth-Century Revolution in Government: A Reappraisal', *Historical Journal*, 1 (1958), 52–67.

care, which almost inevitably resulted from the inadequacy of the inspectorial apparatus.[14]

Proposals for a new department reflected the desire to dissociate the new mental deficiency system from this unpopular legacy of the past, and to give the administration real power to plan and regulate the building of new asylums.[15] The creation of the Board was also linked to a wider vision of the need for a fundamental rationalization of government. Both the Minority and Majority Reports on the Poor Law were in sympathy with the recommendations of the 1908 Royal Commission on the Feeble-Minded for the establishment of a Board of Control to administer the lunacy and mental deficiency systems and to remove the confusion of administrative overlap between the Home Office, the Local Government Board, and the Board of Education. Indeed, the Minority Report saw the Board as a model for a system of specialized departments in a full-scale reorganization and replacement of the poor law.[16]

Despite the new title, the Board of Control continued to encounter many of the same libertarian suspicions which the Commissioners in Lunacy had provoked. Rather than being reassuringly bureaucratic and neutral, the title of the 'Board of Control' had an unfortunately disciplinary tone. Fears about placing control in the hands of unelected experts surfaced in the debates over the 1913 Act and were revived by the press in the early 1920s. Two cases, in particular, inflamed popular anxieties: first, Montagu Lomax's exposé, *Experiences of an Asylum Doctor*, which recounted the horrific conditions which he had encountered when temporarily serving as a superintendent during the First World War; and secondly, the *Harnett* case, in which Commissioner of the Board of Control, C. H. Bond, was personally taken to court by a member of the public for illegal certification.[17] The Board of Control, without its own direct Ministerial representative in Parliament, was attacked in the press for being outside democratic control and unresponsive to personal enquiries from the concerned relatives of the mentally disordered.[18] Descriptions of the Board as, for instance, 'a

[14] Mellett, 'Bureaucracy and Mental Illness', 226–7.

[15] *Roy. Com. Feeb.*, viii, 212–29.

[16] *Report of the Royal Commission on the Poor Laws and the Relief of Distress* (Cd. 4499, London, 1909), 80; there is far more attention to the problem of the feeble-minded in the Minority Report, 'The Break-Up of the Poor Law', 295–308, 499–573.

[17] M. Lomax, *Experiences of an Asylum Doctor* (London, 1921); for *Harnett* v. *Bond* [1924] 2KB 517, PRO MH 58/82–4.

[18] *John Bull* (17 Oct. 1925), 17; (18 Aug. 1925), 12; (3 Sept. 1932), 5.

monstrous instrument of child slavery' forced it into a defensive position and simply reinforced its legalistic emphasis. In such circumstances, it is hardly surprising that it was cautious about supporting the extension of the care in the community. As Secretary of the Board, Percy Barter pointed out:

> In short, if the Board intended to take this line they may provide justification for the attacks which are now being made by some organs of the Press, who say that the Bureaucracy are spreading their net in order to rope into the Mental Deficiency Institutions every child who is in any way sub-normal.[19]

The Board also had to cope with the suspicions of career civil servants within other departments towards new members of the Board, such as Mary Dendy, Ellen Pinsent, and W. H. Dickinson, who had been so intimately involved in the heated and sometimes alarmist campaign for the 1913 Act that they could hardly now be trusted to be disinterested. The practice of appointing people with backgrounds in psychiatric care as Commissioners made the Board something of an anomaly in an era when the cult of the administrative expert was being imposed by Treasury control over the civil service.[20] The value of psychological expertise as a possible tool of administration was dismissed by advocates of the 'administrative mind' who favoured a wholly pragmatic approach.[21] Even the medical practitioners within government were rather suspicious of the Board. For instance, Chief Medical Officer of Health at the Local Government Board, Arthur Newsholme, criticized the 'experts' at the Board for advancing eugenic-inspired policies which ignored the evidence of the latest scientific work on inheritance.[22] The medical men in the Ministry of Health tended to look down on the credentials of psychiatrists, which in medical terms often consisted of little more than experience of managing an institution. Chief Medical Officer, George Newman, viewed the mental hospital system as a 'serious blot on our public health system . . . a sort of Victorian cul de sac of hopeless laissez faire'. And in calling for the

[19] PRO MH 51/374, Barter to Chairman, 21 Nov. 1933, p. 3.

[20] J. R. Greenaway, 'Warren Fisher and the Transformation of the British Treasury, 1919–1939', *Journal of British Studies*, 23 (1984), 125–42.

[21] H. F. Carhill, 'Administrative Habits of Mind', *Public Administration*, 3 (1930), 119–30.

[22] PRO MH 58/85, A. Newsholme to Mr Munro, 22 Sept. 1910; J. S. Davy to Munro, 12 Oct. 1910. On the conservatism and inefficiency of the LGB see Addison Papers, Bodleian Library (ADD), Box 39, twelve-page memo n.d. (*c.*1914), Addison to PM on public health opposition to eugenics: D. Porter, 'Enemies of the Race: Biologism, Environmentalism, and Public Health in Edwardian Britain', *Victorian Studies*, 34 (1991), 159–79.

problem of mental disease to be dealt with instead, 'on the broad lines of Preventive Medicine', it seems that Newman was inclined to move beyond the Board of Control, rather than seeing it as integral to reform.[23]

Although the Board was originally envisaged as the epitome of a new, more specialized, and expert, governmental agency, it was invariably a conservative force, struggling to preserve the cultural and organizational traditions which it had inherited from its predecessor. By incorporating the personnel of the old Commission in Lunacy into its own organization, the Board was also incorporating and perpetuating well-established traditions of administration.[24] Three of the first four legal Commissioners at the Board—Lancelot Shadwell, Arthur Hill Trevor, and B. T. Hodgson—had already served lengthy careers as Commissioners in Lunacy, working their way up through the ranks from being Secretaries to become Commissioners. Similarly, the first medical Commissioners at the Board were far more notable for their administrative experience as Commissioners in Lunacy than for any great expertise in psychiatry: Sydney Coupland does not seem to have had any specialist experience in psychiatric medicine at all; Sir Frederick Needham had practised in psychiatry, but this experience dated back to the very different practices of 1858; and Marriott Cooke, whose medical views were said to be accepted 'without question' at the Board, had served as Medical Superintendent of a large mental hospital, but was not noted for any research or writing on psychiatry.[25] This typifies the tendency for medical Commissioners to be former institutional superintendents, rather than men who were noted for their writing and ideas on psychiatric disorder. The institutional orientation was reinforced by the appointment of Dr Arthur Rotherham, former Superintendent of Darenth Colony, who had been one of the main advocates of the specialized institutional care before the 1908 Royal Commission. And this practice of appointing former superintendents to serve as Commissioners was continued under the 1913 Act, thereby perpetuating the corporate alliance.

The Commissioners had gained their experience in an era when a legalistic inspectorial role was their main function. Needham, for exam-

[23] PRO MH 58/90, Newman to A. Robinson, 17 Feb. 1922.

[24] Mellett, 'Bureaucracy and Mental Illness', 226–9.

[25] Biographical material on Commissioners from: *Journal of Mental Science, Whitaker's Almanack, Dictionary of National Biography (DNB), and Who Was Who (WWW)*, unless otherwise stated. Cooke obituary, *JMS*, 78 (1932), 1–11.

ple, had been a stern opponent of mechanical restraint, but this was hardly a central issue by 1913.[26] The disciplinary demeanour and authoritative physical presence of the Commissioners reflected the importance of establishing an aura of authority in enforcing their regulatory role. Trevor, the legal Commissioner, was described as a 'famous bat' at cricket whose 'sporting and athletic instincts formed an important element in his character and often enabled him to gain a coign of advantage in dealing with patients'.[27] Similarly, Sir Frederick Needham, was portrayed as a medical Commissioner who, 'with his directness of speech, scorn of diplomacy . . . was an official who had to be taken very seriously'.[28] This image was at odds with the reorientation towards a more humane and personal medical regime within public health. The image may have been appropriate in the nineteenth century, but by the interwar years it was detrimental to the Board's relations with the public, the medical profession, and other administrators.[29]

The new requirement under the 1913 Act that at least one Commissioner should be a woman was an attempt to show that the Board would be more responsive to public needs than its predecessor. However, within the civil service this confirmed the low status of the Board's work and raised further suspicions about the presence of 'outsiders' within government, particularly since these women tended to maintain strong links with their former colleagues in voluntary mental welfare.[30]

The first female Commissioner at the Board was Mary Dendy, a leading figure in the movement for permanent segregation of the feeble-minded, the founder of Sandlebridge Colony for the feeble-minded, and the sister of Helen Bosanquet of the Charity Organization Society (COS). Dendy soon became unhappy with the tiring visiting duties as a Commissioner, and missed the daily contact with mental

[26] *JMS* 71 (1925), 1–7. [27] *JMS* 71 (1925), 175.
[28] *JMS* 71 (1925), 1–6.
[29] For example, during discussions over the Mental Treatment Act of 1930 there were serious doubts about whether voluntary mental treatment should be placed under the responsibility of the Board, since this would associate it with the stigma of legal certification and might make it unacceptable to the general public: PRO MH 58/90, Robinson to Minister of Health, 21 Feb. 1925.
[30] On the position of women in the interwar civil service: M. Grant, 'Perceptions of Propaganda in Interwar Britain: The Development of Government Publicity in the Domestic Sphere', unpub. D.Phil. thesis (Oxford, 1988); M. Zimmeck, 'The "New Woman" in the Machinery of Government: A Spanner in the Works?', in R. M. Macleod (ed.), *Government and Expertise: Specialists, Administrators, and Professionals, 1860–1919* (Cambridge, 1988), pp. 185–202; H. Martindale, *Women Servants of the State, 1870–1938* (London, 1938), 119.

defectives in her paternalistic colony. Her discontent in the new role was exacerbated by the difficulties she faced as a woman within the male-dominated service, just as she had earlier experienced prejudice in Manchester local politics.[31] Dendy's successor, Ellen Pinsent, was far happier in the administrative role. She had been the first woman councillor in Birmingham and rose to be the highest-paid woman civil servant of her time. Perhaps because of her background in local government after-care work for the defective, Pinsent seems to have been less aggrieved than Dendy when community care had to be provided as an alternative to lifetime segregation. Unlike Dendy, she thrived in the bureaucratic environment, becoming one of the most dominant figures on the Board of Control during the 1920s.[32]

The presence of these women gave the Board a much broader network of links within civil society. For instance, Ellen Pinsent brought the Board a series of contacts with the intellectual aristocracy of Cambridge. It was a lecture delivered by Pinsent, during the campaign for the 1913 Act, which had stimulated the formation of a local mental welfare association in Cambridge.[33] This group set up special courses to educate Cambridge University women on the problem of mental deficiency. Among these women was Pinsent's daughter, Hester, who subsequently became Secretary of the Cambridge Mental Welfare Association, and who married E. D. Adrian, the Nobel Prize-winning neurologist, who later provided an important stimulus to research in the area as head of the Medical Research Council's psychological section.[34] Pinsent's links with Cambridge mental welfare work brought contact with some influ-

[31] H. McLachlan, *Records of a Family* (Manchester, 1935), 135–84; M. Cruickshank, 'Mary Dendy, 1885–1933, Pioneer of Residential Schools for the Feeble-minded', *Journal of Educational Administration and History*, 8 (1976), 26–9; M. Jackson, 'Institutional Provision for the Feeble-Minded in Edwardian England: Sandlebridge and the Scientific Morality of Permanent Care', in D. Wright and A. Digby (eds.), *From Idiocy to Mental Deficiency: Historical Perspectives on People with Learning Disabilities* (London, 1996), 161–83.

[32] P. Hollis, *Ladies Elect: Women in English Local Government, 1865–1914* (Oxford, 1987), 62, 277, 425; *WWW 1941–50* (Oxford, 1959), 673; G. Sutherland, *Ability, Merit, and Measurement: Mental Testing and English Education, 1880–1940* (Oxford, 1984), 32.

[33] E. Pinsent, 'Social Responsibility and Heredity', paper to Church Congress (Sept. 1910).

[34] 'Close Up', *Trinity Review* (Easter, 1965), 28; *The Times* 23 (June 1966). A. Landsborough Thomson, *Half a Century of Medical Research*, ii (London, 1975), 149–50; J. Woodcock, '75 Years of Voluntary Endeavour: A Review of the Work of the Cambridge Mental Welfare Association, 1908–1983' (privately published, copy at Cambridge County Record Office).

ential families.[35] Most notably, the members of the Darwin family collaborated with the government in funding important research into mental deficiency by C. S. Myers, E. O. Lewis, and Lionel Penrose. The network was linked to the Eugenics Society via the Society's President in the first half of the 1920s, Leonard Darwin. And finally, the network of influence went full circle as Horace Darwin's daughter, Ruth Darwin, moved from Cambridge voluntary mental welfare work to become Pinsent's successor at the Board of Control.[36] It was understandable that other government departments would be suspicious of whether the Board's loyalties lay with the government or with these lobbying networks to which it was connected. In particular, there was concern over the Board's close links to the CAMW, a group which had been formed on the suggestion and with the aid of W. P. Byrne, the first Chairman of the Board, but which also had independent roots within voluntary mental welfare work.[37]

THE ADMINISTRATIVE PROBLEM OF MENTAL DEFICIENCY

Staff and Finance

The dominant government department during the interwar years was the Treasury.[38] With control over distribution of funding and appointment of staff, the Treasury had a decisive influence over the work of a department like the Board. The interwar Treasury has been noted for its fiscal caution and its attempt to create a civil service which was efficiently manned by civil servants who were generalist administrators and under Treasury influence. It was naturally suspicious of the Board

[35] For example, in 1909 the Cambridge COS Association for the Care of the Feeble-Minded Annual Report lists the General Committee as including: Sir Clifford Allbutt, Mr W.. Bateson, Prof. Sir G. Darwin, Mr Horace Darwin, Prof. Inge, Prof. Pigou, Prof. Seward, and Mr Whetham (reports for this and other Cambridge mental welfare groups held at Cambridge County Record Office). For a description of this tightly knit world: Lord Annan, 'The Intellectual Aristocracy', in J. H. Plumb (ed.), *A Tribute to Trevelyan* (London, 1955), 241–87.

[36] Horace Darwin, *DNB, 1922–50*, p. 239. Ruth Darwin, *WWW, 1971–80*, p. 197.

[37] PRO MH 58/85, A. V. Symonds to Mr Lowry, 8 July 1914.

[38] G. C. Peden, 'The Treasury as the Central Department of Government, 1919–1939', *Public Administration*, 61 (1983), 371–85.

of Control, with its unorthodox personnel and its ambition to expand mental hospital provision dramatically. The 1913 Act set the Board on a tight rein of £150,000 expenditure per annum.[39] This sum would have been totally inadequate if all local areas had actually wanted to build hospitals. At a cost of £300 per bed, and even with a local authority contributing equal funding, the construction of a single institution for 1,000 patients would have absorbed the annual total. In the event, this problem did not arise, as the First World War and then the post-war prioritization of council housing diverted energies.[40] In 1919 the limit on spending was raised. However, with a serious financial crisis for the government in 1921, the Treasury refused a further rise, fearing a potential doubling in expenditure if all cases of mental defect were to be dealt with. As the Board of Control pointed out, this was totally inadequate since there were at least 2,000 new cases needing hospital care every year, while only 300 beds a year were being freed through discharges or death.[41]

Despite Treasury pressure to hold down expenditure, the Board of Control succeeded in maintaining a steady, even though inadequate, rise in central expenditure.[42] This is in contrast to central funding of comparable areas, such as services for tuberculosis, venereal disease, lunacy, and maternal and infant welfare, which all peaked in the first years after the First World War and then levelled off or fell.[43] Mental deficiency funding was slower to take off, perhaps because it was less of a political priority, but in the long term it was difficult to avoid expansion after starting from a base of such serious under-provision. The Board defended the area by arguing that any cuts would prove to be a 'false economy', saving money in the short term but in the long term leading to a far higher expenditure as defectives ended up unemployed or as occupants of expensive prison or asylum accommodation.[44] In

[39] PRO MH 51/691, CACMD to Ministry of Reconstruction, 24 Oct. 1917.

[40] Abrams, 'Failure of Social Reform', 43; PRO MH 58/85.

[41] PRO T 161 149/13551, BOC, 17 Nov. 1921.

[42] PRO MH 58/88, 'National Economy Proposals 1921–2', memo, 23 Dec. 1920; MH 51/519, 'Local Authorities' Schemes in the Present Financial Emergency', 24 Nov. 1931.

[43] This is based on analysis of information available in the *Annual Reports of the Chief Medical Officer of the Ministry of Health, 1921–1939*.

[44] PRO T 161 149/13551, J. B. B. (Beresford) to Mr Phillips, 21 Nov. 21; Edward Troup, 8 Dec. 1921.

adopting this position, the Board committed itself to keeping the cost of care in mental deficiency institutions at a bare minimum.[45]

Not only did the Board have to struggle against Treasury parsimony, but it also had to overcome the Ministry of Health which viewed mental deficiency as a low priority compared to other areas under its responsibility, especially in this period of post-war financial stringency. As Newman argued: 'If the times were normal I should venture to advise definite progress all round—but I find it very difficult to see how we can advance with public opinion unfavourable to extensions and without money.'[46] The Board's building schemes had to pass through the Ministry of Health for approval, and were unenthusiastically received when the cost per bed was found to exceed the per capita cost of much needed, and politically vital, working-class housing.[47] In response, the Hedley Committee was set up to investigate whether the cost of building mental institutions could be reduced even further.[48]

Restraints on expenditure also affected the building of special schools. To have provided all mentally defective children with special education would have raised expenditure from £1.3 million to £17 million.[49] The Ray Committee on Local Expenditure of 1932/3 recommended integrating defective children in elementary schools, estimating that this would cost just £12. 12*s.* per child for a year's schooling, compared to £23 in a special school, or £60–£70 in a residential institution.[50] In this economic climate, the building of special schools slowed down over the interwar period, and this placed even greater pressure on services under the Mental Deficiency Act.[51]

Finally, financial restrictions had a direct impact on staffing at the Board of Control, and again this had a debilitating influence on the administration of the Mental Deficiency Act. In 1908, before taking on the additional work of administering the mental deficiency system, Commissioner Dr Mariott Cooke had already expressed his concerns

45 *Report from the Select Committee on Estimates, 1932–33* (v.197), 11–22, 130–1. See Ch. 3.

46 PRO MH 58/90, Newman to Robinson, 17 Feb. 1922, p. 5.

47 PRO MH 51/710, Robinson to Brock, 3 July 1929.

48 PRO MH 58/97, Robinson to Minister of Health, 22 June 1931.

49 PRO Ed 24/1199, Wood to Secretary, 23 Oct. 1925, p. 1.

50 *Committee on Local Expenditure in England and Wales (Ray Committee)* (Cmd. 4200, xiv, 1932–3), 41.

51 A. J. Welshman, 'The School Medical Services in England and Wales, 1907–39', unpub. D.Phil. thesis (Oxford, 1988), 255–311; Sutherland, *Ability, Merit and Measurement*, 61–2.

over the inadequacy of staffing levels.[52] In 1913, an additional Board of four members was established to oversee the preparation of rules and regulations for the new mental deficiency system.[53] In 1914, this new Board was combined with the old Commissioners to form the Board of Control, but this slight expansion in staff hardly compensated for the increase of duties. The Government had to use an emergency war order to appoint additional 'Inspectors'. This action was opposed by members of the Commission, who regarded it as a threat to their own status, and argued that the use of these inexperienced Inspectors would lead to a damaging loss of confidence in the process of inspection among the general public. Nevertheless, the use of Inspectors in the role previously performed by Commissioners was unofficially continued after the war. The Board lobbied the Ministry of Health to reform the situation, but the Ministry refused, cautious about drawing publicity to the fact that the practice had been allowed to develop.[54]

The Commissioners were paid a salary of £1,200–£1,500. Such a salary was essential if qualified medical and legal men were to be attracted at a time of rising professional wages.[55] To attract psychiatric superintendents, the Board also had to compensate for the benefits, such as free housing and a low cost of living, which were associated with control of huge and often economically productive colonies. However, the Treasury was unsympathetic towards such arguments. It regarded the Board as an inefficient agency, overstaffed with highly paid experts, and proposed a central Board of six rather than eight Commissioners, with the others moving to the lower rank of Inspector.[56] The Ministry of Health, also under attack for being spendthrift, defended the size of the Board, which was no larger than in 1913 and was less well staffed than in Scotland, where there were two Commissioners for a population a fifth of the size.[57] The Treasury gave way, but only after it was given

52 *Roy. Com. Feeb.*, viii. 215.

53 F. Robinson, 'Local Administration of the Mental Deficiency Act, 1913', *Public Health*, 29 (1916), 78–81.

54 PRO MH 58/90, Willis to Robinson, 7 Mar. 1922; Robinson, 17 Mar. 1922.

55 Salaries for members of the Board of Control and other civil servants were published annually in *Whitaker's Almanack*. Medical salaries were rising on average by 91 per cent between 1913/14 and 1922/3, and then by 36 per cent up to 1936/8: G. Routh, *Occupational Pay in Great Britain, 1906–60* (Cambridge, 1965), 62; A. Digby and N. Bosanquet, 'Doctors and Patients in an Era of National Health Insurance and Private Practices', *Economic History Review*, 2nd ser., 41 (1988), 74–94.

56 PRO MH 58/90, Grigg, 10 Mar. 1923.

57 PRO MH 58/90, Veale, 22 Feb. 1923; ADD Box 38, 'Organisation and Increase of Medical Staff'.

sanction over future appointments.[58] Thus Treasury parsimony did not manage to reduce the level of staff—this would hardly have been possible at a time of a growing defective and insane population, especially after starting at such a low point of understaffing—but in gaining control over appointments the Treasury was to have a debilitating influence on the Board in the future, restraining its growth and eroding its standards of expertise.

The Board attempted to offset staffing restrictions by appointing retiring Commissioners to nominal positions, where they could continue to contribute their expertise on a voluntary basis. It also used the appointment of women.[59] Only one paid female Commissioner was ever appointed at a time (and this minimum was required by law), suggesting that the incentive of the lower female salary was offset by a concern to guard the status of the position.[60] However, an increasing number of women were employed in the lower paid post as Inspectors, paid on a scale of £400–£650, rather than £500–£800 for a 'medical' (male) inspector.[61] Significantly, although the position of male inspector was abandoned and replaced by a second grade of Commissioners in the 1930s, the position of Women Inspectors was retained.[62]

The Legal Role

In the nineteenth century, the primary role of the Commissioners in Lunacy had been as inspectors to enforce the legal regulations of the lunacy legislation and to protect the liberty, rights, and property of members of the public against wrongful certification. Although the 1913 Act marked a shift towards medical specialization, the legal function of the Commissioners remained paramount, especially in light of

[58] PRO MH 58/90, Grigg, 3 Apr. 1923.

[59] This was part of a wider strategy to economize by using 'paid volunteers' to link up institutions to the net of female-provided after-care and preventive work: ADD Box 39, Miss A. M. Anderson, 'Memorandum on Subsidiary Health and Kindred Services for Women' (March, 1917).

[60] PRO MH 51/677; M. Zimmeck, 'Strategies and Strategems for the Employment of Women in the British Civil Service, 1919–39', *Historical Journal*, 4 (1984), 901–24; M. Zimmeck, 'Marry in Haste, Repent in Leisure: Women, Bureaucracy, and the Post Office, 1870–1920', in M. Savage and A. Witz (eds.), *Gender and Bureaucracy* (Oxford, 1992), 65–93. For the BMA protest about the employment of women: C. Addison, *Four and a Half Years*, i (London, 1934), 25. For an insight into the prejudice suffered by women Commissioners, see the notes on Flora Calder: PRO MH 107/48, Brock to Dr Quine, 31 Mar. 1944.

[61] PRO MH 51/677: summary of salaries, 23 Dec. 1927, p. 1.

[62] PRO MH 58/96.

the acute political sensitivity over the issue of individual rights. Much of the Board's time was taken up monitoring the complex rules and regulations of the legislation, making sure that institutions did not abuse individual rights through the use of such managerial tools as physical restraint and corporal punishment. This produced a mass of paperwork for the Board to process in its London offices.[63] The sheer volume of this work made it impossible for the Board to perform a rigorous monitorial role. It had to deal with the majority of cases on paper, rather than in person, and therefore had to rely on the medical opinion of the doctors based in the localities, many of whom lacked the necessary expertise to diagnose mental deficiency with any accuracy.[64] In doing so, the Board failed to provide the effective lead which was called for, proving too cautious to settle local disputes or to offer binding interpretations of the legislation to confused local authorities.[65] The Board even struggled in its traditional role of regulating the standard of institutional care: with so many institutions, inspectorial visits were too rare to provide close control; many of the institutions were now so large that the inspection could only be cursory in the limited time available; institutional staff were often forewarned about visits; and the Medical Commissioners, often former superintendents themselves, tended to be overly lenient, sympathetic towards the practical difficulties of providing adequate conditions.[66]

By the early 1920s it was clear that the public had little confidence in the system. The Board reacted by setting up the Cobb Committee, appointing two former superintendents, now serving as Commissioners, to lead the investigation.[67] The Committee dismissed all allegations against the Board, but its conclusions were so one-sided that the objectivity of such an internal enquiry was brought into question.[68]

[63] *Report of the Royal Commission on the Law Relating to Mental Illness and Mental Deficiency* (hereafter *Roy. Com. 1954–7*) (Cmnd. 169, 1957), 746. For the Board's sensitivity to prevent corporal punishment: PRO MH 51/490, Board to Agatha Stacey Home, 10 Feb. 1936.

[64] PRO MH 51/470, e.g. Case of Irene Colby 69282/G, Darwin and Thomas, 20 Dec. 1932.

[65] PRO MH 51/410, letter to Oxford City Council, 16 June 1922.

[66] *Report of the Royal Commission on Lunacy and Mental Disorder* (hereafter *Roy. Com. 1926*) (Cmnd. 2700, 1926), 11247–63; *Roy. Com. 1954–7, Minutes of Evidence*, pp. 749–50. David Parfitt, 'A Mental Hospital in 1929', in G. Berrios & H. Freeman (eds.), *150 Years of British Psychiatry*, ii, *The Aftermath* (London, 1996), 465–76.

[67] *Report of the Committee on the Adminstration of Public Mental Hospitals (Cobb Report)* (Cmnd. 1730, 1922).

[68] *Roy. Com. 1926, Minutes of Evidence*, Parker, 11546–7.

Libertarian critics called for the abolition of the London-based Board, and for its replacement by a system of district Commissioners who would be in much closer contact with local problems.[69] Similarly, representatives of local authorities wanted more control in the hands of elected local committees, with the Board of Control's power reduced to an advisory role.[70]

In summary, the Board's legal role was seriously undermined by its centralist structure and inadequate staffing levels. Ironically, despite all its efforts as a legal monitor, it was pilloried by the public as an ineffective regulator of standards, and worse still as actively complicit in the unjust restriction of individual rights. The Board seems to have been caught in an impossible situation: it was still expected to be a legal watchdog by politicians and by the public, but this role was increasingly difficult to perform, both because of the volume of the work and because many of the issues were now medical, rather than strictly legal. Forced into emphasizing its legal role, the Board only attracted further public scrutiny and perpetuated the image of mental health care as an issue of restricting individual rights rather than providing a therapeutic service.

The Medical Role

In 1908, when the Royal Commission on the Feeble-Minded had called for a new central government department to administer mental health care, it had envisaged:

> an active and efficient centre, both for the general protection and supervision of the mad, and also, for the promotion of measures of scientific administration and investigation, which will prevent, as far as may be, the increase of mental disease and defectiveness.[71]

This vision proved difficult to realize.

One way for the Board to provide medical leadership was through its Annual Reports. The Commissioners in Lunacy had used these reports to provide valuable statistical breakdowns. However, faced by an increased burden of work, the Board of Control pared down such information to a few basic tables. It used the reports to encourage efficiency and economy, rather than recommend higher standards or

[69] Ibid., Parker (Chair National Society for Lunacy Reform), 10469–72.
[70] Ibid., Representatives of Mental Hospitals Association, 13435–60.
[71] *Roy. Com. Feeb.*, viii. 280.

new treatments. The publication of medical papers in a supplement to the Reports was obviously an attempt to encourage medical work within institutions, but the content of the papers suggests that this scheme did more to encourage general medical research in mental institutions than it did to stimulate research-led psychiatric therapy.[72]

A number of notable psychiatric figures did join the Board and used their positions to provide some leadership on psychiatric issues. This group included C. H. Bond, W. Rees Thomas, and E. O. Lewis—men who had the respect of the psychiatric community. Bond served as President to the Medico-Psychological Association (MPA), while still a Commissioner, and used his position as Commissioner to direct the profession's development.[73] Rees Thomas used his reports on state institutions for the morally insane in the Board's Annual Reports to disseminate his ideas about the psychiatric, rather than constitutional, basis to such problems.[74] And Lewis used his position as chief investigator of the Interdepartmental Committee on Mental Deficiency—the largest interwar survey of mental deficiency—to extend concern to the dull and backward and the 'social problem group'.

Between 1914 and 1920 the Board's record for encouraging research had been fairly promising. As the only department of government which was exclusively devoted to medical care at this time, and with responsibility for the largest group of the population under state medical care, the Board was ideally placed to benefit from the state's entry into funding scientific research under the 1911 Insurance Act.[75] The Board presented two arguments to justify research on mental disorder: first, it would lead to new therapies and help to prevent mental disorder; and secondly, it would stimulate interest in psychiatry within medicine and help to attract more able candidates into the profession. Funding was requested to set up regional laboratories which would act as foci linking the surrounding mental hospitals and universities. Not only would these centres undertake research on institutional populations, but they would also cast light on prevention within the general population.[76] Particularly impressive was a Manchester scheme which aimed to link up mental hospital, university, and social survey research,

[72] See Ch. 3.
[73] C. H. Bond, 'The Need for Schools of Psychiatry', *JMS* 66 (1920), 10–22.
[74] *Board of Control Annual Report, 1922*, pp. 102–5 (hereafter referred to as *BOC*).
[75] PRO MH 51/78, letter to Lloyd George, 14 Mar. 1914.
[76] Ibid., 'Memorandum by the Commissioners in Lunacy on the Subject of State Aid for Research and Mental Diseases'.

and combined experimental, clinical, and pathological psychology, morbid anatomy, chemical pathology, and anatomical, physiological, and statistical work. With plans to develop mental-testing techniques and investigate inheritance, the scheme placed the problem of mental deficiency near the head of its priorities.[77]

The Board's research schemes were set back by the establishment of the Medical Research Committee (later Council) to control expenditure on research. In 1915 the domineering Secretary of the MRC, Walter Morley Fletcher, warned the Board of Control and Board of Education against independent research.[78] The MRC refused the grant for the ambitious Manchester scheme, seeing its parts as of 'unequal value'. It also refused to fund a psychological laboratory at Cardiff, since an MRC laboratory had already been set up in this area.[79] Funding was continued on research into dysentery within mental institutions, but support reflected concern about dysentery among British troops in the Dardanelles as much as interest in psychiatric medicine.[80] In 1921, to the chagrin of the Board of Control, the grant previously preserved for work in mental institutions was merged with that for general medical research under the MRC.[81] As a piece of rationalization, the move had Treasury approval, but it left the interests of psychiatry, especially an area such as mental deficiency, poorly represented under the biomedically oriented MRC. Thus, over the next two decades, there was hardly any psychiatric research among the numerous projects sponsored by the MRC.[82] The only area of psychiatry which the MRC showed any great interest in funding was research into the genetics of mental disorder.[83] By contrast, it showed little enthusiasm when the Board of Control made an ambitious proposal to establish a network of psychiatric research clinics which would encourage co-operation between local authorities, mental hospitals, and universities, viewing this as an issue of clinical practice rather than scientific research.[84]

[77] PRO MH 51/80, BOC to MRC, 4 Dec. 1914.

[78] J. Austoker, 'Walter Morley Fletcher and the Origins of a Basic Biomedical Research Policy', in J. Austoker and L. Bryder (eds.), *Historical Perspectives on the MRC* (Oxford, 1989), 25.

[79] PRO MH 51/85, Fletcher to BOC, 26 Feb. 1915, pp. 3–4.

[80] PRO MH 51/86, report by Bond on meeting with Dr Fletcher, 22 Sept. 1915.

[81] PRO MH 51/89, Bond, 5 Jan. 1921.

[82] L. Bryder, 'Public Health Research and the MRC', in Austroker and Bryder, *Historical Perspectives*, 69; Landsborough Thomson, *Medical Research*, ii. 27–8.

[83] Landsborough Thomson, *Medical Research*, ii. 149–50.

[84] PRO MH 51/254.

In such circumstances, the Board of Control should not be blamed for failing to provide the leadership in scientific and preventive approaches to mental disorder which had been hoped for. The Board's attempts were thwarted by the low standing of psychiatric research, both within government, and within the medical profession as a whole. Instead, the Board had to attempt to exert its medical influence through encouragement in its Annual Reports or through the recommendations of Departmental or Royal Commissions. This was a far cry from the vision of control by experts which had provoked both fears and hopes earlier in the century.

Interdepartmental Relations

Analysis of the interdepartmental politics of mental deficiency reveals the interwar civil service to have been a complex political arena, divided by different visions of what constituted good government, and certainly not a monolithic whole. With its peculiar character and its unorthodox model for the role of expertise within government, the Board of Control was always at a disadvantage in interdepartmental struggles over policy.

The new Board was given greater powers to regulate and plan than had its predecessors, but in reality building still depended upon the co-operation and capital of the Local Government Board or the Ministry of Health.[85] Neither did the intended rationalization of administrative responsibility take place: at the local level there continued to be a confusing overlap between mental deficiency, education, and poor law authorities, which undermined institutional reclassification and made a co-ordinated ascertainment of defectives almost impossible;[86] while in central government, division of responsibility between the Board of Control, the Ministry of Health, and Board of Education also caused confusion.[87]

The administrative division between the system for adult defectives, under the 1913 Mental Deficiency Act, and that for children, under the 1914 Elementary Education Act, made close co-operation between the Boards of Control and Education crucial. Unfortunately, tensions between the two departments were often acute. The creation of a separate system for children had been justified by the impressive

[85] *Roy. Com. 1954–7*, p. 740; PRO MH 51/520, Barter to Heseltine, 3 Oct. 1931.
[86] *BOC 1922*, pp. 38–42.
[87] PRO MH 51/706, Ed 50/113; Ed 50/121; Sutherland, *Ability, Merit, and Measurement*, 57–96.

expansion in special school provision prior to 1913. However, this came to a virtual halt during the interwar period, rising by only 3,000 from the 14,626 places in 1915.[88] The Board of Control proposed joint schemes to overcome the shortage, but the Board of Education blocked this as an attempt to reverse the split caused by the 1914 Act. A. H. Wood, of the Board of Education's medical department, was typically suspicious when he reported that the Board of Control:

> thought the Act was out of date, and that they proposed so far as possible to evade it in this particular . . . it is very hard to draw the line between low grade educable children, and children who are not educable at all. But there is no doubt that it is the policy of the Board of Control to move this line as high as they can.[89]

The Board of Education viewed the Central Association for Mental Welfare (CAMW) as a mere tool of the Board of Control. So when this pressure group launched a campaign advocating more rigorous notification of mental defectives in schools, the Board of Education interpreted it as a covert attack from the Board of Control, and reacted by defending itself to the Chancellor of the Exchequer.[90] Byrne, Chairman of the Board of Control, was obviously upset at the letter which was sent to the Treasury: 'With considerable vigour [he] said he disagreed with every word of our draft letter.' Newman, in his capacity as head of school medical services, entered the fray, arguing that Byrne did not properly understand the legislation.[91] The dispute probably had at least something to do with Byrne's rather sudden retirement, and it left a legacy of suspicion between the departments which continued to undermine co-operation.[92]

Relations with the medical section within the Board of Education gradually improved, but the legal and educational sections maintained serious reservations. Newman was keen to encourage school medical officers to play a more active role in the ascertainment of defectives, seeing this as one of the areas which would make the school medical services a central part of a public health service.[93] And in the 1920s A. H. Wood, head of the medical branch of the Board of Education, was in favour of extending care beyond mental defectives to the wider popula-

[88] Sutherland, *Ability, Merit and Measurement*, 297.
[89] PRO Ed 50/112, Wood to Newman, 4 Aug. 1920.
[90] PRO Ed 50/112, Leslie Scott (CAMW) to Fisher (President B. of Ed.), 22 Dec. 1920.
[91] Ibid., Barter to Wood, 8 Mar. 1921.
[92] Ibid., memo, Wood to Fisher on meeting with CAMW *c*.1921, and letter, Fisher to Chamberlain, 11 Mar. 1921; Ed 50/266, Duke to Maudsley, 24 Jan. 1936.
[93] PRO Ed 50/112, Newman to Morant, 8 July 1919; Ed 50/116, Pease to Newman 22 May 1914.

tion of the dull and backward.[94] On the other hand, they both wanted control to remain with the school medical officers. Those in the legal and educational sections of the Board of Education were even less sympathetic towards co-operation. The educational branch was generally opposed to the removal of mentally defective children from its own schools, and was suspicious of the Board of Control's plans to place defective children in its own residential colony schools or in occupation centres in the community.[95] It was under pressure from parents, who objected to the stigma of certification.[96] And it was under pressure from special school teachers, who wanted to protect their livelihoods.[97] The legal section was even more reluctant to pass responsibility over children to the Board of Control. It regarded the legal definition of mental deficiency as dangerously imprecise, and recognized that 'ineducability' could only be proven if there had first been a serious attempt to educate a child.[98] It also doubted the legality of the Board of Control's suggestion that merely dull and backward children could be notified to the mental deficiency authorities.[99] In summary, because of a desire to protect its own administrative territory, and because of concern over the rights of defective children and their parents, the Board of Education was often unwilling to pass on mentally defective children to the mental deficiency authorities, and this seriously undermined the attempt to set up a comprehensive system of care.[100] Chairman of the Board of Control, Brock, privately admitted: 'I have always regarded the Board of Education as an exceptionally timid department . . . (but) Publicly departments never criticize each other on the sound principle that 'dog doesn't eat dog.'[101] The public façade of civil service harmony masked rivalries, differences in outlook, and wholescale disagreements over policy between the departments.

[94] PRO Ed 24/1199, Wood to Secretary, 23 Oct. 1925.

[95] PRO MH 51/429, CAMW proposals, Sept. 1925; MH 51/706, Pelham (B. of Ed.) to Home Office, 8 July 1914.

[96] PRO Ed 50/112, H. A. L. Fisher to Austen Chamberlain, 11 Mar. 1921, p. 1.

[97] PRO Ed 50/124, NUT to B. of Ed., 8 Nov. 1929; Turnbull to Maudslay, 4 Nov. 1930; NUT to B. of Ed., 16 Nov. 1934.

[98] PRO Ed 50/113, Barker to Wood, 1 Mar. 1924; PRO Ed 50/121, Barker to Wood, 28 Jan. 1924.

[99] PRO Ed 50/226, L. G. Duke to Maudslay, 24 Jan. 1936, pp. 1–2. (Until the age of 17 'educable' defectives were still the responsibility of the education authorities. Therefore if they left school before this date they could not be legally dealt with or notified to the mental deficiency authorities until aged 17.)

[100] PRO Ed 50/124, 'Office Committee on Mentally Retarded Children; MH 51/374 Brock to Fox, 21 Dec. 1932.

[101] PRO MH 51/374, Brock to Fox, 21 Dec. 1932, pp. 2, 5.

Governmental action to address mental health invariably took second place to the priorities of more powerful departments. Throughout the 1920s, the Ministry of Health prevaricated and delayed reform of mental deficiency, since there seemed to be little point in making any substantial changes until local government and poor law reform had first been tackled.[102] Similarly, in the 1930s, reorganization of the education system to provide special care for the dull and backward was perpetually delayed by the Board of Education, as it would have brought up the more fundamental issue of raising the school leaving age.[103]

Even within the Board of Control, mental deficiency policy generally took second place to mental illness. In the 1920s, the Board's main concern was to introduce voluntary treatment for the mentally ill, and it was wary about jeopardizing this by linking it with the controversial issue of mental deficiency. The issue was effectively shelved during the second half of the 1920s; this inaction was justified by the ongoing investigation of the Interdepartmental (Wood) Committee on Mental Deficiency, which had been set up in 1924. Any mention of mental deficiency was studiously avoided by the 1926 Royal Commission on Mental Illness.[104] And publication of the controversial Wood Report was held back so that it did not interfere with the introduction of voluntary treatment for the mentally ill under the 1930 Mental Treatment Act.[105] The result was an unnecessarily sharp division between policy for mental illness and mental deficiency, and a missed opportunity to discuss reform of the mental deficiency system. The introduction of voluntary treatment for the mentally ill in 1930 made the mental deficiency system even more of an anomaly, trapped in the incarceral mode of the past. Pressure to make space for voluntary curable cases under the 1930 Mental Treatment Act led to an investigation into the possibility of transferring the incurable and bed-bound from mental hospitals to mental deficiency institutions. Administrative priorities were therefore encouraging a reclassification of the mental illness/ deficiency divide, with almost 10,000 chronic cases in mental illness hospitals under consideration for transfer.[106] Although this did not take

[102] PRO Ed 50/112, Fisher to Chamberlain, 11 Mar. 1921, p. 2.
[103] PRO Ed 50/124, Macaulay, 30 June 1931.
[104] *Roy. Com. 1926, Minutes of Evidence, ii*, evidence of CAMW, 21163–5.
[105] PRO MH 58/103, Brock to Minister of Health, 26 July 1929.
[106] PRO MH 51/520, conference of the Minister of Health with the Board of Control, 5 Feb. 1932, Brock, p. 4; MH 51/541, 'Report of Investigation of Drs Wilson and Lewis 1932–3; MH 58/68, Circular 761, May 1931; M. Thomson, '"Though Ever the Subject of Psychological Medicine": Psychiatrists and the Colony Solution for Mental Defectives', in Berrios and Freeman, *150 Years of British Psychiatry*, ii. 138–40.

place, the very fact that it was considered by the government is indicative of the way mental deficiency institutions were regarded as a residual and stagnating area of the welfare system.

THE CHANGING CHARACTER OF THE BOARD: FROM SPECIAL EXPERTISE TO INTEGRATION

The most fundamental organizational issue facing the Board was the position of mental health services in relation to general health care. Integration with the National Health Service in 1946 can be seen as marking the end of a period of relative independence. This gives the interwar years an important comparative significance when set against mental health care under the Welfare State. However, the balance between independence and integration, and between special expertise and general bureaucratic expertise, was already shifting during the interwar period. This transition does not fit into a neat model of modernization. The Board had been set up as part of a broader vision of bureaucratic specialization in 1913, though it drew on the traditions of its predecessors, the Commissioners in Lunacy. During the interwar period this identity came under threat, as there was pressure to develop in two directions: first, away from legalism and towards being a consultative medical body; and secondly, towards integration with the dominant, non-specialist, bureaucratic culture of the civil service. These two trends—the move towards a medical model, on the one hand, and towards a non-specialist bureaucratic model, on the other—were often in tension. Squeezed between the two, the Board's own rather special character was torn in two directions. This can be traced through an analysis of its changing personnel during the interwar period.

The Chairmen

The Chairmen of the Commissioners in Lunacy and the Board of Control were usually laymen rather than doctors. In the nineteenth century, the most noted Chairman had been Lord Shaftesbury, whose position arose from his active lay interest in campaigning for lunacy reform. The first Commissioner of the Board of Control, W. P. Byrne, was a very different type, one of the new generation of post-Macauley–

Trevelyan reform, civil-servant meritocrats.[107] After a successful career in the Home Office, Byrne was brought in to serve on the 1908 Royal Commission on the Feeble-Minded, and then to implement its plans and set up the new administration. His appointment reflected his reputation as a problem-solver, rather than any personal interest or expertise in the field. Thus, as soon as priorities shifted because of the outbreak of war, he was temporarily moved out of the Board to serve as Under Secretary in Ireland.[108]

The war stimulated a debate about reorganizing the civil service on more rational lines.[109] Following on from the ideas already developed by the Minority Report on the Poor Law, the Haldane Commission recommended that there should be a separate department to administer each central function of government. For mental health, this would mean being subsumed under a new health department.[110] The idea of integrating mental with general health care was also encouraged by the wartime shell-shock experience, which showed that even the fittest members of society were prone to mental illness if placed under stress, and that in this sense mental health care, like physical health care, was something needed by all.[111] Therefore, in 1919, with the establishment of a Ministry of Health imminent, the independent status of the Board of Control had to be reconsidered. The Commissioners realized that remaining outside the new health department would reinforce the stigma surrounding mental illness. On the other hand, they were wary about their work and special interests being swamped by the concerns of general health. In the end a compromise solution was reached. The Board was protected on the grounds that it provided a unique combination of legal and medical expertise and experience and could run the services more economically than a new body, but its work was placed under the supervision of the Ministry of Health in the hope that this would help destigmatize mental illness.[112] With the new

[107] J. Pellew, *The Home Office, 1848–1914: From Clerks to Bureaucrats* (London, 1982), 200.

[108] PRO MH 107/47.

[109] ADD Box 38; *Report on the Transfer of Functions of Poor Law Authorities in England and Wales (Maclean Report)* (Cd. 8917, 1918); F. Honigsbaum, *The Division in British Medicine: A History of the Separation of General Practice From Hospital Care, 1911–1968* (London, 1979), 40–1.

[110] PRO MH 79/377, L. G. Brock, notes on Morant's plans for the Ministry of Health; W. H. Greenleaf, *The British Political Tradition*, iii (London, 1988), 170–95.

[111] ADD 38, report of the Reconstruction Committee Sub-Committee on a Ministry of Health, May 1917, p. 6.

[112] PRO MH 51/631, memo, 'Coordination of procedure with Ministry of Health', *c.*1918, and letter to S. W. Harris, 16 July 1918; ADD Box 39, Byrne, 12 Mar. 1919.

Ministry under the influence of Christopher Addison, Robert Morant, and George Newman, medical specialization was the flavour of the day; as such the auguries appeared favourable for the Board of Control.[113]

On Byrne's return, the temporary Chairman, Dr Mariott Cooke, stood down. As described above, friction arose between the Boards of Education and Control over policy for defective children. This argument had a longer history, stretching back to Byrne's period at the Home Office when he had competed with the medical lobbies at both the Local Government Board and the Board of Education for Home Office control over mental defectives.[114] His departure symbolizes the transition from an era in which mental disorder was still dealt with alongside law and order, to one in which it was firmly an issue of health and welfare.

Byrne's replacement, F. W. Willis, had served his career on the Local Government Board and more recently in the National Insurance Commission.[115] As such, he was far more qualified than Byrne to negotiate politically with local authorities and to institute welfare policies. Minister of Health, Alfred Mond, described to his Prime Minister, Lloyd George, the type of new political qualities which were now needed by a Chairman of the Board:

> The main qualifications for the post are those of an experienced administrator capable of managing a comparatively small but still important department of state, well and economically. As Chairman of the Board of Control he must also be a man of judgement and tact, likely to get on well with medical and legal colleagues. It is also very necessary that he should be well acquainted with the machinery of local government and the chief personalities connected with local government.[116]

Ex-businessman as he was, Mond recognized that the Chairman of the Board had to be a skilful manager, negotiating the delicate issues which surrounded mental illness by establishing firm relations with outside professional bodies and local government, and running his own department on an efficient basis.

Willis was succeeded by Laurence Brock, a protégé of civil-service mandarin, Robert Morant.[117] He inherited his mentor's desire to inte-

[113] ADD Box 38. S. Sturdy, 'State Strategies for Health Care and the Development of Clinical Science, 1900–1920', paper given at Wellcome Unit for the History of Medicine, Oxford University, 22 Feb. 1990.

[114] Sutherland, *Ability, Merit, and Measurement*, 39–55.

[115] PRO MH 107/28; *WWW, 1941–50*, p. 1244.

[116] PRO MH 107/28, Mond to PM, 21 June 1921.

[117] V. Markham, 'Robert Morant—Some Personal Reminiscences', *Public Administration*, 28 (1950), 257; Obituary, *Eugenics Review*, 41 (1949), 67, 87; PRO MH 107/4, biographical file.

grate mental health into the mainstream of public health and for government to be placed in the hands of an administrative elite who would concentrate upon policy making rather than day-to-day administration.[118] Prior to joining the Board, Brock had been interviewed by the Royal Commission on Mental Illness on the relationship between the Ministry and the Board. He described it as 'entirely harmonious', with the Ministry having a limited regulatory power over expenditure akin to the Treasury's.[119] This answer was somewhat disingenuous, as the delays caused by having to gain approval were a frequent source of tension.[120] When pressed on the economy of amalgamating the Board with the Ministry, Brock replied that he did see a case for bringing it under the CMO instead of the present cursory overview.[121] Brock's appointment as Chairman in 1928 can therefore be seen as marking a further step towards greater integration of the Board under the Ministry of Health. However, as Brock became involved in work at the Board, he developed close relations with outside groups such as the Eugenics Society, the CAMW, and the RMPA, and therefore he maintained the Board's intermediate and mediating position between central government and outside groups. Moreover, inheriting Morant's enthusiasm for research and preventive medicine, Brock sometimes played an active role in stimulating new initiatives in these areas. As such, though Brock brought the Board under closer bureaucratic control from the Ministry of Health, he also encouraged it to play an independent consultative medical role.

The Commissioners

In the 1950s, the National Council for Civil Liberties (NCCL) argued that during the interwar period the primary function of the Board had shifted from being a legal to a medical one—a shift which was reflected in the changing proportion of medical to legal Commissioners.[122] However, this failed to take into account a concurrent shift of power between medical Commissioners and non-specialist civil servants, which was equally important.

Non-specialist bureaucrats ('experts in administration', rather than law or medicine) were inclined to be very suspicious of the efficiency of

[118] Markham, 'Robert Morant', 260; Stacey, 'Ministry of Health', 85.

[119] *Roy. Com. 1926, Minutes of Evidence*, 1066–222. [120] PRO MH 51/632.

[121] *Roy. Com. 1926, Minutes of Evidence*, 1210.

[122] *Roy. Com. 1954–7, Minutes of Evidence*, 22, evidence of NCCL, pp. 839–40.

the Board of Control. From the early 1920s onwards there was pressure to rationalize the structure of the Board by creating a smaller central Board under tight bureaucratic control and by shifting the Commissioners into a purely inspectorial role:

> What we are trying to do is to get rid of the system where the Board reviewed the papers in the office down to the most meticulous details and to substitute an ordinary office organization where the staff does what it can efficiently do and gets the Board in on the policy and 'expert' questions, the Chairman in effect running the place and the Commissioners being visiting persons first of all and not office men.[123]

Pressure for such a reorganization reflected a concurrent shift of power and influence within the Ministry of Health, away from medical men, such as George Newman, towards administrators, such as Principal Secretary Arthur Robinson, whose administrative conservatism would stifle medical innovation.[124]

The pretext to justify reorganization and tighter supervision over the Board's work was provided in 1930 when the transfer of poor law institutions to the Ministry of Health exposed the appalling conditions under which some defectives were being kept.[125] As a result, the Board was restructured into a smaller central Board of four senior Commissioners, with other Commissioners serving a primarily visiting role. Bond, the senior medical Commissioner, vigorously opposed the reorganization on the grounds that the Board's greatest value lay in its consultative methods.[126] Similarly, the visiting medical Commissioners complained about no longer having access to consult files held by the central Board.[127] And when the government attempted to economize further on standards of colony care in 1930, the medical Commissioners showed signs that they were willing to break official ranks to complain about this.[128]

In 1931, E. J. Fitzgerald, the Assistant Medical Officer at Winwick Mental Hospital, publicly complained that none of the eight new Commissioners appointed that year were serving members of mental or mental deficiency hospitals, and that three were in fact retiring naval officers. He viewed these appointments as an affront to the psychiatric profession, which would seriously undermine its morale.[129] Brock

[123] PRO MH 58/90, Robinson to Minister, 15 Mar. 1923.
[124] Stacey, 'Ministry of Health', 129–43.
[125] PRO MH 58/99, Newman to Brock, 2 June 1930. [126] PRO MH 58/96.
[127] PRO MH 51/517, meeting of circuit leaders and Board, 18 June 1940.
[128] PRO MH 58/97, Newman to Heseltine, 11 June 1931.
[129] PRO MH 51/516, Fitzgerald to Sir Robert Young, 10 Mar. 1931.

defended the appointments, stressing the value of their 'personal qualities; good manners, social experience, and a good presence'. Privately, however, he admitted that the real problem was that the Treasury refused to provide a salary adequate to attract 'men of real standing', and had forced this policy upon the Board.[130] Despite this private sympathy, Brock requested the Lancashire Mental Hospitals Board to deliver an official reprimand to Fitzgerald for writing about the issue to the press.[131] The episode highlights the tensions which were emerging in the alliance between the Board and the psychiatric profession by the 1930s, suggesting that the model of an interwar shift from a legal to medical mode is too simplistic, since it failed to take into account the concurrent trend towards tighter administrative control.

STATESMEN IN DISGUISE?

During the interwar period there was serious disquiet that, under democratic pressure to introduce more and more complex systems of legislation, Parliament was left with insufficient time or expertise to monitor seriously the details of new laws. As a consequence, effective control to shape policy was being delegated into the hands of the civil servants who administered it—a 'new despotism'.[132] However, such criticism is hardly compatible with the opposite complaint, voiced by many historians, that the interwar civil service was overcautious and destructively negative, especially when it is compared to the innovative and interventionist Edwardian period, under Morant, Beveridge, and Llewellyn Smith, which preceded it, or the watershed formation of the Welfare State which followed.[133] The incompatibility of these assess-

[130] PRO MH 51/516, Brock, 23 Feb. 1931; Brock to Leggett, 20 Mar. 1931.

[131] PRO MH 51/516, Brock to Sir George Etherton, 27 Apr. 1931.

[132] Lord Hewart, *The New Despotism*; C. K. Allen, *Bureaucracy Triumphant* (London, 1931), 22; G. K. Fry, *Statesmen in Disguise* (London, 1969); T. Taylor, 'An Early Arrival of the Fascist Mentality: Robert Morant's Rise to Power', *Journal of Educational Administration and History*, 17 (1985), 48–62.

[133] Though the two criticisms were combined in Harold Laski's definition of bureaucracy as: 'the term usually applied to a system of government the control of which is so completely in the hands of officials that their power jeopardizes the liberties of ordinary citizens. The characteristics of such a regime . . . are a passion for routine in administration, the sacrifice of flexibility to rule, delay in the making of decisions and a refusal to embark upon experiment. In extreme cases the members of a bureaucracy may become a hereditary caste manipulating the government to their own advantage.': F. R. A. Seligman (ed.), *Encyclopaedia of the Social Sciences* (New York, 1930), 70.

ments suggests that they may tell us as much about the different expectations of the critics as they do about the interwar civil service. As Rodney Lowe has argued, the apparently opposite critiques are both reflections of the fact that during the interwar period the civil service was rigorously restrained by political influences from both above and below. In Lowe's view it is unfair to be overly critical about caution and inaction, for the interwar civil service was faced by a much greater challenge of 'adjusting to democracy' than any of its forebears; attacks on the 'new despotism' were as much a sign of this determination to assert democratic regulation, as they were a sign of civil servants out of control.[134]

On the whole, this analysis of the Board of Control supports Lowe's argument. The Board's independence was checked by pressures from above: political caution and concern about individual liberties meant that Parliament was reluctant to act on the recommendations of the Board; and within the civil service itself, the Board was restrained by the tight fiscal control from the Treasury and the Ministry of Health. At the same time, the Board was restrained from below because of its own concerns about public opinion, as well as those of the Ministry of Health and Board of Education. The introduction of the idea of a 'special public'—one whose interest in the problem was permanent, rather than fleeting like that of the general public—adds an extra dimension to this equation.[135] The Board was undoubtedly influenced at times by the views of these pressure groups, in particular by the psychiatric profession and voluntary mental welfare groups. This might be considered further evidence of governmental attention to the public; it certainly does not support the view of the civil service as a closed environment.[136] On the other hand, critics might view it as evidence of the Board acting as an undemocratic front for outside pressure. A problem with both positions is that they view the civil service as the passive object for political pressure from outside groups.[137] The close affiliation between the Board's members and outside groups shows that there are serious problems in assuming that there was always a strict separation between public and private, or governmental and non-

[134] Lowe, *Adjusting to Democracy*, 44–7.

[135] H. E. Dale, *The Higher Civil Service of Great Britain* (Oxford, 1941), 178–90.

[136] Stacey, 'Ministry of Health', 141.

[137] For an examination of the interaction between government and pressure groups during this period: G. Wootton, *Pressure Groups in Britain, 1720–1970* (London, 1975). Greenleaf, *British Political Tradition*, iii. 338–501.

governmental spheres.[138] In fact, the Board could be an active agent of pressure itself, protecting its territory and role, spreading the ideas of its expert members, and coercing and influencing, rather than being influenced by, outside pressure groups. This active and independent role does not justify the epithet of 'statesmen in disguise' being attributed to the Board; the political, financial, and administrative constraints on the Board surely make this description misleading. However, in order to overcome these constraints, it was sometimes necessary for the Board to take initiatives which were not strictly under the control of Parliament. In this sense, the Board, and more generally the new welfare bureaucracies of the twentieth century, did have the potential to produce a 'new despotism', though paradoxically this was an integral part of adjusting to democracy.[139]

There were a number of ways in which the Board of Control could exert its influence on policy, but it is difficult to delineate the extent to which such action went beyond the purely administrative, which was expected of civil servants, to become independently political. Through its Annual Reports, the Board could encourage local authorities to adopt non-statutory services. And to encourage local activity, it published figures of the number of reported mental defectives in each local area, even though such figures were unofficial—a strategy which was later pointed out to be an abuse of official power.[140] In the hope of stimulating local authorities into action, the Board's Annual Reports sometimes adopted language which was so alarmist that it might have come straight from the pages of the Eugenics Review; for instance, the 1927 Annual Report claimed that:

> mental deficiency is the fundamental cause of much pauperism, and of a great deal of petty and serious crime. It is also the fundamental cause of much inebriety. Mental deficiency is intimately allied with insanity. . . . Therefore mentally defective parents must create centres of degeneracy and disease which welfare cannot reach.[141]

Likewise, the Annual Reports contained numerous case histories which graphically pointed out the eugenic danger of leaving feeble-minded women at large.[142] The Board even took the bold step of using its

[138] C. Maier (ed.), *Changing Boundaries of the Political: Essays on the Evolving Balance Between the State and Society, Public and Private in Europe* (Cambridge, 1987).

[139] Extensive analysis of the relationship between bureaucracy and democracy has been stimulated by the work of Max Weber: e.g. D. Beetham, *Max Weber and the Theory of Modern Politics* (Oxford, 1974), pp. 63–89; M. Albrow, *Bureaucracy* (London, 1970), 47–9.

[140] PRO MH 51/374, Barter 16 Feb. 1934.

[141] *BOC 1927*, p. 39.

[142] *BOC 1925*, pp. 59–60.

Annual Reports to recommend legislation to prohibit marriage of mental defectives.[143]

A less visible form of influence was the effective delegation of legislative power to the Board in its ability to change the rules and regulations of the Act through administrative orders to local authorities.[144] By using these circulars to mark out certain types of defective as in 'urgent' need of care (such as the young, the high-grade, and the morally defective), the Board had a crucial influence in determining which defectives were placed in institutions and which were left in the community.[145] Other circulars set down the standards for mental hospitals,[146] or pressurized local authorities to be more active.[147] Such independent action upset the Treasury and raised the wrath of other spending departments when it interfered with their duties.[148] However, criticism of the Board's use of such 'delegated legislation' should be balanced by realizing that it had little choice, since the often outdated rules and regulations it had to work with were not reformed systematically until 1948. In 1935, there had been a move to revise them, provoked by the establishment of the Donoughmore Committee to investigate the accusations of delegated legislation made in *The New Despotism*, but action was deferred because of ministerial and parliamentary reluctance to address the time-consuming and controversial issue of mental deficiency.[149] In these circumstances the Board's use of circulars to adjust rules was understandable and was excused by the Minister of Health as preferable to the problems of having to draft and pass new legislation.[150]

A third way in which the civil service could exert influence over policy was in appointing personnel to investigative committees. Indeed, the general preference of interwar governments for Departmental and Interdepartmental Committees, rather than Royal Commissions, was because they had greater control in appointing the Committees and could therefore be more confident about the eventual findings.[151] For

143 *BOC 1927*, p. 48.
144 *Roy. Com. 1954–7*, *Minutes of Evidence*, 22, evidence of NCCL, p. 803.
145 PRO MH 51/240, circular 591.
146 PRO MH 51/239, circular 557A, 1919.
147 PRO Ed 50/124 circular 1341, 1931.
148 PRO MH 58/81 Treasury to BOC, 12 Dec. 1921.
149 PRO MH 58/66; Greenleaf, *British Political Tradition*, iii. pp. 541–2.
150 PRO MH 58/66 Young to Sec., 25 Jan. 1934.
151 PRO MH 58/103, Hilton Young (Minister of Health) to Sec., 15 Jan. 1932; H. Mcd. Clokie and J. W. Robinson, *Royal Commissions of Inquiry: The Significance of Investigations in British Politics* (London, 1937), 191–7.

example, great care was taken over appointing the members of the Interdepartmental Committee on Sterilization, and through the skilful influence of Brock, the Committee's Chairman, it was able to produce a clear statement in favour of voluntary sterilization for defectives.[152] Similarly, considerable effort went into making sure that those who were appointed to the Departmental Committee on Colonies, which was to investigate whether the cost of institutional care could be further reduced, were all 'rigid economists'.[153]

Despite these examples of ways in which civil servants could exert some independent influence over mental deficiency policy, on substantive issues their personal views and biases were largely stifled by political restraints. For instance, as early as 1923 Chairman of the Board, Willis, and Chief Medical Officer of the Ministry of Health, Newman, had expressed interest in the use of sterilization for mental defectives, but they dropped the issue when it became apparent that legislation would be needed.[154] In the 1930s the Board's next Chairman, Brock, was also personally converted to support sterilization, and it was he who extended the remit of the Sterilization Committee's recommendations to include the physically handicapped, the blind, and the hereditarily mentally ill.[155] However, Brock's influence, and the sympathy of Robinson and Hilton Young, Permanent Secretary and Minister of Health respectively, could not overcome the political barriers to legislation.[156] Moreover, because of the culture of political caution within the civil service, the Brock Committee presented its findings in a manner which others felt seriously underplayed its backing for sterilization.[157]

Ultimately, the realization of influence depended upon the response in the localities. The Board could coerce by threatening to remove funding, but it could not force developments. When it did place heavy pressure on local authorities, they responded by complaining to the Treasury, who sided with the localities and restrained the Board.[158]

[152] MH 79/292, Brock to Robinson, 16 Mar. 1932; PRO MH 51/210, Brock, 6 July 1933, 16 Aug. 1933; PRO MH 58/104B, Fisher to Hilton Young, 5 Feb. 1934.

[153] PRO MH 51/710; MH 58/97, Brock, 4 Nov. 1929; *Colonies for Mental Defectives: Report of the Departmental Committee (Hedley Report)* (London, 1931).

[154] PRO MH 79/292: Willis to Robinson, 7 Feb. 1923; Newman, 12 Feb. 1923; Robinson, 12 Feb. 1923; Willis to Robinson, 1 Mar. 1923.

[155] PRO MH 51/210, Brock to Committee, 6 July 1933: MH 79/292, Brock to Robinson, 16 Mar. 1932.

[156] PRO MH 58/103: Hilton Young, 15 Jan. 1932; Robinson, 18 Apr. 1929.

[157] PRO MH 51/210, Tredgold, 26 Sept. 1933; Adams, 29 Sept. 1933.

[158] PRO MH 51/520, Brock to Robinson, 15 Dec. 1931; Brock to CCA, 17 May 1932; report of Public Health and Housing Committee. 10 Feb. 1932; *BOC 1932*, pp. 12–13.

Because of the continued inadequacy of local authority provision, the Board went beyond its official remit to encourage the delegation of administrative responsibility to non-statutory bodies, using voluntary groups to pioneer new services.[159] Similarly, because of the limitations of inspection, responsibility for the standards and running of the colony system was effectively delegated to the psychiatric profession. Paradoxically, therefore, delegation of administration was often the result of attempts to overcome the weakness of central government, rather than a reflection of 'ministerial despotism'.

CONCLUSION

A number of points can be made about the literature on the civil service in the interwar years from the perspective of the Board of Control. First, the distinctive character of the Board indicates that there were exceptions to the growing dominance of the Treasury style over the interwar civil service. The Board cannot be seen as a closed environment; its members' links to the outside were vital if the administrative system was to hold together. Moreover, this analysis shows that the civil service could be a conduit and a battleground for ideas. The interdepartmental and intradepartmental politics which resulted indicates the danger of the bland term 'official opinion', and should lead us to question the caricature of the interwar civil service as a monolithic body, united by social background and unreceptive to pressure from without. Secondly, any criticism of lack of innovation must take into account the limitations under which the Board had to work.[160] The Treasury, the Board of Education, and the Ministry of Health all played a part in inhibiting Board of Control initiative.[161] Thirdly, the history of the Board is valuable for uncovering the tensions which existed between different concepts of expertise within the civil service.[162] This is apparent within the Board of Control, and also in the internal politics of the Ministry of Health and the Board of Education. The gradual loss of power by the medical and legal experts of the Board of Control was an indication of the failure of the visions of the Minority Poor Law

[159] M. Rooff, *Voluntary Societies and Social Policy* (London, 1957), 136–46. See Ch. 4.
[160] Davidson and Lowe, 'Bureaucracy and Innovation'.
[161] McDonald, 'The Geddes Committee'.
[162] Macleod, *Government and Expertise*, 20.

Report and Haldane Committee for the transfer of government into the hands of specialists. The administrators who took control instead were heavily constrained by their political awareness. Thus, the administrative stagnation which often resulted was integrally linked to the process of adjusting to democracy, and the attempt to overcome this through using administrative tools to shape policy was largely understandable and certainly did not amount to ministerial despotism. As this analysis of the Board of Control suggests, the administrative and political spheres cannot be clearly divided in interwar Britain: the attempt to make the problem of mental deficiency a purely technical one—to place power into the hands of the 'experts'—was always in tension with constraints from above and below.[163]

An awareness of the special character of the Board and its low status is vital to understanding how the problem of mental deficiency evolved: it caused the bifurcation between children and adults which was crucial in provoking the continuing concern of social hygienists and eugenicists; it lay behind the limited realization of segregation which kept the 'problem' alive during the interwar years, and led to consideration of the alternative solutions of sterilization and community care; it resulted in the incorporation of voluntary and professional groups into an informal extrastatutory administration;[164] and finally, it left much of the responsibility for instituting policy with the local authorities.[165] Therefore, although analysis of central government politics is crucial in understanding the construction of policy to deal with the problem of mental deficiency, for a full picture we have to go beyond the strict confines of central government to analyse its interaction with professional, voluntary, and local authority groups.

[163] For futher discussion of this theme: P. Starr and E. Immergut, 'Health Care and the Boundaries of the Politics', in Maier (ed.), *Changing Boundaries*, 221–54.

[164] See Chs. 3–5. [165] See Ch. 6.

3
The Colony Solution

INTRODUCTION

The role of the medical profession and the institutional site of its practice—the hospital—have dominated accounts of the history of medicine. Early histories tended to portray doctors as heroic and self-less pioneers, struggling against lay ignorance and suspicion. However, the tone of recent literature has been far more critical: the definition of the medical profession is recognized as being problematic, and it is no longer automatically accepted that doctors possessed a unique or accurate understanding of illness and health, or that their motives in advancing medicine were wholly disinterested.[1] Nowhere has the role of the profession been more controversial than in the history of psychiatry, since it is here that the profession's claims to be able to treat and cure illness have been viewed with the greatest scepticism. In the historiography of British psychiatry, the leading critic of the psychiatric profession has been Andrew Scull, who has argued that the emergence and expansion of asylums was driven by the ambition and self-interest of the psychiatric profession, rather than the needs of the patients. Although Scull acknowledges that psychiatrists may have had humanitarian and idealistic ambitions in the early days of the asylum,[2] he argues that by the second half of the century the failure of psychiatry to cure mental illness was apparent and that psychiatrists presided over mass 'dumping grounds' for the unwanted and chronically ill. Despite this, psychiatrists maintained and expanded their roles, adopting hereditarian theories of degeneration to explain their inability to cure the increasing numbers of mentally ill.[3] Although Scull's account ends before the

[1] For instance, E. Freidson, *Professional Dominance: The Social Structure of Medical Care* (Chicago, 1970).

[2] A. Scull (ed.), *The Asylum as Utopia: W. A. F. Browne and the Mid-Nineteenth-Century Consolidation of Psychiatry* (London, 1991), pp. xxxiv–xxxix.

[3] A. Scull, *Museums of Madness: The Social Organisation of Insanity in Nineteenth-Century England* (London, 1979); for an overview of the debate, see A. Scull, *Social Order/Mental Disorder: Anglo-American Psychiatry in Historial Perspective* (London, 1989), 1–30.

passage of the 1913 Mental Deficiency Act, it is easy to see how the Act could be interpreted as a further stage of psychiatric imperialism: providing a further outlet for professional expansion, creating a new class of deviants in need of control, and adopting hereditarian arguments to highlight the urgency of the problem and the need for medical management. Scull's work has provoked considerable criticism from those who are impressed by the commitment of Victorian society to asylum building in comparison to that of the late twentieth century, and who attribute the shortcomings of the asylum to external faults such as inadequate resources and an unreformed legal framework.[4] In the light of later work on admittance and discharge, Scull's claims that asylums were 'dumping grounds' for the unwanted, and that the insane entered never to leave asylums, are both in need of revision.[5] On the other hand, his view that the institutional system was in a mounting crisis by the end of the nineteenth century—overcrowded and filling with chronic and incurable cases—is still largely accepted.

The motivations of the psychiatric profession have come under renewed scrutiny because of growing awareness of its involvement in sterilization and euthanasia in Nazi Germany. Historians are divided as to whether this involvement was a consequence of unscrupulous professional self-interest in the search for employment and status within the Nazi regime, was driven by the terrible logic of a murderous eugenic science, or was the response to a financial and curative crisis of institutional care.[6] This debate on Nazi psychiatry raises obvious comparative questions about the motivations and attitudes of psychiatrists elsewhere during the interwar period, particularly those involved in the care of mental defectives, who were one of the first groups targeted for 'euthanasia'.[7]

[4] G. Grob, *Mental Illness and American Society* 1875–1940 (Princeton, 1983); G. Grob, *From Asylum to Community: Mental Health Policy in Modern America* (Princeton, 1991); G. Grob, 'Marxian Analysis and Mental Illness', *History of Psychiatry*, 1 (1990), 223–32; G. E. Berrios and H. Freeman, 'Introduction', in id. (eds.), *150 Years of British Psychiatry*, i, *1841–1991* (London, 1991).

[5] J. Walton, 'Lunacy in the Industrial Revolution: A Study of Asylum Admissions in Lancashire, 1848–50', *Social History*, 13 (1979), 1–22; L. J. Ray, 'Models of Madness in Victorian Asylum Practice', *European Journal of Sociology*, 22 (1981), 229–64; A. Scull, '*Museums of Madness* Revisited', *Social History of Medicine*, 6 (1993), 3–23.

[6] P. Weindling, *Health, Race, and German Politics between National Unification and Nazism 1870–1945* (Cambridge, 1989); M. Burleigh, 'Psychiatry, German Society, and the Nazi "Euthanasia' Programme", *Social History of Medicine*, 7 (1994), 213–28; M. Burleigh, '"Euthanasia" in the Third Reich: Some Recent Literature', *Social History of Medicine*, 4 (1991), 317–28.

[7] M. Thomson and P. Weindling, 'Sterilisationspolitik in Grosbritannien und Deutschland' (The Politics of Sterilisation in Britain and Germany), in F. W. Kersting, K. Teppe, and B. Walter (eds.), *Nach Hadamar* (Paderborn, Germany, 1993), 137–49.

As these ongoing debates indicate, determining professional motivations is far from straightforward. There are at least three reasons for this. First, the boundaries of professional status are often unclear: professionalization is a process in which professional identity is constantly being reconstituted as a result of negotiation, debate, and conflict between competing factions, and, as such, professions rarely speak with a singular voice.[8] Secondly, there is always a danger of retrospectively judging action as self-interested, simply because it turned out to be of long-term benefit to a profession, whereas at the time the notion of what was in a profession's best interests was heavily contested and was modified by a complex of motivations. For instance, the desire for a respectable and disinterested professional identity often offset economic and monopolistic interests. And thirdly, there is the problem of balancing a profession's public explanations of its actions with those motivations which, for one reason or another, remained unstated or unrecorded. Bearing these points in mind, in this chapter I shall consider the role of the psychiatric profession in the development of the colony solution for mental defectives. In doing so, my aim is twofold. First, I shall assess the significance of professional self-interest in the formation of social policy. Secondly, this chapter will be the first of three in which I shall examine the influence of what Harold Perkin has termed 'professional society' in the development of the problem of mental deficiency.[9]

This concept of 'professional society' goes well beyond the rather narrow perspective of studying professional self-interest to include the broader attitudes and ideologies which motivated action. Perkin's use of the term 'professional society' embraces a wide range of social groups, from professional associations, to occupational groups, voluntary organizations, and expert bodies, who all advanced a vision of organizing society according to what he terms the 'professional ideal'. Of course, invariably the recommendations of 'professional society' were also in the self-interest of the groups concerned, but the policies were advanced as disinterested 'solutions' to social problems, making society fairer (more meritocratic), more efficient, and paving the way towards a

[8] This account will question the use of 'professional' as an ideal category, and instead focus upon the process of professionalization in which 'professionalism' was used as a strategy to claim control over new areas of the welfare complex. See E. Friedson, 'The Theory of Professions: The State of the Art', in R. Dingwall (ed.), *The Sociology of the Professions* (London, 1983), 19–37.

[9] H. Perkin, *The Rise of Professional Society: England since 1880* (London, 1989).

new social citizenship based on equality of opportunity, which found its ultimate realization in the Welfare State.[10] Perkin argues that because it led to social efficiency, the 'professional ideal' became the dominant force in twentieth-century society, overcoming class divisions, and restructuring society into a system of parallel professional hierarchies competing for the resources of the state.[11] As such, the rise of professional society played a key role in mediating Britain's relatively peaceful and welfarist adjustment to democracy.[12] We can examine this thesis through the role of 'professional society' in dealing with the problem of mental deficiency, beginning in this chapter with the psychiatric profession. Three issues seem particularly pertinent. First, to what extent were psychiatrists guided by the 'professional ideal' of meritocracy and social efficiency, rather than by self-interest? Secondly, to what extent was the 'professional ideal' becoming the dominant organizing force in society: did this professional idealism unite the psychiatric profession, and how far did it also influence the professionalization of nursing and thereby overcome class and gender divisions between the two groups? Finally, how compatible was the vision of expert control of society with democratic government, what were the tensions between the two, and how did the psychiatric profession react to this?

THE EMERGENCE OF THE COLONY SOLUTION

In the context of its own time the 1913 Mental Deficiency Act was in many respects a progressive piece of social legislation.[13] The idea of the colony was crucial in the construction of this progressive image, since it associated the Act with a fashionable tool of social engineering, and obscured its reliance on the type of warehouse-like institutions which catered for the mentally ill and were broadly condemned as ineffective and inhumane. The census of 1911 saw the highest ever proportion of the British population in institutions.[14] The Mental Deficiency Act was the last major attempt to solve a social problem by locking it away from the rest of society. It can be seen as a turning point—the culmination

[10] H. Perkin, *The Rise of Professional Society: England since 1880* (London, 1989), 331–58.
[11] Ibid. 1–26. [12] Ibid. 217. [13] Ch. 1.
[14] J. Harris, *Private Lives, Public Spirits: A Social History of Britain, 1870–1915* (Oxford, 1993), 218.

and most mature expression of the institutional era, but one which also contained the seeds of decline.

Confidence in the institution had already seen its troughs in the nineteenth century, but at each point it had been revived by enthusiasm for new institutional designs and the undiminished willingness of the Victorians to express their beneficence through monumental building. However, the idealism and pride was difficult to maintain, as institutions were continuously forced to expand beyond their capacities, became overcrowded, and lost the close contact between staff and residents which was essential for 'moral therapy'.[15] In an attempt to produce an illusion of intimacy, and also to aid management of the growing institutions, new architectural designs separated patients into 'wings', 'pavilions', and 'villas'.[16] The colony was the last and the fullest development of such attempts to rescue the institution from becoming a warehouse.

The use of the term 'colony' for a site of care for the mentally disabled, and the allied idea that provision of a healthy environment would aid cure, can both be traced back to 1839 when the Swiss doctor, Guggenbühl, attempted to cure cretins by relocating them from the unhealthy air of Alpine valleys to the restorative environment of the mountains.[17] In the second half of the nineteenth century the term began to be used to describe mental hospitals which, instead of being barrack-like 'blocks', were built in the 'congregate' style, consisting of a collection of smaller 'cottages' or 'villas' with shared central administrative facilities. The colony was designed to address both the practical and the therapeutic problems of institutional care. Practically, the design addressed the problem of institutional overcrowding, as colonies would be able to expand, villa by villa, to potentially huge scale, gradually and non-disruptively. Therapeutically, the villas were to be small enough to maintain an intimate atmosphere, and were grouped together to create a kind of village community, which could help to socialize the patients; there was a sense that mental problems were a product of civilization, industrialization, and urbanization, and could be

[15] P. McCandless, 'Build! Build!: The Controversy Over the Care of the Chronically Insane in England, 1850–70', *Bulletin of the History of Medicine*, 53 (1979), 553–74. On 'moral therapy': R. Porter, *Mind-Forg'd Manacles: A History of Madness in England from the Restoration to the Regency* (London, 1990), 222–8.

[16] J. Taylor, *The Hospital and Asylum Architecture in England, 1840–1914* (London, 1991), 133–61; H. Burdett, *Hospitals and Asylums of the World*, ii (London, 1893).

[17] D. Pick, *Faces of Degeneration: A European Disorder, c.1848–c.1918* (Cambridge, 1989), 45–7.

solved by recreating a rural idyll.[18] The countryside location provided space and cheap land for the dispersed design, and farmland on which patients could be set to work cultivating their own food; at the same time, the fresh food, exercise, and fresh air seemed to offer therapeutic advantages for the residents. Thus, the colony design aimed to be both more economical and efficient, and more humane and therapeutic than earlier institutions[19]—the two aims were integrally linked in the progressive vision of social policy.

In the gloomy backwater of mental health care, the great appeal of the colony was that it associated the area with a social engineering tool which was being presented quite seriously by leading reformers as a solution to a host of the nation's major social problems.[20] The attraction, aims, and limitations of the mental deficiency colony can only be fully appreciated if they are placed in the context of this broader contemporary movement. The colony seemed to provide a new form of institutional care which was both more humane and more efficient than the closed asylum, workhouse, or hospital. As such, it had great potential for those groups who could not justifiably be placed in a closed institution (either in terms of equity or cost) but who still needed assistance and/or discipline. As early as the 1880s Charles Booth was proposing the colony as an ideal tool to excise the unreformable and incapable from the broader community, both to protect society from their polluting influence, and for their own good.[21] Such proposals continued into the first decade of the next century when the Report of the Royal Commission on the Poor Law recommended compulsory detention in a colony, lasting from six months to three years, to discipline those who wilfully refused to maintain themselves or their families, or whose poverty was the outcome of persistent drinking, gambling, or idleness.[22] However, this disciplinary function was given a progressive gloss by the adoption of medical language, with the Report describing the colony as serving a 'therapeutic

[18] *Roy. Com. Feeb.*, viii. 77–8, 237–41, 297–301, 317–21. On the idealization of the countryside: R. Williams, *The Country and the City* (London, 1973).

[19] This is consistent with a broader trend in which hospital design became less monumental and paid more consideration to the needs of the patients: A. Forty, 'The Modern Hospital in England and France: The Social and Medical Uses of Architecture', in A. King (ed.), *Buildings and Society: Essays on the Social Development of the Built Environment* (London, 1980), 61–93.

[20] *Report of the Royal Commission on the Poor Laws and Relief of Distress* (PP, Cd. 4499, 1909), pp. 427–30, 638, 1207.

[21] J. Brown, 'Charles Booth and the Labour Colonies', *Economic History Review*, 2nd ser., 21 (1968), 49–70.

[22] *Roy. Com. Poor Law*, 638.

role' and as reforming the 'morbid state of mind' of malingerers;[23] discipline had been recast as a tool of welfare rather than punishment.[24] Because of its reformist credentials and its status as a tool of discipline and re-education rather than punishment, the colony could also be advanced as a tool to deal with members of the population who were merely 'socially inefficient'. For instance, recognizing that unemployment could be caused by lack of work rather than malingering, there were proposals to place the unemployed in rural colonies where they would help resettle the depopulated land and where the fresh air and exercise would restore them to physical vigour.[25] And at the other end of the spectrum, but sharing many of the same ideological roots, there was the practice of sending out groups of orphans to populate Britain's overseas colonies.[26] As these examples suggest, the attraction of the colony was not simply that it provided new institutional solutions to discrete social problems, but also that it was a salve to broader social anxieties about rural depopulation, urban squalor, and racial degeneration.[27]

Although colonies for the sick and handicapped placed greater emphasis on medical therapy, their regimes were still disciplinary in style. The evolution of tuberculosis colonies epitomizes this and provides a striking analogy to the shift from mental asylums to colonies. At the turn of the century tuberculosis had been tackled by confining the infected in sanatoria, thereby containing the spread of the disease and assisting recovery through a healthy environment and rest. The problem with this strategy was that discharged patients often relapsed when they returned to the stress of industrial work in the community. To overcome this, treatment was to be followed by an extended 'hardening-up' period in a colony, where residents would perform light outside work (viewed as far less stressful than industrial work). Having benefited from the fresh air, strengthening experience, and training, they would be better prepared for return to the outside world. In reality, such farm work turned out to have little long-term value, since discharged patients were forced to return to their unhealthy industrial

[23] *Roy. Com. Poor Law* (Minority Report), 1207.

[24] For a full discussion of shifting strategies of discipline at this time: D. Garland, *Punishment and Welfare: A History of Penal Strategies* (Aldershot, 1985).

[25] *Roy. Com. Poor Law*, 427–30.

[26] W. and V. W. Carlile, *The Continental Outcast: Land Colonies and Poor Law Relief* (London, 1907).

[27] For further exploration along these lines: G. S. Jones, *Outcast London: A Study in the Relationship between Classes in Victorian Society* (London, 1976), 281–315.

jobs to obtain reasonable wages, and therefore often relapsed, became unemployed, and perhaps spread the infection. The solution was to build colonies where patients and their families would willingly settle after treatment, establishing their own separate sheltered communities.[28] Infection would thereby be contained, relapse prevented, and appropriate work would be provided for these weakened individuals. From a eugenic perspective, any hereditary predisposition to tuberculosis would be contained within the colony. The colony was also presented as a humanitarian advance: these nervous and weakly individuals would be placed in a community of similar types, an environment which would be far less stressful than life in the outside community. Segregation was thus justified as a therapeutic solution, in the interests of patients, their families, and the wider community.

Despite being a highly interventionist policy, the colony solution could attract the support of those who otherwise wanted to restrain the advance of the state. The first reason for this was that the intervention of the colony was set apart from the community, away from areas of population, and within its own walled confines. The second reason was that the colony intervened only with the incapable or irresponsible—those who could not be relied upon or treated as independent citizens. And finally, removal of these problem groups would leave the broader community to function more smoothly and with less need for social intervention, while the threat of the colony would discourage social inefficiency.

For all the interest it attracted, the colony only had limited success. Colonies were not established in any systematic sense for those who were expected to accept such a life voluntarily.[29] The main use of the colony was to provide a new design and ethos for the housing of groups who were already provided with institutional care, who had no choice over the matter, and for whom the colony seemed a progressive step away from enclosed institutions. The one notable experiment in using the colony to segregate a new group was the Mental Deficiency Act of 1913, with its extension of care to the feeble-minded. The feeble-minded provided an ideal target for the colony solution. Segregation

[28] Sir G. Woodhead and P. Varrier-Jones, *Industrial Colonies and Settlements for the Consumptive* (Cambridge, 1920), 99–115; L. Bryder, *Below the Magic Mountain* (Oxford, 1989), 46–70.

[29] Several voluntary colonies were established to provide work for unemployed townsmen. These included the LCC farm colony at Hollesley Bay: T. Hancock Nunn evidence to *Roy. Com. Poor Law*, 712–13. By contrast, compulsory colonies for the unemployed were established on the continent: *Roy. Com. Poor Law*, 428.

from the outside community could be justified for eugenic reasons and because the moral instability and social inefficiency of these individuals was a threat to the community. At the same time, the colony promised a sheltered environment in which defectives could be protected from the community and would be better able to contribute to their own support, and a rural and relatively free environment which made lifetime segregation more acceptable than in traditional mental hospitals. Finally, the fact that mental deficiency colonies, like those for tuberculosis, were to be controlled by doctors was crucial in making them publicly acceptable, since it obscured their disciplinary function and associated them with the progressive and expert ethos of medicine.

The 1908 Royal Commission on the Feeble-Minded undertook an extensive survey of possible ways to care for the mentally defective, and was particularly interested in the idea of the colony. By this time, a number of different types of mental institution had adopted the title of 'colony'. The most famous of these was Gheel colony in Belgium, which had attracted considerable attention in Britain since the mid-nineteenth century. In the middle ages, Gheel, the location of the shrine of Saint Dymphna, became a pilgrimage site for the sick and insane in search of cure. Many of these visitors remained close by, boarding out with the villagers. By the mid-nineteenth century, over a thousand mentally ill people were living with families in the surrounding area; by contrast, the central institution held only sixty patients. The colony was therefore not an institution as such, rather it encompassed the whole village and the surrounding farms—an area of fourteen by eight miles.[30] The Gheel 'family colony' attracted a few British advocates, such as William Bucknill who experimented with a 'cottage system' in the 1860s, but it was generally seen as unique and not easily transferable.[31] British psychiatrists condemned the lack of medical control, the unhygienic conditions, and poor diet of the peasant homes, and the pecuniary motivations of the hosts.[32] Commentators drew parallels between Gheel and the practice of 'boarding-out' the insane in the Scottish 'colonies of the north'.[33] In fact, this practice of boarding-out was not novel; for

[30] W. Parry-Jones, 'The Model of Gheel Lunatic Asylum and its Influence on the Nineteenth-Century Asylum System in Britain', in A. Scull (ed.), *Madhouses, Mad-Doctors and Madmen: The Social History of Psychiatry in the Victorian Era* (London, 1981), 200–4.

[31] Ibid. 208–9.

[32] Ibid. 204–6.

[33] G. Gibson, 'The Boarding Out System in Scotland', *JMS* 71 (1925), 253–63; Parry-Jones, 'Model of Gheel'.

centuries it had also been a practice of poor law authorities in England. What made the Scottish practice stand out was that it remained more common, perhaps because there was greater demand from farmers to supplement their incomes, and it was concentrated in certain villages which therefore became known as colonies. The reaction of doctors to these colonies, with their lack of medical control, was again largely negative.[34] The Royal Commission on the Feeble-Minded accepted medical opinion and rejected the 'family colony' model of Gheel and Scotland, preferring models which confined the residents within institutions.

A more important influence than Gheel was the colony model developed in the United States. The Royal Commission undertook a special survey of the American mental deficiency colonies, and devoted a whole volume of its report to this subject. The American model was of interest because it had developed in reaction to a similar set of therapeutic and economic problems to that encountered in Britain.[35] The key to the system was the development of satellite colonies, still controlled by the central institution, but located up to a hundred miles away, where high-grade defectives were sent to work on farms or even in towns. These colonies provided a solution to the problem of overcrowding in the host institutions, a more economic form of care in which the defectives contributed to their own self-support, and the prospect of retraining defectives for eventual life back in the community.[36] The great advantage of such an open colony model was that it could dramatically increase the numbers of defectives under care; as one witness reporting on European colonies pointed out, for every one defective in the host institution, up to three might be boarded out in the surrounding environs.[37] Finally, the Commissioners also considered a range of institutional models already in operation in Britain, including systems of small voluntary-run hostels and refuges for feeble-minded girls,[38] and the integrated system of specialized institutions for each grade of mental defective provided under the poor law by the Metropolitan Asylums Board in London.[39]

[34] Parry-Jones, 'Model of Gheel', 209–10.

[35] J. Trent, Jr., *Inventing the Feeble-Mind: A History of Retardation in the United States* (London, 1995), 60–130.

[36] S. P. Davies, *Social Control of the Mentally Deficient* (London, 1930), 216–90; *Roy. Com. Feeb.*, vii.

[37] *Roy. Com. Feeb.*, i. 521.

[38] Mrs Bramwell Booth, *Roy. Com. Feeb.*, ii. 174–80.

[39] Dr A. Downes (Senior Medical Officer LGB for London), *Roy. Com. Feeb.*, i, qu. 1784–1981; D. Cochrane, '"Humane, Economical and Medically Wise": The LCC as Administrators of Victorian Lunacy Policy', in W. Bynum *et al.* (eds.), *Anatomy of Madness*, iii. (London, 1988), 247–71; see Ch. 6.

Rather than following those colony models, which utilized satellite hostels and boarding-out, the Commission came down heavily in favour of a closed industrial villa colony—an institutional model which took on some of the style of the colony but did not open its walls to the outside world, and was not that radical an advance on the villa design which was already in practice.[40] The villa system offered scale, economy, and expandability, and could integrate the colony practice of employing patients in farm and industrial work to reduce costs and occupy the residents. However, a system of boarding-out would have been even cheaper, and therefore the choice of a closed rather than open colony model was not based on economy alone. The closed model was chosen because this was the only design which could both guarantee medical control over care and secure segregation of the residents. Although shaped by broader currents in social policy, the recommendations of 1908 should also be seen as a victory for medical opinion. It was a choice against alternatives which threatened medical control: against boarding-out on poor law relief, against placing defectives with lay persons in the community, against the lay management of voluntary hostels, and against poor law control of institutions.[41] Because of the Report's recommendation that institutions should contain mixed grades of defectives, rather than specializing after a process of administrative pre-classification, doctors were able to claim that only they could act as managers, for the key to management was their unique ability to classify defectives into different groups so that care could be specialized and the institution could be run on efficiency and 'modern' lines.[42] To understand the willingness of doctors to argue their case on lines of economy and accept the role of custodians, it is necessary to consider the status of the psychiatric profession at this time.

THE PSYCHIATRIC PROFESSION AND THE COLONY SOLUTION

At the start of the twentieth century, psychiatric medicine was a poorly developed and low-prestige field which was hardly recognized in the

[40] *Roy. Com. Feeb.*, vii. 235–7.

[41] Miss M. H. Mason, *Roy. Com. Feeb.*, i, qu. 1982–2077.

[42] F. W. Mott, *Roy. Com. Feeb.*, i, qu. 8015–8093. For the challenge from the poor law: PRO MH 58/85, West Bromwich Union Guardians of the Poor, 13 June 1912.

mainstream of general medicine.[43] Most of those working in the field had received no special training or qualification in psychiatry. Nevertheless, from the mid-nineteenth century onwards we can speak of a psychiatric profession, since the dramatic expansion of mental hospital work created a group of doctors with a common set of interests which marked them off from the rest of the medical profession. A sense of corporate solidarity was maintained by the Medico-Psychological Association (MPA), a body dominated by asylum superintendents and issues of institutional management, and by its *Journal of Mental Science (JMS)*. However, if we are to talk of a psychiatric profession in the early twentieth century, it must also be recognized that it was far from being united or clearly defined: the loyalties of doctors in this field were divided between their psychiatric and general medical work; there were clashes of interest between superintendents and their assistant asylum medical officers, who often lacked medical qualifications and rarely belonged to the rather exclusive MPA; and there was often a gulf of opinion between doctors in public mental hospitals and the small but growing number in private and clinic-based psychiatric practice.

The predominant and depressing image of the late nineteenth-century mental asylum was of a huge custodial fortress filled with the incurably insane.[44] Because of the stigma surrounding mental illness and the requirement that the mad had to be certified insane before they could enter an asylum, people were unlikely to come forward for treatment except as a last resort—generally too late for there to be much chance of cure.[45] This was not an attractive area of work for doctors to enter, unless compensated by the power and prestige associated with being manager of a large institution.[46] No such attractions existed for junior medical officers, who were poorly paid and suffered the added deterrence of restrictions on their freedom to marry. Since assistant officers had little chance of promotion, they quickly moved on to more prestigious fields, giving them little time to gain experience or a commitment to the work.[47]

[43] A. Walk, 'The Royal College of Psychiatrists', *St Bartholomew's Hospital Journal*, 77 (May 1973), 135–9.

[44] Scull, *Museums of Madness*, 186–219, 250–3.

[45] K. Jones, *Mental Health and Social Policy 1845–1959* (London, 1960), 153; C. Unsworth, *The Politics of Mental Health Legislation* (Oxford, 1986),19.

[46] B. Pierce, 'The Status of British Psychiatry and of Medical Officers', *JMS* 59 (1913), 699–701; 'Report of the Committee Regarding Status of British Psychiatry' and Discussion, *JMS* 60 (1914), 667–93.

[47] J. B. Spence, 'Assistant Medical Officers in Asylums', *JMS* 59 (1913), 263–73.

In the first two decades of the century, psychiatrists made a concerted attempt to raise the status of their profession: they developed a specialist psychiatric diploma, which was introduced in 1910;[48] they lobbied successfully for the 1909 Asylums Officers Superannuation Act, which improved the pay and conditions of employment for junior medical officers;[49] and they continued to support reform of the rigid legalism of the 1890 Lunacy Act (eventually achieving this with the 1930 Mental Treatment Act).[50] The profession's support for the 1913 Mental Deficiency Act should be placed in the context of this broader platform of reform. The Act promised to remove defectives from lunatic asylums and thereby alleviate the twin problems of overcrowding and low cure rates.[51] The new system offered the prospect of new posts to alleviate the bottleneck of promotion within the profession.

However, it was not altogether clear that mental defectives—by definition incurable—needed medical care.[52] In this respect, the colony was an important tool in legitimating the profession's control over the new system of care. To understand this we need to look at the origins of institutional care for the defective. The first idiot asylums, established in the mid-nineteenth century, had an idealistic mission of educating and training idiots for life in the outside world. This vision was stimulated by the work of the French doctor and Saint-Simonian, Edouard Séguin, who aimed at the cure of idiots through an education of the senses.[53] His educational theory was based on the idea that stimulation of the muscles and senses could gradually reactivate the higher thought processes:

> sensation perceived like a notion, notion fecundated to an idea realized in life itself. . . From collecting the sparse powers of muscles and nerves disconnected by the absence of will, to the gathering of faculties in the act of thinking, our progress has been a constant ascension on the steps leading from isolation to sociability.[54]

[48] C. H. Bond, 'The Need for Schools of Psychiatry', *JMS* 66 (1920), 14–15; T. S. Clouston, 'The Diploma in Psychiatry', *JMS* 57 (1911), 207–18.

[49] Asylum Officers Superannuation Act 1909 (9 Edw. VII, c. 48).

[50] Unsworth, *Mental Health Legislation*, 119–27.

[51] McCandless, 'Build! Build!', 553–74.

[52] See the debate over whether mental deficiency should be placed under the asylums department of local authorities: *JMS* 60 (1914), 338–40, 651–3.

[53] Trent, *Inventing the Feeble Mind*, 40–59.

[54] E. Séguin, *Idiocy and its Treatment by the Physiological Method* (New York, 1866), 209. (French edn., *Traitement Moral, Hygiene, et Education des Idiots* (Paris, 1846), tr. into English several times before Séguin's official translation after his exile to the United States.)

This philosophy was adopted by Britain's first idiot asylums, such as the Royal Earlswood Idiot Asylum whose charter set out its aims as:

> not merely to take the idiot and imbecile under its care, but especially, by the skilful and earnest application of the best means in his education, to improve his bodily powers, and prepare him, as far as possible, for the duties and enjoyments of life.[55]

However, this educational mission proved to be more difficult than had been hoped. As the asylums expanded and became filled with low-grade and older cases, there was a loss of confidence in the educability of defectives, and training degenerated into a tool of management.[56] Despite this, the idea of 'the education of the senses' had a long-term importance, for it helped to give training and employment of defectives a therapeutic and hence a medical authority. Recognizing this, psychiatrists began to advocate hospital designs which would enhance the educational function of the institution. They therefore became more sympathetic towards 'congregate' designs and began to advocate colonies as long as they were enclosed and under medical control.

The shift towards colonies began in London where the vast scale, expense, and seemingly inexorable growth of asylums spurred on the search for alternatives. The movement was led by psychiatrist T. Knowles Stansfield, who had been an advocate of the colony since visiting the German colony of Alt Scherbitz in the early 1890s.[57] He was also influenced by a visit to the United States, where he was particularly impressed by the villa design of the asylum at Toledo, Ohio. Here, the chronically ill were housed in small cottages distributed around lakes and baseball fields, promoting a communal way of life which aimed to recreate an idealized village. The design appealed to Stansfield because it offered what seemed a more humanitarian form of long-term care, yet it also had the practical advantages of being cheap to build, easy to administer, and flexible enough to allow internal specialization and expansion.[58] Although he initially struggled to convince the local authorities of these advantages, Stansfield found allies in the architect, William Charles Clifford Smith, and the designer, Gregory Hine. Together they constructed a series of huge

[55] *Roy Com. Feeb.*, viii. 157. Wright, 'The National Asylum for Idiots, Earlswood'.

[56] Barrett, 'From Education to Segregation'.

[57] Cochrane, '"Humane, Economical, and Medically Wise"', 247–71.

[58] T. E. K. Stansfield, 'The Villa or Colony Sytem for the Care and Treatment of Mental Disease', *JMS* 60 (1914), 30–7.

epileptic and insane asylums in London around the turn of the century.[59] A second important influence on the development of the colony in London was the construction of a series of institutions by the Metropolitan Asylums Board (MAB) in the second half of the nineteenth century. The MAB demonstrated how the villa design could be used to construct huge institutions for mental defectives. Darenth Colony, under E. D. Sherlock, with its farms and industrial workshops, provided a model of what was possible if defectives were set to work in a supervised environment.[60]

Superintendents, such as Sherlock, C. H. Bond, and A. Rotherham (the latter two subsequently appointed as Commissioners of the Board of Control), supported a system of specialized mental deficiency institutions under medical control, rather than small lay-run homes.[61] Behind the scenes, the MPA added its support and helped to organize political support for the legislation.[62] Because of their heavy dependence on public asylums for employment, psychiatrists were willing to back state expansion in this field, even though the medical profession as a whole vigorously opposed state employment in general medicine.[63] At the same time, the profession managed to ensure that its private work and control of the charitable Royal institutions could continue under the new legislation. The profession had reason to be optimistic about its prospects under the Act, especially after it had secured the continuation of its old corporate links to the Commissioners in Lunacy—now renamed as the Board of Control.[64]

The colony seemed to have provided a solution to a series of the profession's problems, and can, as such, be viewed as self-interested. However, psychiatric support for the colony was far more complex than this: the colony seemed to be an efficient, economic, yet humane solution for lifetime confinement, and through its regime of work, exercise, and country life it seemed to offer the institution renewed therapeutic credibility. In summary, psychiatrists turned to the colony, not simply out of self-interest, but also because it appeared to align the

[59] Obituary, *JMS* 85 (1939), 1129–39; Cochrane, '"Humane, Economical, and Medically Wise"'.

[60] E. B. Sherlock, *The Feeble-minded* (London, 1911).

[61] D. Turner, 'Notes about Institutions for Defectives', *Studies in Mental Inefficiency*, 1 (1920), 32–8; A. Rotherham, 'The Case for the Large Mixed Institution for Defectives', *CAMW Conference Report 1920*, pp. 59–64.

[62] T. B. Hyslop, 'The Mental Deficiency Bill 1912' and Discussion, *JMS* 58 (1912), 548–97.

[63] F. Honigsbaum, *The Division in British Medicine: A History of the Separation of General Practice from Hospital Care, 1911–1968* (London, 1979).

[64] *JMS* 61 (1915), 267.

profession with progressive social policy—Perkin's 'professional ideal'. In retrospect, much of this confidence was misplaced.

PSYCHIATRISTS UNDER THE MENTAL DEFICIENCY ACT

Although most institutional psychiatrists probably did view the colony as offering new opportunities for the profession in 1913, subsequent developments highlighted the problem of tying the profession to a system in which cost-efficiency was the main priority. Divisions were exposed within the profession. Mental deficiency work, which had seemed such an innovative and progressive new venture in 1913, rapidly became an unfashionable and rather neglected backwater.

The 'shell-shock' episode during the First World War is often seen as a turning point towards a more progressive, psychotherapeutic, and sympathetic psychiatry.[65] In fact, it had little impact on psychiatry within public mental institutions.[66] Because of the time-consuming nature of psychotherapy, and since it was less effective for the more serious type of psychotic cases who filled institutions, most superintendents dismissed its value as a practical therapeutic tool.[67] Therefore, the superintendent-dominated MPA generally remained divorced from the growth of more experimental wings of the profession,[68] and continued to have a low status within the medical profession.[69] This therapeutic caution was mirrored by a retreat from the overtly political role which the MPA had played in lobbying for the 1913 Act. In an attempt to align itself with the altruistic and disinterested image of general medicine, the MPA presented itself as being above politics or personal interests. Despite misgivings at the results, it began to leave its medico-political role to the British Medical Association (BMA).[70] Instead, the MPA

[65] J. R. Lord, *JMS* 66 (1920), 69; M. Stone, 'Shellshock and the Psychiatrists', in Bynum *et al.*, *The Anatomy of Madness*, ii. 242–71.

[66] H. Merskey, 'Shell-Shock', in Berrios and Freeman, *150 Years of Psychiatry*, 245–67.

[67] For example, Armstrong Jones, *JMS* 67 (1921), 105–07. M. Clark, 'The Rejection of Psychological Approaches to Mental Disorder in Late Nineteenth-Century British Psychiatry', in Scull (ed.), *Madhouses, Mad-Doctors, and Madmen*, 271–312.

[68] Walk, 'The Royal College of Psychiatrists', 132.

[69] 'The Reproach of Psychiatry in England', *The Lancet* (6 Mar. 1920); *BMJ* (1922), 443.

[70] A. W. Pierce, *JMS* 81 (1935), 729. Unsworth, *Politics of Mental Health*, 205–15, argues that the RMPA in fact followed a cautious line over the extension to voluntary treatment.

advanced its interests through professional strategies. For instance, it created stricter educational qualifications, both to bar infiltrators from the profession,[71] and to enhance the scientific credentials of psychiatry within the medical profession and thereby attract better candidates.[72] This strategy was reflected in the *JMS*, under its new editor J. R. Lord, which became far more scientific in its orientation, with an increased volume of research papers expanding the journal from an average size of 536 pages in 1920, to 722 in 1924. The *JMS* also began to reach out to a wider medical readership, as reflected by a rise in subscriptions from 3,600 in 1920, to 4,100 by 1924. As Lord wrote:

> The index finger of the time was pointing very definitely to the need for the Association to extend its sphere of activities, especially in the direction of scientific psychiatry, in which junior members of the speciality were showing the deepest interest.[73]

Mental deficiency work benefited little from these post-1913 developments. Prior to the war, the new mental deficiency system may have offered a potential outlet for surplus psychiatrists, but because of war fatalities and retirements the number of psychiatrists in the MPA fell from 747 in 1913 to 661 in 1919.[74] In such circumstances it was more difficult than ever to find qualified men to fill mental deficiency posts.[75] This situation was not helped by the neglect of mental deficiency in the syllabus of new psychiatric courses. For example, in Part I of the 1920 Maudsley course on psychiatry there were no lectures on mental deficiency, and in Part II only two out of the forty-eight talks addressed the topic. By 1922, even these lectures had disappeared.[76] Those doctors already in post were so closely tied to their isolated institutions that they tended to be divorced from developments in public health and community care, and had limited opportunities to take advantage of new outside training courses.[77]

[71] A postgraduate committee was set up: 'Model Scheme for Diploma in Psychological Medicine', *JMS* 69 (1923), 534–7.

[72] 'The Reproach of Psychiatry in England', *The Lancet* (6 Mar. 1920), 19.

[73] J. R. Lord, *JMS* 71 (1925), 801–4.

[74] These figures are derived from the annual membership lists of the MPA published in the *JMS*.

[75] *BOC 1929*, pp. 12–13.

[76] *JMS* 66 (1920), 193–4; *JMS* 68 (1922), 201–2. Through an analysis of the *JMS* contents for 1923–34, compared to 1935–49, it has been argued that during the first period, research was dominated by an interest in the physical basis of mental problems: G. Baruch and A. Treacher, *Psychiatry Observed* (London, 1978), 33.

[77] Set up in co-operation with the London School of Hygiene and Tropical Medicine as part of Diploma in Public Health: M. Rooff, *Voluntary Societies and Social Policy* (London, 1957), 129–30.

The low status of mental deficiency work can be demonstrated through a prosopographical analysis of members of the RMPA for the year 1937.[78] Of the 874 members of the RMPA in 1937, ninety-five either worked in the mental deficiency field or were involved through a scientific or administrative interest. Judged in terms of education, only 35.8 per cent of these mental deficiency psychiatrists (MDPs)[79] had specialist psychiatric qualifications, compared to 48.1 per cent in the Association as a whole. The MDP group was dominated by superintendents: it included forty psychiatrists who were currently running public institutions, two running private homes, and eight retired superintendents, but only nineteen junior officers. The orientation was overwhelmingly towards institutional care, with only three MDPs working in the community, just twelve holding a Diploma in Public Health, and only one working in a local government department. The majority of the MDP superintendents had attained their positions at the end of careers outside mental deficiency work: fourteen had begun their careers in mental hospitals (significantly, this group formed the majority of the sixteen MDPs who had published scientific literature), five came from clinical backgrounds in specialist or general hospitals, five came from Ministry of Pensions or military hospitals, and eight came from local government backgrounds. It cannot have helped morale among junior officers in mental deficiency institutions that these elite positions tended to go to doctors who had made their reputations elsewhere. In fact, only three of the mental deficiency superintendents of 1937 had worked their way up within the field. Prestigious membership of the Royal Colleges was confined to MDPs, such as A. F. Tredgold, R. L. Langdon-Down, and R. G. Gordon, who had backgrounds in private practice rather than in public institutions. The low status of mental deficiency work was also reflected in the limited number of MDPs on special committees of the MPA: five

[78] In 1926, as a sign of the rising status of psychiatry, the MPA was renamed the Royal Medico-Psychological Association (RMPA). On mental deficiency work within the MPA: K. Day and J. Jancar, 'Mental Handicap and the Royal Medico-Psychological Association: A Historical Association, 1841–1991', in Berrios and Freeman (eds.), *150 Years of British Psychiatry*, i. 268–78.

[79] Each year the *JMS* published membership lists and committee lists. The mental deficiency group (MDPs) has been created by including those in the mental deficiency committee (JMS, 83 (1937), pp. xxii–xliv), and others who were officers in mental deficiency institutions. This second category is based on an analysis of the *BOC Annual Reports*, Part II, which include lists of institutions and their superintendents, and on information given in *The Medical Register*.

served on the parliamentary committee of thirty, seven sat on the education committee of forty-three, only one sat on the examination committee for the certificate in psychological medicine, and none at all served on the nursing committee of twenty or on the library committee of nine. MDPs who did serve on committees were usually those, such as T. Saxty Good, W. F. Menzies, and G. de Montjoie Rudolf, whose interest in mental deficiency arose from their research rather than any involvement in institutional care of defectives. The main exception to this pattern of low status was F. D. Turner, the Superintendent of the Royal Eastern Counties Institution for Mental Defectives at Colchester, who rose to become President of the RMPA in 1933; but Turner was the only psychiatrist from such a background who was given this honour.[80] The insecurity of MDPs was captured by the complaint of E. J. Fitzgerald in 1935:

> [mental hospital doctors] tended to look down professionally on the mental deficiency people. Those who entered upon mental deficiency work were often regarded as those who had gone into a sterile world, where there was little to be done on the lines of curative medicine.[81]

With mental deficiency work having such low status within the profession, and with MDPs having so little influence, it became even harder to realize the vision of the colony as a therapeutically progressive system of care.

A QUANTITATIVE ASSESSMENT OF THE COLONY SOLUTION

Judged simply in terms of the total of defectives placed under institutional care, the colony solution would appear to have been a qualified success: the number rose fairly steadily over the whole period from 13,842 in 1921 to 46,054 by 1938;[82] though it fell far below the level of 94,000 somewhat unrealistically recommended by the Wood Report of 1929.[83] A rather more complex picture emerges if the total figure is

[80] Presidential address. 'Mental Deficiency', *JMS* 79 (1933), 564–77; Obituary, *The Lancet* (9 Nov. 1957), 955; F. D. Turner, 'Forty Years in an Institution', *Mental Health*, 7 (1946), 1–5.

[81] *JMS* 81 (1935), 731.

[82] Figures derived from the *BOC Annual Reports*.

[83] *Report of the Interdepartmental Committee on Mental Deficiency, 1925–1929* (*Wood Report*) (London, 1929), pt. III, p. 42.

subdivided into different types of institutions, and provision in local authority and state institutions is compared to that in the types of institution which they had been intended to replace. Local authority institutions did take up an increasing proportion of the total, rising from 8,513 beds in 1921 to 34,290 by 1938. Likewise, the number in state institutions for violent or criminal defectives at Rampton and Moss Side expanded rapidly in the first years of the Act, from 323 in 1921 to 1,043 by 1930; though their growth then slowed in the 1930s, only rising to 1,475 by 1938, as the Treasury became suspicious that local authorities were using this option to avoid the cost of care in their own institutions.[84] Contrary to intentions, the number of defectives still in poor law institutions—permitted under Section 37 of the Mental Deficiency Act only because of the lack of alternatives—continued and rose from 4,052 in 1921 to 9,219 by 1930. After this date, the Local Government Act placed these institutions under the county borough and county council authorities, but defectives continued to be sent to these old poor law institutions: despite the Act having been in operation for over two decades, 9,445 defectives were still in residence in 1938. The number of those accommodated in profit-making private certified houses was never substantial, since new licences were not given after the 1913 Act, but it did remain at between 200 and 300 residents throughout the period. The number of defectives in approved houses under charitable and religious control—accommodation which local authorities were permitted to use under the Act—rose, but only slowly, from 286 in 1921 to 671 by 1938. Combining the totals for certified houses and approved homes, the number of those outside direct statutory control only rose from about 500 to 800. In terms of total accommodation, therefore, there was a shift towards direct local authority provision, as intended.

An alternative way of examining the success of the Act's institutional plan is to consider how many local authorities were directly providing their own institutional care. In this respect, achievements were disappointing for the first two decades, with less than thirty authorities establishing their own institutions. The situation improved after 1929, though this was partly an illusion since it reflected the transfer of poor law institutions to local authorities under the Local Government Act. The number of local authorities providing their own institutions or being involved in joint schemes

[84] PRO MH 51/496, Treasury memo. 7 Jan. 1926; *BOC 1926*, pp. 36–7.

levelled off in the mid-1930s at about eighty authorities, even though the national financial situation was improving. This suggests that a maximum had been reached; for the remaining forty authorities, because of small size or awkward geography, it was not feasible to build their own institutions, and instead it was more convenient to use the facilities of others on a contractual basis.

Although the interwar years did see considerable expansion in institutional care, there was no wholesale adoption of the colony model. This can be examined through a breakdown of the listing of all mental deficiency institutions in 1937.[85] The fact that only 18 institutions had adopted the title of 'colony' suggests that the colony model had not become dominant. In fact few institutions were purpose built as colonies. Most were still dominated by central block-style buildings built in the nineteenth century,[86] or were developed from old poor law buildings or country houses.[87] Similarly, the intention that local authority certified institutions should be the main form of care was still some way from being fully realized: there were 129 certified local authority institutions taking 32,653 patients and two state institutions taking 1,445, but there were also 114 institutions registered under Section 37 of the Act (old poor law institutions) with 9,301 patients, four certified houses with 201 patients, and forty approved homes with 663 residents. Therefore, although local authority institutions contained the majority of defectives by 1937, they were still outnumbered by other types of institutions. Moreover, only sixty-nine of the certified institutions were actually directly owned or managed by the local authorities, the others being backed by charities and religious denominations.

If institutions are judged according to size, the majority again failed to live up to the colony ideal. Of the 129 certified institutions, just seven held over 1,000 patients, nine held between 500 and 1,000, and nineteen held 300–499.[88] To this one could only add the Section 37 institutions at Leavesden and Caterham, which each held over a 1,000 patients. The majority of Section 37 institutions held under 50 patients, a scale which was far too small for the internal specialization of care recommended by

[85] *BOC 1937*, pt. II, pp. 560–601.
[86] PRO MH 51/453, 694/1351—Royal Eastern Counties Institution.
[87] PRO MH 51/453, 1580/1351—Great Barr Colony; 724/1351—Coleshill Hall Certified Institution.
[88] Figures derived from *BOC 1938*, pt. II, pp. 560–601.

the Board of Control.[89] Moreover, although the ideal of 1908 had been for mixed ability institutions, the majority of these Section 37 institutions took only medium to low-grade active types; high-grade cases needed too much control, and bed-ridden cases needed too much care, whereas medium-grade active types could prove stable and willing workers, valuable to the regime and easy to care for. The only certified house of substantial size, holding 150 residents, was Normansfield, run by the specialist in mental deficiency R. L. Langdon Down. The approved homes were all small, none holding over fifty residents. The majority tended to be narrowly selective in the type of cases which they took, discriminating according to creed, sex, and age, and refusing admittance of defectives who were too seriously handicapped or had behavioural problems. They enabled middle-class families to obtain private care for family members, but they contributed little to the wider problem of public provision.

A striking indication of the limited success of psychiatrists in realizing the colony ideal is that most institutions remained under non-medical management. In 1937, only twenty-three of the certified and state institutions were under medical heads; though doctors also ran five of the voluntary institutions, all of significant size, including three of the prestigious Royal Institutions. All of these medical superintendents were men. This reflected male predominance in the profession as a whole, with only forty-two of the MPA's 503 members being women.[90] By contrast, the remaining 103 institutions, including forty-nine statutory hospitals, were all run by lay managers, just twenty-two of whom were men.[91]

In summary, if mental deficiency institutions are assessed in terms of who controlled and managed them, their scale, and their design, they clearly failed to live up to the colony ideal of 1908–13. However, this failure should not be confused with a failure of the idea of the colony solution: the progressive ethos and style of the colony model, however illusory and rare in reality, was still vital in making the

[89] *BOC 1926*, pp. 76–8.

[90] Women's Medical Federation, *JMS* 75, (1929), 351. For the tensions: 'Women Doctors in Asylums', The Times (7 Mar. 1922). On women within the profession: E. Showalter, *The Female Malady: Women, Madness and English Culture, 1830–1980* (London, 1987), 53–4; C. Mackenzie, 'Women and Psychiatric Professionalization, 1780–1914', in London Feminist History Collective (eds.), *The Sexual Dynamics of History* (London, 1983), 107–19.

[91] The Board of Control continuously complained about the lack of medical superintendents in mental defective institutions: *BOC 1929*, pp. 12–13; *BOC 1937*, p. 5.

segregative system politically acceptable and in legitimating the claim of psychiatrists to control what, in reality, was largely a custodial system.

A QUALITATIVE ASSESSMENT OF THE COLONY SOLUTION

Nursing Care

In Perkin's analysis of 'Professional Society', efficiency lies at the core of professionals' claims to the management of society. As we have seen, psychiatrists did indeed use arguments about efficiency to maintain and expand their role in mental deficiency. However, professionalization also entailed the promotion of higher standards of expertise and care. Therefore, there was an inherent tension over professional attitudes towards efficiency, when the main measure of this was economy. The tensions between professionalization, efficiency, and economy can be explored through the roles played by psychiatrists and nurses in colony provision. Consideration of these two groups also provides an opportunity to consider the extent to which common interests and the 'professional ideal' were able to establish 'hierarchical professional alliances' which transcended divisions running along lines of class and gender.[92]

Since the colony solution had been premised on the idea that it would provide a less expensive and more efficient form of segregation than earlier institutional designs, standards of care were acutely vulnerable in the economizing environment of interwar social policy.[93] Revelations that the cost of building an institution was more per person than that of working-class housing panicked the Government into appointing the Departmental Committee on Mental Defective Colonies (Hedley Committee) to investigate whether institutions could be built more cheaply. The Report recommended increasing the size of institutions from the level set down in the guidelines of 1919, of 320 patients with

[92] Perkin, *Rise of Professional Society*, 2–9. Similarly, a recent study argues that professional consensus and medical modernization were dominant forces behind the emergence of the British health-care system: D. Fox, *Health Policies and Health Politics: The British and American Experience, 1911–1965* (Princeton, 1986).

[93] *BOC 1922*, p. 72; *BOC 1931*, pp. 8, 13.

the potential to expand to 640,[94] to the economy-oriented ideal of over 1,000 and up to 2,000.[95] It was argued that because of its villa design the colony would still be able to maintain a community spirit among this huge population.[96] However, the sixty-bed villas were hardly 'home-like'; in reality, they were barrack blocks on a smaller scale. Most witnesses had argued that a change from two to three rows of beds in the narrow wards would destroy any sense of domesticity. Nevertheless, the plan was approved on the grounds that it would save £200 per villa. Similarly, the Report recommended that villas should be semi-detached rather than detached, with a saving of £600 for every two villas which were combined,[97] and that they should be placed closer together so that money could be saved on amenities, such as roads, heating, lighting, water, and drainage.[98] Instances where economy was not given priority were rare and tended to involve security issues, as in the decision to build two-storey villas, rather than to use a cheaper bungalow design.[99] The Report also recommended that cheaper materials and temporary structures should not be adopted—though simply because it was viewed as uneconomic in the long run.[100] Altogether, the proposed savings brought the estimated cost of building defective colonies below that of mental hospitals, at £475 per bed, or isolation hospitals, at £770 per bed. However, it was not possible to bring the cost per bed down below £300 towards the £100 figure which had been hoped for, and therefore attacks on the extravagance of colony expenditure continued.[101]

The provision of nursing staff raised further conflicts between standards and costs. Since doctors, in reality, had little daily contact with patients, standards of care rested heavily on the quality and number of the nurses and attendants. In the nineteenth century, mental hospitals had been staffed by male 'keepers' or 'attendants', rather than 'nurses'.[102] The status and quality of this work was low, and it was often carried on over generations as a family trade.[103] There were two possible strategies available to improve standards of nursing care: professionalization and medicalization through the introduction of training,

[94] PRO MH 51/239, Board of Control circular 557A 1919.

[95] *Report of the Departmental Committee on Colonies for Mental Defective (Hedley Report)* (London, 1931), 13–14.

[96] Ibid. 13.

[97] Ibid. 29.

[98] Ibid. 40.

[99] Ibid. 22.

[100] Ibid., vi. 44–5.

[101] H. F. Stephens, 'Cheaper Institutional Care', *Mental Welfare*, 13 (1933), 8.

[102] L. D. Smith, 'Behind Closed Doors: Lunatic Asylum Keepers, 1800–1860', *Social History of Medicine*, 1 (1988), 301–27.

[103] M. Carpenter, *Working for Health: The History of COHSE* (London, 1988), 30.

education, and professional organization; or unionization to improve working conditions. The tension between and within these two strategies highlights the continuation of class and gender divisions alongside the 'professional ideal'.

The MPA supported the reform of nursing on the grounds that this would help to medicalize and destigmatize mental care, but also hoped thereby to offset unionization, which threatened to weaken its control over nurses. Following the example of professionalization in general nursing, the MPA established a mental nursing examination,[104] and founded the Asylum Workers' Association, which included nurses but was dominated by doctors and propounded an ethos of professionalism against the threat of unionization.[105] Motivated by a desire for improved medical standards, doctors also tried to encourage women trained in general nursing to enter asylum work.[106] In reaction, male attendants founded their own Asylum Workers' Union (NAWU), which excluded superintendents from membership.[107] Thus, what seems to have been developing at the turn of the century was a professional alliance across gender to counter a class threat: male doctors allying with professionalizing female nurses against militant male attendants.

However, this professional alliance between doctors and female nurses was only temporary, soon cracking along lines of gender. In the first decades of the century, the MPA actively debated the relative merits of female and male nurses. Female nurses were regarded as having the advantage of natural mothering instincts and as having a beneficial influence in stimulating the chivalry of male patients.[108] On the other hand, the arguments against women proved to be numerous and more compelling: it was argued that men were better suited to the physical and mental hardships of the job,[109] that female probationers would be unable to cope with the trauma of the work,[110] and that the authority of superintendents might be undermined by female managers brought in for the female nurses.[111]

[104] R. Dingwall, C. Webster, and A. M. Rafferty, *Introduction to the Social History of Nursing* (London, 1988), 128.

[105] *JMS* 65 (1919), 124; obituary of founder, the psychiatrist G. E. Shuttleworth, *Mental Welfare* 9 (1928), 72–3.

[106] G. Robertson, *JMS* 62 (1916), 351–62; Carpenter, *Working For Health*, 31–2.

[107] F. Adams, 'From Association to Union: Professional Organisation of Asylum Attendants, 1869–1919', *British Journal of Sociology*, 20 (1969), 11–26.

[108] Wolsley Lewis, *JMS* 62 (1916), 451.

[109] Brander, *JMS* 62 (1916), 451.

[110] B. Abel-Smith, *A History of the Nursing Profession* (London, 1960), 274.

[111] MPA, *Handbook for Attendants on the Insane*, 6th edn. (London, 1912), 346.

Allying against the joint threat of female carers, superintendents and male attendants both supported the merits of the MPA certificate in mental nursing against that of the General Nursing Council (GNC) of the female nurses. Through its regulation of this certificate, the MPA was able to strengthen its control over nursing.[112] In doing so, it was willing to compromise on the professional ideal: popularity of the certificate was ensured by making it easier to obtain than that of the GNC—it was said that even illiterates could pass the exams.[113] The MPA certificate helped to perpetuate the attitude that it was more important for mental nurses to be able to organize patients in work and sporting activities than to have an understanding of psychiatry or general medicine.[114]

The NAWU deployed the MPA qualification alongside unionization to protect male dominance in mental nursing. Despite approximately equal numbers of male and female nurses in mental hospitals, men outnumbered women by 13,000 to 6,000 within the union.[115] NAWU annual conferences passed resolutions to exclude female nurses from male wards on the basis that it was dangerous for women, embarrassing for the male patients,[116] and undermined the self-respect of the male staff.[117] Arguing that women tended to enter mental nursing for temporary employment when no better alternative was available, rather than as a career, male nurses were able to secure better training, wages, and working conditions on the grounds that they were more career-minded.[118] In effect, the male nurses of the NAWU were beginning to appropriate the same strategy of professionalization which it had earlier seen as a threat when in the hands of the GNC female nurses.

Despite the developing alliance of professional and gender interests between the NAWU and MPA, there were continuing tensions across

[112] Dingwall *et al.*, *Social History of Nursing*, 132–3.

[113] Ibid. 133; P. Nolan, 'Psychiatric Nursing Past and Present: The Nurses' Viewpoint', unpub. Ph.D. thesis (University of Bath, 1989), 184–5.

[114] Board of Control, *The Nursing Services in Mental Hospitals* (London, 1925), 46–7.

[115] *MHIWJ* 27 (Feb. 1938), 4.

[116] For example: Annual Conference Report, MHIWJ 25 (suppl.) (Aug. 1936), 22, resolution from Whalley Branch; *MHIWJ* 26 (July 1937), 27–9; 'Women Nurses in Peril?', *Daily Herald* (9 July 1925). Similarly, Revd Bancroft stressed the dirty and violent nature of work: *Report and Special Report from the Select Committee on Asylum Officers' Employment, Pensions, and Superannuation Bill* (London, 1911). This can be contrasted to the views of female nurses who stressed the moral nature of the work: *Board of Control, Nursing Services in Mental Hospitals*, 31. See Nolan, 'Psychiatric Nursing', 209–40, for oral interviews indicating gendered differences in the way nurses viewed their work.

[117] Board of Control, *The Nursing Services*, Mr Hathrill, 23–4.

[118] Statement of MHIWU to Interdepartmental Committee on Nursing Services, *MHIWJ*, 27 (Feb. 1938), 6–7.

lines of class. This came to the fore over the issue of employing defectives within institutions. The NAWU objected to the practice, on the grounds that it exploited the patients and undercut the outside labour market.[119] Superintendents countered this, arguing that employment served a therapeutic role of occupational therapy.[120] At first, nurses resisted this argument, reasoning that if work really was therapy, then they should be paid extra if they were expected to train defectives. However, once threatened by the introduction of specialists in training and occupational therapy, the union backed down.[121] Nurses in mental deficiency institutions, where work was such an integral part of the regime, were at the forefront of this movement to adopt occupational therapy as part of nursing expertise.[122]

More fundamentally, the huge gulf in power between superintendents and nurses makes it difficult to view their relationship as a professional alliance. The blatant disparity in pay and living conditions, and the often autocratic control of superintendents over their fiefdoms, resulted in a constant threat of class tension between the superintendents and the militant union.[123] Nurses had reason to be aggrieved when the 'extravagance' of colony care was attributed to their rising wages and shorter hours.[124] As a participant at the 1936 NAWU Annual Conference argued, if cuts had to be made, nurses were more indispensable than doctors:

> I wonder whether it is necessary to have a medical superintendent at £1,500 a year, plus emoluments, several assistants with DPMs and the whole hierarchy of administrators to look after people whose chances of recovery, for all practical purposes, amount to nil. Would it not be possible to devise some plan for accommodating such types in colonies or elsewhere, with skilled nurses to look after them and with a medical supervisor if necessary?[125]

[119] *MHIWJ* 26 (Nov. 1937); similarly *MHIWJ* 24 (Apr. 1935), 18.

[120] A. M. McCutcheon, 'Institutional Treatment of Mental Defectives with Special Reference to Occupation', *JMS* 71, (1925), 694–703.

[121] Resolution at Annual Conference, *MHIWJ* 25 (suppl.) (Aug. 1936), 20. The Board of Control argued that craft work did not deserve extra pay: *BOC 1934*, p. 6. For the problems experienced by specialists: A. Raven, 'The Training for Occupational Therapy', *Mental Welfare*, 11 (1930), 34; N. Ross, 'Occupational Therapy', *Mental Welfare*, 17 (1936), 12–15.

[122] *MHIWJ* 24 (Jan. 1935), 19; *MHIWJ* 24 (Feb. 1935), 18 (contrast to opposition from mental hospital nurse, p. 20).

[123] Mental nursing was the first area of the health services to see serious labour disputes in the 1950s: Carpenter, *Working for Health*, 264–400.

[124] *BOC 1921*, p. 5.

[125] 'Annual Conference Report', *MHIWJ* 25 (suppl.) (Aug. 1936), 21.

Unsurprisingly, this radical and anti-psychiatric alternative did not materialize; the fact that it was voiced, however, indicates the inherent tension which lay beneath the surface of the professional alliance.

If standards of care were to be improved it was essential to secure a high ratio of nurses to patients. Because of interwar unemployment there was less difficulty than before attracting candidates to the low-status area of mental nursing. By 1924, the nurse to patient ratio had risen to 1:10—the level recommended by the Board of Control. However, financial cut-backs reduced this ratio to 1:11 on the male, and 1:11.8 on the female side, by 1939.[126] Because of the poor conditions of work, the recovery of the labour market had a less beneficial effect than in other areas of nursing:[127] between 1933 and 1937, nursing employment rose by just 6 per cent in mental hospitals, compared to 12.4 per cent in voluntary hospitals, and 18.4 per cent in local authority general hospitals.[128]

Following the First World War, the NAWU had won concessions over shorter hours, reducing the working week of some nurses from 60–66 hours to 50—significantly shorter than that of probationer nurses in general hospitals.[129] On the other hand, in rural areas, where mental deficiency colonies were typically located, lack of union organization placed employees in a far weaker position.[130] Moreover, shorter hours were often an illusion, as a high proportion of mental nurses had to reside in their remote institutional workplaces: in 1937 just 10.7 per cent of female nurses lived outside their institutional workplace, 25.4 per cent lived in rooms adjoining the wards, 12.4 per cent lived elsewhere in the hospital, and 50.1 per cent lived in the hospital grounds. Men were less tied to living in institutions, with 56.8 per cent living outside, although 20.6 per cent still lived in rooms adjoining the wards.[131] Since they remained at work during their free time and were used as back-up staff to cover the very low level of night staff employed, the relatively short working week of mental nurses was something of an illusion.[132]

It proved difficult to attract high-quality recruits to a career with

[126] Dingwall *et al.*, *Social History of Nursing*, 133. [127] Ibid. 130, 133.

[128] Abel-Smith, *History of the Nursing Profession*, 270, app. on the Athlone Report into conditions of employment for nursing in 1937.

[129] Carpenter, *Working for Health*, 81–127.

[130] Dingwall *et. al.*, *Social History of Nursing*, 133.

[131] Abel-Smith, *History of the Nursing Profession*, 275.

[132] *Hedley Report*, p. 26.

such poor working conditions. Nurses were often poorly educated Irish immigrants, who had a tradition of working in mental hospitals, or ex-soldiers, who were used to a militaristic and institutionalized regime.[133] In 1937, 92 per cent of the recruits to mental hospitals had only elementary education, compared to 25 per cent in voluntary hospitals and 71 per cent in local authority general hospitals. Young women entering nursing as a career tended to choose the general hospital service. In 1937, just 28 per cent of female recruits to mental hospitals were below the age of 18, compared to 94 per cent in voluntary, and 69 per cent in local authority general hospitals.[134] Once young women had entered training in general nursing they were unlikely to move to a mental hospital, unless it was to a position of authority. Instead, recruits were usually older women who had lost their previous jobs, who wanted to return to employment after rearing children, or who had been pushed into employment by temporary financial difficulties. The resulting high turnover rate of staff further undermined standards of care.

Working conditions in mental deficiency institutions were even worse than in mental illness hospitals. With only the largest of the mental deficiency institutions having their own branches of the NAWU (eight had their own branches in 1935),[135] the union's pay-negotiating board had little power. A low proportion of mental deficiency institutions were recognized as RMPA training schools, with the consequence that by 1934 just 30 per cent of mental deficiency staff were trained compared to 52.4 per cent in mental hospitals.[136] Promotions generally went to outsiders who had qualifications in general nursing.[137] In 1938, the *MHIW Journal* summed up the problems of mental deficiency work:

> the patients [are] out of sight and out of public mind. This applies equally to the staffs, whose conditions are as much in need of improvement as are the care and treatment of the patients. Very few of these institutions provide facilities for the training of their staffs in nursing. . . . Nursing certificates being unobtainable, the staffs have to remain on a lower scale of pay with very little prospects for the future. Long hours of duty with unnecessary restrictions are the common lot of the male and female staffs.

Despite the relatively strong union, the prospects for the future of

[133] Nolan, 'Psychiatric Nursing', 130–86.

[134] Abel-Smith, *History of the Nursing Profession*, 274.

[135] *MHIWJ* 24 (Apr. 1935), p. iii (seven of the eight branches were in London).

[136] Carpenter, *Working For Health*, 120; J. T. Waite, 'Mental Deficiency Institution Staffs', *MHIWJ* 27 (1938), 24.

[137] *MHIWJ* 28 (June/July 1939), letter of complaint, 49.

mental nursing were bleak. With the birth-rate peaking in 1920, a future decline in the population of potential nursing recruits seemed inevitable. As employment opportunities improved in the late 1930s it became harder than ever to attract the career-minded to mental nursing.

Occupational Therapy and the Work Regime

Part of the progressive gloss surrounding the colony was the idea that it would involve patients in useful activity. The Board of Control encouraged institutions to occupy mental defectives in work activities, and it even criticized private homes for failing to actively employ their paying residents.[138] However, driven by the exigencies of economy, it was easy for the occupational regime to become a form of exploitation. Employment of defectives within institutions was justified as serving a useful managerial role, providing an outlet for the energies of the patients, and distracting malicious defectives from disruptive behaviour.[139] By the 1930s, employment of residents was also being justified as training them for a return to outside life.[140] In reality, the colony had become dependent upon the labour of its residents. Not only was patient labour used on colony farms, and in industrial workshops and institutional laundries, but it helped to compensate for scarcity of nursing and auxiliary staff. Manpower shortages and the departure of staff to the armed forces in the First World War forced institutions to employ high-grade defectives.[141] Much of the work undertaken by these mental defectives was 'dirty' work—the cleaning of the institution and of other defectives. By the 1930s, it was becoming officially recognized that high-grade defectives sometimes also served in caring roles: 'there is no objection to the creation of a class of paid patients who would be given definite responsibility and a regular wage, but who would not rank as staff, and who would not be left in sole control of patients.'[142] Inevitably, these two criteria of pay and supervision were both open to abuse. As late as 1951, high-grade patients were left alone to supervise the lower grades, and it was acknowledged that 'without the assistance of high-grade and not so high-grade patients it would be impossible to carry on the nursing functions of the colony'.[143] This practice was not confined to mental deficiency institutions: defectives were also valued as supplementary

[138] *BOC 1919*, p. 73. [139] *BOC 1934*, p. 5. [140] *BOC 1932*, p. 66.

[141] GLRO, Caterham Hospital Records, Report Book of the Superintendent, 2 June 1916. [142] PRO MH 51/457, Brock to Barter Nov. 1934.

[143] PRO MH 51/503, 22 Feb. 1951, Barter, p. 1.

staff in public assistance institutions and mental hospitals, so much so that transfer to mental deficiency colonies was often resisted.[144]

Superintendents argued that colonies did not exploit defectives. Sherlock, Superintendent of Darenth colony and open advocate of employing defectives within institutions, argued that the expense of extra staff, materials, and facilities demanded in such a regime offset any potential profits. There was, anyway, only a limited market for such low-quality handicraft. And there was the added problem of resentment among the outside labour force against being undercut by 'forced' labour.[145] The production and sale of goods was defended on the grounds that it aided the management of otherwise bored defectives.[146] Patients were said to show more enthusiasm and satisfaction if their handiwork was sold.[147] Aware that lay managers might claim to be more efficient supervisors of such an occupational regime, psychiatrists asserted that their medical expertise was essential in assessing the capacities of defectives for employment:

> The main purpose of the colony is to provide work of such faculties as the inmates are possessed of and therefore it becomes necessary to ascertain, as far as is practicable in regard to any individual patient what capacity for remunerative employment lies latent in his abnormally constituted mental machinery.[148]

In reality, it was impossible for psychiatrists effectively to monitor the therapeutic role of the occupational regime within their huge institutions.[149] Occupation of patients was therefore far more often a tool of management and a source of institutional manpower than a form of therapy.

Sherlock's claim that Darenth and other mental deficiency institutions did not make a profit was probably true, but the evolution of the Darenth regime suggests that therapy and training did become sacri-

[144] PRO MH 58/94, 15 Feb. 1934, Young to Caswell; PRO MH 51/451, 9 July 1934, Brock.

[145] E. B. Sherlock, 'On the Earning Capacity of Mentally Defective Persons', *MAB Annual Report 1914*, pp. 246–7. This may be supported by the slightly higher running cost per patient in mental deficiency institutions than in mental hospitals: in 1934, 28s 0.1d per week, compared to 25s 7.1d: *BOC 1934*, p. 69.

[146] *MAB Annual Report 1928–9*, p. 42.

[147] *MAB Annual Report 1914*, p. 246.

[148] *MAB Annual Report 1923–4*, p. 198.

[149] R. Ström-Olsen, 'The Therapeutic Value in Mental Illness of Physical Fitness Through Exercise', *Mental Welfare*, 18 (1937), 1–10; McCutcheon, 'Institutional Treatment of Mental Defectives', 694–703, 847–50; 'Occupation Therapy and "Action Therapy"' *JMS* 79 (1933), 758–63.

ficed as the occupation of patients was reoriented to maximize income. For example, tailoring at Darenth had originally been introduced as a useful form of education and manual stimulation; reflecting its non-profit-making intent, it had been taught on scrap material. However, by the early 1920s, instead of learning tailoring skills, defectives were being employed in mass production of face cloths.[150] Subsequently, the Metropolitan Asylums Board began contracting out all its hospital needlework to Darenth Colony. The defectives were crowded into industrial needle-rooms, and the productive capacity of the colony was increased from 2,000 to 8,000 articles per week. Although Darenth's reports placed great emphasis on the training of patients in skilled trades, such as bookbinding and printing, most of the defectives were employed in unskilled work. By 1925, this move to factory-style production had boosted Darenth's annual income to £74,837.[151] Farming, originally justified as providing defectives with fresh air, exercise, and fresh produce, was also overtaken by economic rationales. For instance, the LCC's Brunswick House Institution produced all of its own food supplies and was still able to send annually 3 tons of potatoes, 2,100 eggs, and 600 lbs of rhubarb to other LCC institutions.[152] The impressive achievements of institutions such as Darenth and Brunswick House made the productive, industrial regime a model for institutions in the rest of the country.

Medical Care

At the start of the century, the psychiatric profession had supported the colony, both as a source of employment and because of the progressive ethos which surrounded the policy. In the long term, the colony did provide some opportunities for professional expansion, but association with the huge, non-therapeutic, and work-centred regimes was not so clearly in the interest of the profession; there were fault lines within the 'professional ideal', since efficiency was in constant tension with medical standards. In the 1930s, divisions emerged in the profession over how to react to this dilemma.

For reasons of economy, the often huge mental deficiency institutions were expected to function with only one medical superintendent.[153]

150 *MAB Annual Report 1923–4*, pp. 156–9.
151 *MAB Annual Report 1928–9*, pp. 157–9.
152 *LCC Annual Report 1933*, vi. 42–7.
153 *Hedley Report*, p. 14.

The local power and material benefits associated with these 'fiefdoms' compensated superintendents for the low status of the work. However, junior medical officers were frustrated, as there was little prospect of promotion and no compensation in terms of satisfaction over rates of cure or involvement with innovative treatments. Unsurprisingly, tensions developed between the younger generation of assistant medical officers and their seniors.[154]

The Board of Control attempted to attract young doctors to work in mental hospitals through the prospect of research opportunities.[155] In fact, much of this research bore little therapeutic relation to the problem of mental deficiency and was criticized on these grounds by members of the older generation of psychiatrists.[156] The residents of institutions were often used as controlled and passive populations for medical investigations that were of little direct benefit to the defectives themselves. For instance, the patients at one colony were used as guinea pigs to test the Flexner vaccine against dysentery.[157] Elsewhere, defectives were used in experiments with the Wassermann test for detecting syphilis; this involved uninhibited use of the lumbar puncture which was painful and potentially dangerous, and even when the disease was detected the defectives were not necessarily treated.[158]

Another area of potential research was the heredity of mental deficiency. In the 1920s the caution of the profession had led to a downgrading of such work.[159] In the 1930s, stimulated by the sterilization debate and by government interest in the genetic basis of mental disease and deficiency, more sophisticated work began to emerge which compared institutional populations to their families and control groups outside.[160] These control studies offset the limitations of earlier research which had concentrated solely on those inside institutions, but such work was expensive if it was to be performed on an adequate scale and demanded rare genetic expertise. Although the genetics of

[154] This was perhaps reflected in the development of separate annual dinners: *JMS* 69 (1923), 392.

[155] *BOC 1929*, pp. 12–13.

[156] H. Yellowlees, 'Modern Psychiatry and Mental Hospitals', *JMS* 81 (1935), 250–5.

[157] K. C. L. Paddle, 'The Control of Dysentry by Prophylactic Innoculation', *JMS* 84 (1938), 134–9.

[158] T. Lindsay, 'Mental Deficiency Practice at Caterham', *JMS* 80 (1934), 402–8; K. C. L. Paddle, 'Lumbar Puncture and the Cerebro-Spinal Fluid in 2,000 Cases of Mental Deficiency', *JMS* 80 (1934), 674–80.

[159] Dr Pickworth, *JMS* 80 (1934), 173.

[160] E. Slater, 'The Inheritance of Manic Depressive Insanity and its Relation to Mental Defect', *JMS* 82 (1936), 626–34; T. A. Munro, 'Consanguinity and Mental Disorder', *JMS* 84 (1938), 708–14; L. Penrose, 'Some General Problems in Mental Deficiency', *JMS* 84 (1938), 693–707.

mental deficiency attracted a few talented scientists, such as Lionel Penrose, to the site of the colony, such figures were the exceptions.

After the introduction of voluntary and temporary treatment in mental hospitals in 1930, there was a renewed therapeutic optimism among those treating the mentally ill. This made mental deficiency work, by contrast, seem even less attractive. In 1920 a commentator on mental deficiency had predicted that:

> Time alone will show whether ultimately it will overshadow lunacy, but there is no doubt that activity in this direction will lessen lunacy operations, especially as regards the chronic insane. The time may come when lunacy under a more appropriate name will be largely a routine hospital matter like other diseases, the residuum insane being housed in suitable homes, and with the degenerate, mental deficient and criminal ceasing administratively to be primarily of medical concern, though ever the subject of psychological medicine.[161]

Towards the end of the 1930s this prospect was being realized. Mental deficiency colonies were becoming the solution to the psychiatric profession's perennial problem of how to deal with the chronic and the behaviourally disruptive. With many of these difficult cases removed, the mental illness hospital could align itself with the curative model of general medicine, whereas the defective colony assumed the role of institutional 'dumping ground'.

Diet was one of the few areas where medical improvement could be demonstrably achieved. Dietary expertise was presented as vital in improving physical health and an important tool in managing the differing mental states of the patients.[162] Judged in terms of the death rate, medical attention to diet seemed to be a success. For example, the death rate in London had been about 10 per cent before the First World War, and rose to the appalling level of 23 per cent during the war, due to a combination of disease and nutritional deficiency.[163] By 1930 the national rate had fallen and levelled off at 1.3 per cent.[164] However, partly because of this increased life-expectancy, the number of elderly patients increased: the proportion of people aged 65 or over in mental institutions rose from 14.8 per cent of men and 19.7 per cent of women

[161] *JMS* 66 (1920), 289. [162] R. Armstrong Jones, *JMS*, 67 (1921), 146–8.

[163] G. Ayers, *England's First State Hospitals and the Metropolitan Asylums Board, 1867–1930* (London, 1971), app. 3, table C; GLRO, Report Book of Superintendent Caterham, 4–5 Apr., 8 Feb. 1918; E. D. Sherlock, *Minds in Arrear* (London, 1932), 130.

[164] PRO MH 51/520, 5 Feb. 1932, conference between Minister of Health and Board of Control.

in 1938, to 19.1 per cent of men and 27.6 per cent of women in 1948.[165] Better medical care had merely highlighted the therapeutic failure of mental deficiency institutions.

In the 1930s, faced with a shortage of accommodation, a low turnover of patients, and rising demand, some institutions reacted by moving towards the more open models of the colony which had been considered but rejected in 1908. F. D Turner, Superintendent of the Royal Eastern Counties Colony, called for institutions to become 'flowing lakes' rather than 'stagnant water . . . always taking in, always sending out, but sending out only—and this is a most important point—to other smaller lakes, each of which shall be fed from it'.[166] By 1928, there were 1,146 defectives in the community on temporary licence from institutions. Institutional regimes were adapted to provide a pathway back to the community. For instance, at the Manor Institution, defectives were 'socialized' in scout and guide groups and by a system of rewards which led from 'parole' to wander freely within the grounds, to temporary leave, and eventually to discharge.[167] The division between the colony and the community was also broken down through the development of hostels, outside the walls of the institution.[168] Such hostels provided stepping stones to test whether defectives were ready for discharge. Finally, institutions began to use psychiatric social workers to visit patients and maintain links to their families.[169]

It is easy to exaggerate the success of these schemes. The system of licensing was frequently abused, with licences constantly renewed rather than used as a test for discharge.[170] Consequently, discharge rates climbed only slowly to 451 in 1937. In 1938, a scandal arising from cases unfairly kept on licence forced the Board of Control into a

[165] *Annual Report of the Board of Control to the Lord Chancellor for 1950*, p. 4.

[166] F. D. Turner, 'Aims of a Mental Deficiency Institution', *JMS* 74 (1928), 469 (465–73).

[167] The Superintendent, 'The Manor Institution Epsom: Some Comments on its First Two Years', *Studies in Mental Inefficiency*, 5 (1924), 25–32; GLRO LCC MIN/1225 Manor Sub.Com. Papers 1932–3, report of Superintendent.

[168] W. Gibson, 'The Hostel Method for Feeble-minded Young Men and Women', *Mental Welfare*, 11 (1930), 75–7.

[169] J. R. Lord, '"After-care" and Other Aspects of Social Services as an Adjunct to Mental Treatment', *JMS* 76 (1930), 622–31; J. R. Lord, *Social Workers and the Insane* (London, 1923); Unfortunately for developments in this area, Lord died in 1931: obituary *JMS* 77 (1931), 669–79. In the 1930s the main RMPA advocate of this type of service was W. J. T. Kimber: 'Social Values in Mental Hospital Practice', *JMS* 85 (1939), 29–44.

[170] NCCL, *50,000 Outside the Law*; *Roy. Com. 1954–7: Minutes of Evidence* (London, 1954), 825–9.

reassessment of all cases, and this boosted the discharges to 698. But this was still totally inadequate to alleviate the continued rising intake.[171] Medical superintendents remained cautious, if not unenthusiastic, about a shift to the community. Even Turner's model stressed sending out to 'small lakes' only—that is, to annexes of the institution, rather than into the community itself. It was a strategy to extend, rather than draw back, the institution.[172] E. O. Wildensköv, a Danish mental deficiency superintendent, pointed out that the number of defectives actually discharged in Britain was very low and that the criterion for discharge—ability to support oneself in the community—indicated that these individuals should probably not have been in a mental deficiency institution in the first place.[173] Behind Wildensköv's arguments lay his view that a discharge system could only be effective in collaboration with a sterilization programme. In the 1930s the MPA had considered sterilization as another potential solution to its problems but had refrained from giving it official support, despite considerable support from its membership.[174] Wielding influence as President of the MPA at this date, Turner was at the forefront of the opposition to sterilization. He argued that very few discharged mental defectives went on to bear children—though, as Wildensköv pointed out, this was partly because few were released in the first place. In fact, the policies of licence and discharge were held back by continued fears about the sexual behaviour of released defectives and so did little to alleviate the pressure on accommodation.[175]

British superintendents seem to have been less enthusiastic about community care than many of their foreign counterparts, perhaps because their practice tended to be more tightly confined to institutions.[176] Drawing on American experiments, J. R. Lord began to advocate greater use of psychiatric social workers, but he wanted them to be under the control of the superintendent; saving the institution through

[171] BOC circular, no. 835, May 1937; *BOC 1937*, p. 55.

[172] Turner, 'Mental Deficiency Institution', 469.

[173] H. O. Wildensköv, 'Denmark's Care of Mental Defectives', *Mental Welfare*, 18 (1937), 37–44.

[174] For psychiatrists' position on sterilization, see Ch. 5.

[175] PRO MH 51/437; *BOC 1929*, pp. 65–72; *BOC 1937*, p. 9; *MHIWJ* 26 (Mar. 1937), 4; *MHIWJ* 27 (June–July 1938), 3.

[176] Wildensköv, 'Denmark's Care'; L. Beach, 'Beyond Hospital Gates', *Mental Welfare*, 16 (1935), 49–59; G. Grob, *Mental Illness and American Society*; S. C. Brown, 'A Visit to Gheel Colony Belgium', *Mental Welfare*, 14 (1933), 3–7; Trent, *Inventing the Feeble Mind*, pp. 184–224.

reform, rather than replacing it.[177] In Lord's view, a total reorientation to community care would see: 'psychiatry going back to a situation of isolation from general medicine and the social services'.[178] Most psychiatric superintendents were isolated from developments in community care, and were far less sympathetic than Lord. Significantly, there was scarcely any mention of community care for defectives in the *JMS*, even though some coverage was given to the Mental After Care Association's limited work with the mentally ill.[179] Constantly tied to their isolated and closed institutions, mental deficiency superintendents had limited opportunities to develop close contact with outside community care groups.[180] Perhaps hardened by the regimental and dehumanizing nature of institutional life, superintendents were often dismissive about the rights of individual 'inmates' or their suitability for return to the community. Sherlock, Superintendent of Darenth Colony, scorned the complex legal safeguards of the Mental Deficiency Act as a waste of time:

> Common sense would suggest that if a patient has not sufficient intelligence to ask for discharge, he is better off without it, and that it is unnecessary to provide the costly apparatus for reopening periodically the question whether or not he should remain under control.[181]

The American expert in the field, E. A. Doll, was shocked by the conservative nature of institutional care in Britain (in contrast to his enthusiasm for Britain's community care and voluntary work):[182]

> Leadership in mental deficiency in England seems more definitely restricted to the medical profession than is the case in the United States. This is particularly true in the institutional field. In this regard I sensed an atmosphere of conservative administrative policy, apparently reflecting traditional social attitudes

[177] J. R. Lord, 'American Psychiatry', *JMS* 76 (1930), 456–95.

[178] Ibid. 463.

[179] *JMS* 60 (1934), 343–4. C. H. Bond, 'After-Care in Cases of Mental Disorder, and the Desirability of its More Extended Scope', *JMS* 59 (1913), 274–83. For more on the MACA, see Ch. 4.

[180] Just D. Turner and F. Corner were members of the CAMW. Equally, few members of the MPA were members of the Eugenics Society (see Ch. 5): Maurice Craig, the leader of the National Council for Mental Hygiene, was not an MPA member, while C. P. Blacker, Secretary of the Eugenics Society and psychiatrist, was only made a member in 1934.

[181] *MAB Annual Report 1926–7*, p. 289; Sherlock, *Minds in Arrear*, 115–17.

[182] E. A. Doll, 'Community Control of Mental Deficiency in the USA', *Mental Welfare*, 14 (1933), 65–6; Grob, *Mental Illness and American Society*.

> . . . practice lags rather seriously behind ideas and a conventional inertia cools the ardour of enthusiastic noviciates.[183]

It seems that British mental deficiency superintendents, their work unrespected within the profession as a whole, turned inwards to seek status through efficient and economic management as overlords of huge institutional domains. By and large, this had a detrimental effect on the development and standard of care and hindered the breakdown of barriers between institutions and the community.

CONCLUSION

Despite developments in the 1930s, the British colony system continued to be a primarily segregative regime divorced from the outside community. The progressive elements of the colony regime, such as the villa system, training, and occupational activity, became oriented towards management and economy, rather than the development of a therapeutic community. Administrative practice blunted humanity, and underresourcing and professional self-interests blocked alternatives.

In the late nineteenth century, psychiatrists had turned towards the colony, partly because they began to view it as in their professional self-interest, but also because the colony was attractive as an efficient and progressive social policy. The colony was surrounded by a powerful humanitarian and therapeutic ethos. Moreover, it associated the profession with the development of a new tool of social engineering which found much broader favour among progressives. The colony appeared to provide a scientific and politically acceptable way to segregate off the socially inefficient so as to facilitate social development in the rest of society and thereby ensure a harmonious adjustment to democracy. Bearing in mind these broader ideological influences, it is unfair to attribute the involvement of psychiatrists in the development of the colony to professional self-interest alone.[184]

In many respects, the emergence of the colony solution appears to

[183] E. A. Doll, 'Impressions of Mental Deficiency Work in England', *Mental Welfare*, 15 (1934), 18.

[184] On the problems of interpreting the gap between rhetoric and reality: Scull, *Social Order/Mental Disorder*, 256–60; D. Garland, *Punishment and Modern Society* (Oxford, 1990), 196–8.

epitomize the growing influence of the 'professional ideal' in the sphere of social policy. However, the colony also needs to be placed in the context of the lay culture of the time. For instance, the colony regime was shaped by popular attitudes towards mental deficiency, by ideas about the qualities of the countryside compared to the town, by contemporary moral concerns about sex, and by attitudes towards the reformative value of work.[185] It is also apparent that there could be fundamental tensions and divisions over how to balance professional self-interests with the broader 'professional ideal'. It proved impossible to satisfy simultaneously the demand for efficiency and economy on the one hand, and for high quality care and therapy on the other. Both psychiatrists, and the mental nurses who attempted to follow them in a strategy of professionalization, ran into the same dilemma: higher wages and higher standards of care made it difficult to satisfy a demand for efficiency which was assessed in terms of economy. As a result, there was constant competition for resources between and within professional groups, professional alliances were frequently shattered by divisions which ran along lines of class and gender, and there was a constant threat of care being shifted to more efficient and cheaper lay management. Thus examination of the colony solution suggests that 'professional society' was neither a united nor dominant force in interwar Britain. If professionals were to secure positions within the welfare system they often had to compromise on the 'professional ideal'.

[185] For the need to look at the broader cultural roots of policies: M. Wiener, *Reconstructing the Criminal: Culture, Law and Policy in England, 1830–1914* (Cambridge, 1990), 8.

4
The Community Care Solution

INTRODUCTION

As has been described in the two previous chapters, there were serious shortcomings in administrative and institutional provision under the 1913 Mental Deficiency Act. In this chapter I shall explain how these problems were partly overcome through utilizing the voluntary sector to provide care in the community. The chapter will serve a number of additional purposes: first, I shall explore the contribution of the voluntary sector in the construction of policy; secondly, I shall continue to examine the roles of professionalization and gender in welfare provision; thirdly, I shall consider the process of adjusting to democracy; and finally, I shall indicate the existence of community care in interwar Britain, well before the period in which its origins have normally been located.[1]

Most histories of social welfare are heavily state-centred, with scarcely any mention of the voluntary sector, except to point out its shortcomings; when voluntarism is discussed it is usually as a temporary stage on the route to state services.[2] However, writing in an era in which there is less confidence of an onward march of the state and in which power is 'contracted out' to the private and voluntary sectors, recent historians have become more alert to the ephemeral nature of the boundary between the public and private, the statutory and the volun-

[1] R. Titmuss, 'Community Care—Fact or Fiction', in R. Farndale and H. Freeman (eds.), *Trends in the Mental Health Service* (London, 1963), 221–5, said he was unable to locate the origins of the term. M. Bulmer, *The Social Basis of Community Care* (London, 1987), 11, continues to argue that the first official use was in 1957. J. Busfield, *Managing Madness: Changing Ideas and Practice* (London, 1986), 338, like a number of other accounts, does point out the development of community care for the mentally ill in the 1930s but continues to ignore the earlier development of community care for defectives. For a more detailed discussion see: M. Thomson, 'Community Care and Control of Mental Defectives in Interwar Britain', in P. Horden and R. Smith (eds.), *The Locus of Care* (forthcoming).

[2] D. Fraser, *The Evolution of the British Welfare State*, 2nd edn. (London, 1984), 124–32.

tary, and the political and 'non-political'.[3] This situation has stimulated a reappraisal of the role of the voluntary sector in the last century: Prochaska, though working mainly on the nineteenth century, has pointed to the continuing rich vein of philanthropy in the twentieth century; Finlayson, reviving the interwar notion of a 'new philanthropy', has written of 'a moving frontier' between voluntarism and the state, in which the voluntary role continued to be far more important than is reflected in traditional histories of the emergence of the Welfare State.[4]

The view that the voluntary sector did continue to play a significant role in the emerging Welfare State is generally shared by the growing body of historians who have been drawn to the study of women's roles in the formation of social policy. These historians depict women as using the voluntary sector to take advantage of male and state neglect of certain welfare areas, as colonizing these areas, and as developing 'maternalist' policies which, depending on the strength of the state, provided a ladder to enter the public sphere of the state itself.[5] However, Lewis has recently presented some valuable reservations to this emerging orthodoxy: in the British case the policies advanced by women were not necessarily 'maternalist'; the influence of women's class backgrounds was probably as important as their gender; and women's success in the voluntary sector rarely provided a stepping stone into the public sphere of the state in the twentieth century.[6] This latter point is largely supported by Hollis's detailed study of women in nineteenth-century local government, which emphasizes

[3] C. Maier (ed.), *Changing Boundaries of the Political: Essays on the Evolving Balance Between the State and Society, Public and Private in Europe* (Cambridge, 1987).

[4] F. Prochaska, *The Voluntary Impulse in Modern Britain: Continuity and Change, 1880–1987* (London, 1988), 1–2; F. Prochaska, 'Philanthropy', in F. M. L. Thompson (ed.), *The Cambridge Social History of Britain, 1750–1950*, iii. 357–93; G. Finlayson, 'A Moving Frontier: Voluntarism and the State in British Social Welfare, 1911–1949', *Twentieth Century British History*, 1 (1990), 183–206; G. Finlayson, *Citizen, State, and Social Welfare in Britain, 1830–1990* (Oxford, 1990), 201–86. This view adopts the earlier work of E. Macadam, *The New Philanthropy: A Study in the Relations between the Statutory and Voluntary Social Services* (London, 1934); and H. Mess, *Voluntary Social Services Since 1918* (London, 1948).

[5] S. Koven and S. Michel, 'Womanly Duties: Maternalist Politics and the Origins of Welfare States in France, Germany, Great Britain and the United States, 1880–1920', American Historical Review, 95/4 (1990), 1076–1108; S. Koven and S. Michel (eds.), *Mothers of the New World: Maternalist Politics and the Origins of Welfare States* (London, 1993).

[6] J. Lewis, 'Gender, the Family, and Women's Agency in the Building of "Welfare States": the British Case', *Social History*, 19 (1994), 37–55.

the importance of women voluntarists in developing new areas of social provision, but concludes on a pessimistic note regarding women's failure to attain powerful positions in government in the twentieth century.[7] This chapter will examine these debates from the fresh perspective of mental deficiency—an area of policy where women did play an important role, though one which has thus far received little attention in the historical literature.[8]

THE EMERGENCE OF A VOLUNTARY-STATUTORY ALLIANCE

Care of the mentally defective had been one of the many areas innovated by voluntary organizations in the nineteenth century. In the mid-nineteenth century, philanthropists opened the first idiot schools and asylums. Later in the century, charities and female volunteers established hostels for the feeble-minded and after-care services for special-school children.[9] By the end of the century, voluntary organizations were lobbying for the state to assist in providing care for the mentally defective.[10] Even the Charity Organization Society (COS), often seen as the archetype of anti-statist voluntarism, accepted the state's role.[11] In fact, the COS led the way, organizing the first major survey of the scale of the feeble-minded problem in 1893.[12] Whereas such groups generally opposed state intervention on the grounds that it undermined individual self-help, an exception was made for those, such as the defective, who could not help themselves. If the poor law was to continue to be an effective deterrent against improvidence, groups

[7] P. Hollis, *Ladies Elect: Women in English Local Government, 1865–1914* (Oxford, 1987), 471–83.

[8] The one account was written in the 1950s as part of a broader survey of the way voluntary activity was adapting to playing a role alongside that of the state: M. Rooff, *Voluntary Societies and Social Policy* (London, 1957), 79–122.

[9] See Introd.

[10] Rooff, *Voluntary Societies*, 98–104.

[11] C. S. Loch, 'Charity Organization and the Feeble-minded', *Charity Organization Review*, 33 (1913), 6–8. This was part of a wider transition: M. J. Moore, 'Social Work and Social Welfare: The Organization of Philanthropic Resources in Britain, 1900–1914', *Journal of British Studies*, 16 (1977), 85.

[12] Charity Organization Society, *The Feeble-Minded Child and Adult: A Report on the Investigation of the Physical and Mental Condition of 50,000 School Children, with Suggestions for the Better Education and Care of the Feeble-Minded Children and Adults* (London, 1893).

such as the defective, who could not be blamed for their circumstances, had to be removed.[13] Helen Bosanquet, a leading figure in the COS and sister of Mary Dendy, the pioneer in providing feeble-minded colonies, argued: 'We cannot hope to make them good citizens, to make them clever men and women, but we aim at keeping them innocent.'[14] The term 'innocent' is significant: defectives were not viewed as citizens or even as potential citizens in need of reform, but instead, care was provided as a paternal act and because of moral and eugenic concern to maintain the 'innocence' of the feeble-minded. Such voluntarists looked to the state not to replace their own efforts, but to bolster schemes like Mary Dendy's Sandlebridge Colony, and in return, they were willing to accept that the state would have greater power to regulate the care of defectives in the future.[15]

Voluntary organizations played an active role in lobbying the government to act on the 1908 Report of the Royal Commission on the Care and Control for the Feeble-Minded, and they accepted the state's increased role under the 1913 Act; if anything, they tended to be critical that the Act was too limited in its scope. The voluntary–statutory alliance was eased by the inclusion of voluntary-run institutions under the 1913 Act, and by the willingness of the voluntary sector peacefully to cede the principal role in the institutional sector to the state.[16] Although they pioneered care in the community, voluntary groups continued to view segregation as the optimal form of care. In fact, local voluntary groups were often established in an attempt to fund care in state institutions,[17] and only in the absence of such accommodation did they advocate community care as an alternative.[18] Even Miss Darnell, Secretary of the London Association for the Care of the Mentally Defective (LAMW), one of the largest local providers of community care in the country, acknowledged:

[13] C. S. Loch, 'The Development of Charity Organization', *Charity Organization Review*, 86 (1904), 79–80.

[14] National Association for Promoting the Welfare of the Feeble-Minded, *Report of the Conference with Poor Law Guardians* (1901), 10. On the COS support for legislation on the feeble-minded: Hollis, *Ladies Elect*, 23–4.

[15] Ibid. 15; National Association for the Feeble-Minded, *Annual Conference Report* (Brighton, 1911), 16.

[16] *Report of the Feversham Committee: The Voluntary Mental Health Services* (London, 1939), 128.

[17] *Annual Report of the Cambridge Voluntary Association for Mental Deficiency for 1912.*

[18] Report from Oxford Association, *CAMW Annual Report 1925–6.*

We know quite well that at present, owing to shortage of residential accommodation, it is not a matter of who will benefit by treatment in Institutions, so much as of whom we can safely leave even for a time, without such safeguard.[19]

Central government, in turn, welcomed and facilitated the continuation of a prominent voluntary role: the 1913 Act was designed to protect charitable institutions; leaders of the voluntary movement, Mary Dendy and Ellen Pinsent, were made Commissioners on the Board of Control, followed later by Ruth Darwin; and one of the first actions of the new Chairman of the Board of Control, W. P. Byrne, was to aid the establishment of a new national voluntary organization, the Central Association for the Care of Mental Defectives (CAMW).[20] The statutory–voluntary alliance was cemented in the constitution of the CAMW which ensured that about half of its executive committee were central and local government representatives.[21] Local branches of the CAMW adopted a similar practice of co-opting representatives from local government. In fact, examining the membership of the CAMW the very notion of a clear division between voluntary and statutory status is brought into question: 'volunteers' sometimes also served as co-opted members of local government; while the statutory representatives on CAMW committees might themselves serve in local government on a voluntary basis, and they or members of their families might also be voluntary subscribers to the CAMW.

Without a local administration of its own, the Board of Control used the CAMW and its national network of branches to reach out to local authorities and pressurize them into action. From its first Annual Reports onwards, the Board encouraged local authorities to delegate administrative roles to the voluntary associations, arguing that this was cheaper and that voluntary groups could provide desirable services which were outside the statutory remit of the 1913 Act.[22] By 1923, with little prospect of a rapid expansion in colony provision, the Board

[19] Miss Darnell, *CAMW Conference Report 1926*, p. 48. The Association changed its name to the London Association for Mental Welfare at the end of the 1920s, and the abbreviation LAMW is used for the group throughout.

[20] PRO MH 58/85, A. V. Symonds to Mr Lowry, 8 July, 1914; *Report of the National Association for the Welfare of the Feeble-Minded for 1913*, pp. 12–14. The title of the group was the Central Association for the Care of Mental Defectives until 1921 and thereafter the Central Association for Mental Welfare. The abbreviation CAMW is used when referring to the group generally.

[21] *CACMD Annual Report 1915–16*, p. 11.

[22] *BOC 1915*, p. 42.

was outlining an ambitious plan for community care and pointing out the superior performance of those local authorities aided by voluntary associations.[23] In the same year, speaking to the City of Leeds Voluntary Committee for the Care of the Mentally Defective, Commissioner Ellen Pinsent admitted that provision of colonies had been a 'great disappointment' and expressed the Board's interest in the city's schemes for care, or as she put it 'control', in the community:

> You believe that for a small number of cases, what has been called 'Community Control,' as distinguished from 'institutional control,' will prove sufficient protection and enable these specially selected cases to lead happy and useful lives, and it is in view of this that we are so immensely interested in the experiment you are starting in Leeds.[24]

After the publication of the Wood Report in 1929, with its inflated estimate of the defective population, and after the financial crisis of the early 1930s, the Board's support for community care became even more forthright.[25] In its 1929 Annual Report it advocated that, 'Day centres should form an integral part of each Authority's scheme for the care and supervision of defectives living in the community'.[26] The following year, the term 'community care' itself was used: 'It is clear that society cannot afford to segregate defectives who are in other respects fit for community care merely to prevent them from marrying.'[27] Whereas in 1927 the Board had described supervision as 'better than nothing',[28] by 1935 institutional care was being termed 'a last resort'.[29] In fact, by the late 1930s some local authorities had up to 18 per cent of their institutional populations on licence,[30] and by 1939 there were about 40,000 defectives under statutory supervision, a further 26,000 under voluntary supervision, 4,500 under guardianship orders, and 3,000 on licence from institutions. In total, the number of those under the various types of community care was almost twice that in institutions.

The government turned to voluntary organization not simply because of its own limitations in providing institutional care and administration, but also because it needed non-governmental allies to undertake propaganda and public education—work which was seen

[23] *BOC 1923*, pp. 41–60.
[24] City of Leeds Voluntary Committee for the Care of the Mentally Defective, *Address by Mrs Ellen Pinsent: on the occasion of the official opening of the new workshop and occupation centre for feeble-minded young men* (Leeds, 1923), 5.
[25] *BOC 1930*, p. 59.
[26] *BOC 1929*, p. 61.
[27] *BOC 1930*, p. 71.
[28] *BOC 1927*, p. 50.
[29] *BOC 1934*, p. 11.
[30] *BOC 1935*, p. 59.

as inappropriate for supposedly disinterested civil servants.[31] Bureaucrats excused the government for the limitations of services, on the grounds that voluntary groups had to convert popular opinion first: 'It is idle to pretend that this particular service is one of the most popular of health services . . . Politicians of all kinds, national politicians and local politicians are alike in that none of us can run very far ahead of the crowd.'[32] Career civil servants at the Board openly admitted their personal distaste for active involvement in care and welcomed delegation of responsibility to voluntary groups and to women members of their department. Chairman of the Board, Brock, confessed: 'I am very glad that Mrs Pinsent, to whom defectives owe so much, can enthuse over jolly little Mongols. I know others who find defectives much less attractive.'[33] Thus the success of the interwar voluntary sector in providing a supplement to statutory services was partly based on well-established 'maternalist' ideas about there being an essential difference between the sexes, with women uniquely qualified to serve caring roles and break down public–private barriers.

The relationship which rapidly developed between the CAMW and the Board of Control can be seen as one of symbiosis, with the Board attempting to use the CAMW to overcome statutory limitations, and the CAMW using this to gain a privileged position of influence at the local and central levels. The nature of the relationship was captured in the description of the CAMW President in 1913:

> Under this system the Board of Control would in effect retain very real control over the whole of the voluntary organization, but, conversely, the voluntary organization would gladly submit to that control in consideration of the official recognition of its work and the effective power it would obtain thereby.[34]

As late as 1930, the Board of Control reiterated this view when it defended the CAMW against funding cuts and closer monitoring:

> It is in fact a mistake to speak of these relations simply in terms of an inspection. As between the Board and the Central Office of the CAMW in London, there is frequent consultation. As regards the local branches, I do not think the Board

[31] M. L. Grant, 'Perceptions of Propaganda in Interwar Britain: The Development of Government Publicity in the Domestic Sphere', unpub. D.Phil. thesis (Oxford, 1988), 422.

[32] S. Lawrence, *Report of the Proceedings of the Second Conference on Mental Health* (London, 1931), 4.

[33] L. Brock, 'Mental Health Workers: An Appreciation', *Mental Welfare*, 12 (1931), 65 (65–7).

[34] L. Scott, *Medical Officer*, 10 (1913), 241.

would be prepared to say that the bulk of inspection should necessarily be done by the Local Authority. The Central Association will often be doing pioneer work on which the Board can judge the value far more effectively than the Local Authority.[35]

The CAMW's intermediate status is reflected in its dependence on a mix of independent and state funding. In 1919 subscriptions raised £234 and donations peaked at £515, in total almost double the statutory grant of £450. These voluntary donations were essential in setting up local groups and vital in supporting the national CAMW during its financially precarious early years.[36] Through such donations, single individuals and families could gain considerable influence, as was most notably the case with the Darwin family and the MP, Leslie Scott, who became the CAMW's first President.[37] However, by 1920, the central government grant had increased to £2,000 and had become vital in enabling the CAMW actively to encourage the establishment of local associations by sending out representatives to train and organize new members.[38]

In summary, in terms of their common aims, overlap of personnel, financial relationship, and functional interdependence, a close alliance did develop between the Board of Control and the CAMW in the 1920s. With the reservation that co-operation in this particular area of welfare had earlier roots, the increased intimacy of the relationship and the blurring of the division between statutory and voluntary spheres suggests that what we might term a 'new philanthropy' was emerging.

INTERWAR COMMUNITY CARE

The scale of the mentally defective population under the various types of care in the community, and the fact that the government itself was using the term by the 1930s, suggest that one can indeed speak of there being a 'community care' policy in the interwar period. The most common category of community care for defectives was what was termed 'supervision'. Before being placed under supervision, defectives had to be 'ascertained', which meant that they had to be brought to the

35 PRO MH 58/70, Barter to Heseltine, 1 Sept. 1930.

36 *CACMD Annual Report 1915–16*, pp. 14–15.

37 Subscription and donation lists in *CAMW Annual Reports*.

38 Rooff, *Voluntary Societies*, 124.

attention of the mental deficiency authorities by teachers, doctors, parents, or workhouse or penal authorities, and then had to be proved to be mentally defective from birth or early age and to be in need of care or control for social or moral reasons. Once ascertained, the defective was 'subject to be dealt with' and would either be placed in an institution or under some form of community care, usually supervision. Supervision was chosen if the home situation of the defective was deemed to be adequate in terms of care and control, though it was sometimes chosen by default since there was so little space available in institutions. Supervised defectives were visited by social workers at regular periods. There was wide variation in the timing of these visits, though the Board of Control recommended that visits be 'at least once a month and more often if the defective is likely to get into difficulties'.[39] Visits to the home continued even after a defective had been removed to an institution: the 'case-history' remained on file and visitors continued to visit the family, partly to collect contributions towards the cost of care, but also to monitor any improvements in home conditions which might justify a return to the community.

It is difficult to reconstruct the nature of the supervisory visits. The reports from the visitors concentrated on evaluating the standard of home care, the economic status of the family, the physical and moral environment of the home, the hereditary history of the family, and the behaviour of the defective. As this suggests, the primary aim of visits was to monitor whether home care and control were adequate.[40] There is little record of what sort of care, if any, was provided under supervision. The Board of Control advocated that 'success depends almost entirely on the personality of the supervisor. Unless supervisors have the social qualities which will enable them to become the trusted friends of the families they visit, very little good will be effected.'[41] Such intimacy was unlikely, given the inspectorial style of supervision and the social distance between visitors and clients—though, to families in distress with the problems of caring, official recognition and advice from those who had been trained in the area and brought with them a scientific explanation of mental deficiency may have been of comfort.

Where one can be more certain about the limitations of care is in the provision of material welfare; here, one would expect at least some documentation, and since there is no such record it can be assumed

[39] *Board of Control Annual Report for 1923*, p. 47.

[40] For examples, see Ch. 7.

[41] *BOC 1923*, p. 47.

that care stopped short at donation of food, clothes, or financial assistance. In fact, visitors often performed the roles of 'rent collectors', visiting families whose members were already in institutions and threatening them with fines if they failed to make contributions towards support.[42] Indirectly, however, some of the supplementary services linked to supervision did provide material assistance to defectives and their families. For instance, voluntary groups attempted to place defectives in employment, seeking out sympathetic employers who would provide sheltered work, and they set up occupation centres in the community. These centres provided some supplementary income to aid family care, relieved families from their caring roles, diverted the energies of defectives from potential misconduct, and aimed to train the defectives for non-sheltered employment. The number of defectives registered at the centres peaked in 1939 at about 4,200.[43] They were more common in towns, where the density of population made it viable for a sufficient number of defectives to visit a centre daily or several times a week. In rural areas, different types of community care developed: for instance, home teachers visited families who lived too far from a special school for their defective children to attend on a daily basis.

Although supervision was the most common form of community care, a series of other forms also emerged over the interwar period. Guardianship never reached the scale of supervision, rising steadily but reaching only about 5,000 by the end of the 1930s. Under guardianship, ascertained defectives were placed under the legal responsibility of a named guardian who was paid a fee and, in return, was to provide a suitable home for the defective.[44] The third main form of community care was licensing. Under this system, defectives were granted extended leave from institutions on the condition that they kept to the regulations of a licence. The licensed defective was monitored by supervisory visits, and if the conditions of the licence were obeyed it was rescinded after a set period.[45]

The role of the licence in bridging the gap between institution and community, like that of supervision carrying the threat of institutionalization into the community, indicates the close relation between community and institutional care; the two were not diametrically opposed, but rather they were overlapping strategies linked in a common aim of

[42] See Ch. 7. [43] Rooff, *Voluntary Societies*, 126.

[44] For a discussion of this type of care under the 1913 Act: *BOC 1923*, pp. 50–4.

[45] *BOC 1932*, pp. 64–5.

care and control and developing in tandem.[46] Ideas of community were present in institutional care, where the colony aimed to establish a 'community' of defectives, in 'villages' of 'villas' or 'cottages'.[47] Likewise, community care adopted institutional models, most notably in the case of occupation centres, which translated the location of the workshop from the colony to the community.

In summary, interwar institutional and community care did not emerge in opposition to one another: they were intimately linked both strategically and ideologically. The community care provided for mental defectives in interwar Britain was not that of an ideal model—if such a model exists—of care by the community and cut off from the system of institutional care. However, it was a formalized system of care, provided by the state, and located in the community. Therefore, under any reasonable definition, it surely qualifies as constituting a community-care policy. This suggests that we need to reassess our ideas about the history of community care—a history which has exclusively focused on the period after the 1950s—and should pay more attention to continuities stretching back to the interwar period.

CHALLENGES AND PROBLEMS

Tensions within the Voluntary–Statutory Alliance

In the early years of the Act, local authorities were generally enthusiastic about assistance from voluntary groups.[48] The main exception was that some local medical officers of health objected on the grounds that it fragmented public health services.[49] However, local authorities became less willing to devolve work to independent voluntary groups as the fees for such work increased, particularly after the Local Government Act enforced five-year grants upon local authorities and thereby made the scale of devolved expenditure more visible. In response, local

[46] For a more detailed exploration of this interaction see: M. Thomson, 'Sterilisation, Segregation, and Community Care: Ideology and the Problem of Mental Deficency in Inter-War Britain', *History of Psychiatry*, 3 (1992), 473–98; and A. Scull, 'The Asylum as Community or the Community as Asylum', in id., *Social Order/Mental Disorder* (London, 1989), 300–30.

[47] See Ch. 3.

[48] F. Robinson, 'Local Administration of the Mental Deficiency Act 1913', *Public Health*, 29 (1916), 78–93.

[49] *Public Health*, 37 (1924), 301–2.

authorities began to take direct control of the services which had been introduced by the voluntary groups.[50] In the 1930s, the take-over of voluntary work accelerated in response to increased pressure for rationalization of social services. This shift in attitudes was evident in the 1937 survey on social services by Political and Economic Planning (PEP), which criticized the inefficiency, poor organization, and small scale of voluntary organizations.[51]

In his survey of voluntary activity of 1948, Mess argued that there were also growing ideological objections to philanthropy:

> To large numbers of the working classes, to the majority (probably) of active trade unionists and active members of the Labour Party, to many others who are strongly democratic in outlook, the voluntary charitable society is anathema.[52]

Since volunteers tended to be middle-class women who could afford to give up their time unpaid, it was inevitable that in providing welfare for a predominantly working-class clientele there would be an element of paternalism. It is difficult to establish, though, whether this really did make voluntary services unpopular among working-class clients (or at least, any more unpopular than state services), especially since voluntary groups argued exactly the opposite.[53]

In response to criticisms of its work and to the take-over of its services, the voluntary sector had to defend its continued relevance. There were of course some voluntarists who reacted by attacking the advance of the state and any voluntary-statutory alliance.[54] But the predominant attitude was to see new legislation, such as the Mental Deficiency Act, as opening up new opportunities for voluntary activity.[55] Voluntary groups argued that they still had a series of advantages over statutory service in this area: at little or no cost, they could provide

[50] S. Wormald, 'Mental Deficiency Under the Local Government Bill, 1928', *Mental Welfare*, 10 (1929), 1–4; PRO MH 58/70.

[51] PEP, *Report on the British Health Services* (London, 1937), 400–1; Mess, *Voluntary Social Services*, 207–8. This was part of a broader vogue for 'planning': A. Marwick, 'Middle Opinion in the Thirties: Planning, Progress and Political Agreement', *English Historical Review*, 79 (1964), 285–98. On PEP: J. Pinder (ed.), *Fifty Years of Political and Economic Planning* (London, 1981).

[52] Mess, *Voluntary Social Services*, 209.

[53] On working-class suspicions of state services: H. Pelling, 'The Working Class and the Welfare State', in id., *Popular Politics and Society in Late Victorian Britain* (London, 1968).

[54] 'Voluntary or Official', by a 'Wayfaring Man and Fool', *Charity Organization Review*, 36 (1914), 393–9.

[55] e.g. A. T. Thompson, 'National and Municipal Demands of Local and Voluntary Work', *Charity Organization Review*, 36 (1914), 47.

local authorities with an organized labour force to perform the duties of ascertainment and supervision of defectives;[56] emerging out of existing philanthropic societies, they already had local experience in dealing with the problem section of the population from which defectives were likely to come;[57] and because of their voluntary status and mainly female membership, they were better suited to intruding into the private sphere of the household.[58]

In the voluntary sector there was considerable dissatisfaction at the limited achievements of local government under the 1913 Act.[59] The CAMW presented itself as providing essential co-ordination for the chaotic and divided administrative apparatus of the Act. Local voluntary groups often provided the one forum which could bring together representatives of mental deficiency, poor law, and education authorities. Not confined to working within municipal boundaries, they could link up neighbouring local authorities and encourage co-ordinated projects.[60] And not bound by the limits of local authority responsibility, they were able to extend 'preventive' supervision on a voluntary basis to cover non-certifiable but 'borderline' cases.[61] Because of the voluntary nature of its work, the CAMW was also able to reach out to middle-class clients, whose defective dependants were unlikely to be considered in need of statutory certification, and who were unlikely to turn to the public health authorities for aid and advice. The needs of such direct applicants—peaking at over 1,000 in 1929[62]—stimulated the CAMW to extend its services beyond the bare minimum of care.[63] With freedom to experiment and to reach beyond the strict confines of the Act, it became the leading innovator in mental deficiency provision, developing a series of new services which were subsequently adopted by local authorities.[64]

[56] Miss Fox, *CAMW Conference 1918*, 25–6.

[57] Miss Darnell, *CAMW Conference 1926*; Hollis, *Ladies Elect*, 462.

[58] *CACMD Annual Report 1915–16*, p. 14; *BOC 1922*, p. 46; *BOC 1930*, p. 30. For the way that 'separate spheres' shaped voluntary work: F. Prochaska, *Women and Philanthropy in Nineteenth Century England* (Oxford, 1980); R. Walton, *Women in Social Work* (London, 1975).

[59] E. Fox, 'National and Municipal Demands for Local and Voluntary Work', *Charity Organization Review*, 36 (1914), 62; E. Fox, 'The Mental Deficiency Act', *Charity Organization Quarterly*, 8 (1924), 136.

[60] 'Mental Defectives under the Poor Law', *Studies in Mental Inefficiency*, 4 (1923), 35–8.

[61] 'The Mentally Defective and the Community', *Studies in Mental Inefficiency*, 4 (1923), 71–9; *BOC 1922*, p. 47.

[62] Rooff, *Voluntary Societies*, 167–8.

[63] *CACMD Annual Report 1920–1*, p. 22.

[64] Rooff, *Voluntary Societies*, 125–9.

The Challenge of Professionalization

The second challenge faced by voluntary groups was the internal division over the relative merits of voluntary, statutory, and professional status that resulted from the process of professionalization. The founding of specialist voluntary groups to provide services for the mentally defective can itself be viewed as a step in the process of professionalization. In the late-nineteenth century the dominant model for the organization of voluntary work had been that of a single Charity Organization Society, which co-ordinated all social work in any one area and thereby dealt with all the members of a family, even though they may have had very different problems as individuals. Although there was resistance to replacing this model,[65] it was recognized that this was necessary if voluntary groups were to provide the type of specialized services which could compete with those of the state.[66] Before long, however, since specialization of voluntary work led to a greater emphasis on expert knowledge and to delivery of services rather than mere relief, the status of social workers began to be elevated beyond that of mere charitable volunteers. Through placing volunteers alongside representatives of statutory authorities, the CAMW helped to further break down amateurism, while the development of training courses, employment of trained personnel in full-time jobs, and publication of a specialist journal fostered a growing sense of professional identity among its employees.[67]

Although many of the characteristics of professionalization were already emerging in the 1920s, it seems that this was achieved without causing a fundamental division within the CAMW. To understand why this was the case we need to examine the role of its Honorary Secretary, Evelyn Fox, who attempted to negotiate an identity for the CAMW that lay between the poles of voluntary and statutory, and amateur and professional. Born in Ireland in 1874, Fox was said to have inherited a forceful and combative personality from her ancestor, the eighteenth-century politician, Charles James Fox. During the interwar period she was, as one admirer described, like 'the Florence Nightingale of the mental health movement'. It is perhaps significant that Fox, herself, did

65 T. Hancock Nunn, *Charity Organization Review*, 36 (1914), 58.

66 'National and Municipal Demands for Local and Voluntary Work', *Charity Organization Review*, 36 (1914), 51–3.

67 Hargrove, 'Work for the Mentally Defective', *Women's Employment* (26 Feb. 1925), 5; Rooff, Voluntary Societies, 129–30.

not have strong roots in voluntary activity. She seems to have been a relatively late and sudden convert to the cause of mental deficiency, her attention only drawn to the issue at the age of 32 in response to the Royal Commission on the Feeble-Minded. In reaction she joined the Women's University Settlement in Southwark where she received some basic training in social work practice. Thus she had relatively little experience when she set up the CAMW in 1913 with, it is said, just a typewriter and £10.[68]

The fact that Fox had only entered voluntary work at a late stage helps to explain why she was so open to the CAMW adopting a close alliance with the state. At the same time, with only limited training, she was not wholly part of the new generation of interwar social workers who embraced a more stridently professional model of psychiatric social work. Thus, although she considered case work to be a skill, she felt that this rested on experience, local knowledge, and sympathy, rather than psychological expertise:

> The social adaptation of a defective, the study of his own limitations, the correlation of that study with his environment, his home conditions, the further consideration of his employability, the finding of a suitable environment for him if he can remain in the community in perhaps better surroundings than are his at the moment—this is not work that can be carried on in a few moments or by one or two casual visits. It requires the most careful investigation and study, and it requires also that fundamental, real knowledge of the social conditions of the area in which the defective is living.[69]

During the interwar period, Fox's pragmatic vision of social work probably remained dominant among the voluntary foot soldiers of the CAMW, but it did meet a growing challenge from a younger generation who were influenced by the model of professionalized psychiatric social work developing in the United States, and with it a more strident claim to professional status. It was this 'psychiatric deluge' of the early 1930s which proved difficult to contain within the confines of the community care consensus of the early 1920s.[70] Although psychological clinics had already been set up in England, the introduction of child guidance clinics from the United States in the late 1920s was an important

[68] Obituary, *Social Service*, 28 (1955), 172–5; *DNB*, *1951–60*, pp. 372–3; *Mental Health*, 15 (suppl.) (1955), pp. 2–3.

[69] E. Fox, *CAMW Conference Report 1929*, p. 72.

[70] K. Woodroofe, *From Charity to Social Work in England and the United States* (London, 1962), 125.

turning point.[71] In response to a visit from Mrs Esther Strachey, wife of the journalist and future Labour MP Evelyn St Loe Strachey, the North American Commonwealth Fund agreed to support a trial child-guidance clinic in London and to fund British social workers to visit the United States for training.[72] The scheme, established in 1927, continued over the rest of the interwar period.[73] Fox, who envisioned her own CAMW as reaching out to the 'borderline' of mental health, was appointed as the first Secretary of the Child Guidance Council. However, before long she was coming into conflict with the model of work which was being imposed by the Commonwealth Fund. In particular, there was disagreement over two issues: first, over whether mental deficiency should be part of child guidance work—a view which Fox obviously endorsed; and secondly, over the degree to which mental welfare workers needed to be trained—with the Commonwealth Fund backing the creation of a university psychology course, whereas Fox felt that long training courses were impractical and not as important as everyday experience. As a result of these differences, Fox resigned from the Child Guidance Council in 1930. During the 1930s, the Commonwealth Fund directed its support at expanding provision of child guidance clinics and establishing a three-year training course at the London School of Economics in psychiatric social work. With the departure of Fox, child guidance work was distanced from mental deficiency work, and a new generation of leaders, such as Noel Hunneybun and Doris Robinson, established their own professional organization, the Association of Psychiatric Social Workers.[74]

In the 1930s there were signs of increasing tension between the older generation of volunteers and the new generation of trained social workers. This was already apparent in 1931 when the LCC took control of the community care work, previously performed by the London Voluntary Association for Mental Welfare (LAMW). In one sense, the change was superficial, since the council agreed to employ the LAMW visitors

[71] C. Burt, 'Psychology and Social Work', *Charity Organization Review*, NS 43 (1918), 5–9, 51–60 suggests that the impact of psychology was earlier than maintained by Woodroofe, *From Charity to Social Work*, 125. For a summary of the debate: N. Timms, *Psychiatric Social Work in Great Britain, 1939–1962* (London, 1964), 15–21.

[72] A. Harvey and S. Abrams, *For the Welfare of Mankind: The Commonwealth Fund and American Medicine* (Baltimore, 1986).

[73] D. Thom, 'Wishes, Anxieties, Play, and Gestures: Child Guidance in Inter-War England', in R. Cooter (ed.), *In the Name of the Child* (London, 1992), 200–19.

[74] Boxes 1–2, Commonwealth Fund Archive (CF), Rockefeller Archive Center, North Tarrytown, New York (RAC).

on its own staff. However, the shift from voluntary to statutory status probably also had an impact on the style of the work. The LCC's requirement that the LAMW should drop its links to the network of other voluntary groups which fed it with information suggests that there was a hardening of the division between the statutory and voluntary sectors.[75] The COS vocalized the concern of those who felt there was still a place for independent voluntary work, condemning the LCC for 'that spirit of jealous exclusiveness which has done more than any of the other many causes to destroy voluntary social service on this island'.[76] By contrast, the social workers taken on by the council do not seem to have objected; in fact, there are signs that they became critical of the moralistic visiting style of their voluntary partners.[77]

The great problem for those who wanted to professionalize this area of social work was that local and central government had been attracted to community care because it could be provided by a plentiful and voluntary workforce, with little expense to the state.[78] Brock, the Chairman of the Board of Control, continued to support the view that social work was better performed by female voluntary workers than trained professionals:

> it may be found that personality is a good deal more important than any elaborate technical training, and the voluntary worker may be more successful precisely because she is a voluntary worker, because she comes as a friend simply with a desire to help and not in the execution of a job which it is her profession to carry out.[79]

Although the alliance between the Board of Control and the CAMW had been constructed on such a basis, Brock's assumptions about social work provoked objections from a new generation within the CAMW. Leading this more self-confident membership was Letitia Fairfield, sister of the feminist journalist and author Rebecca West, and herself a leading activist for women's rights in the Medical Women's Federation. Fairfield, who was qualified to practice in both law and medicine, and was medical secretary of the CAMW, attacked the double standards which underpinned male attitudes towards social work:

[75] GLRO LCC MIN/731 Report of the MD Acts Sub. Com., 16 June 1931.

[76] *Charity Organization Quarterly*, NS 6 (1932), 134–5.

[77] London County Council, *Handbook on Mental Health Social Work* (London, 1947), 29.

[78] *BOC 1931*, pp. 13–14.

[79] L. Brock, *CAMW Conference Report 1930*, p. 69.

It is a very strange thing that men constantly think that women are using tact and personality when in fact they are using sheer knowledge and experience . . . The important thing for us to realise is that this is an extremely difficult and highly technical job. Anyone who undertakes it will have to learn it in order to be of any real value to the doctor or the patient.[80]

Fairfield's professional orientation led her to adopt an increasingly critical attitude towards the political positions of the CAMW. Believing that social workers should follow the lead of the medical profession in aiming to distance the mental health system from the stigma of legal certification, she objected to the CAMW's attempt to extend the definition of certifiable defect in 1926.[81] Then, in 1934, she resigned from the CAMW over its decision to join the campaign for voluntary sterilization.[82] This event and her subsequent appointment as Chair of the Association of Mental Health Workers symbolized the emerging division between the old generation of voluntary activists and the new generation of professional social workers.[83]

Although professionalization encouraged the gradual elevation of psychiatric expertise over older moral and religious sources of authority,[84] the impact of psychological theory should not be exaggerated. From the writing of the advocates of the new psychiatric social work, it would seem that psychoanalytical thinking encouraged visitors to be more self-conscious about their own role in the process of observation, and, as a consequence, to be rather more sympathetic towards their clients.[85] However, even Sybil Clement Brown, who ran the training programme at the LSE, admitted that a psychological approach had always been an intrinsic part of case work and that it was as much an art based on practical experience as it was a science.[86] In reality, psychology often simply provided a new language to support the traditional aims of

[80] L. Fairfield, *CAMW Conference Report 1930*, pp. 93–4.

[81] *CAMW Conference Report 1926*, p. 93.

[82] *Mental Welfare*, 15 (1934), 47; EUG C 103, Fairfield to Blacker, 16 Mar. 1932: Fairfield's opposition cannot simply be attributed to her conversion to Catholicism, for she was also a supporter of birth control.

[83] *Mental Welfare*, 16 (1935), 37–41.

[84] N. and J. Parry, 'Social Work, Professionalism, and the State', in N. Parry, M. Rustin, and C. Satyamurti (eds.), *Social Work, Welfare and the State* (London, 1979), pp. 33–4. For an indication of the content of the new psychiatric social work training courses: History of the London School of Economics Papers (LSE) 27/3 (vi) (Papers Set I, 1921–30).

[85] S. C. Brown, 'Methods of Social Case Workers', in F. C. Bartlett *et al.* (eds.), *The Study of Society* (London, 1939), 379–401; id., 'Family Case Work and Mental Health', *Charity Organization Quarterly*, NS 13 (1939), 45–50.

[86] Brown, 'Family Case Work', 40–50.

social work. Thus mothers continued to be the focus of social workers, only now they were scrutinized for 'emotional neglect' of their children.[87] Even the conservative COS was able to appropriate the new rhetoric to support its call for self-help and independence. There was a suspicion about what was seen as the pretentious use of psychological language, but a feeling that when the rhetoric was stripped away it was mainly plain common sense. As one social worker reasoned:

> Not that I think we can or should dabble in psychology, but I think the specialists can and should translate some of the essentials of their complicated knowledge into terms of everyday life and make it available for use. For it is no secret cult. . . . So far as I can see, behind the most complicated psychological discussion one always finds the same basic simplicity—a balance between power and individuality with love and relationships.[88]

When psychological language was adopted, it tended to be the psychology of humans struggling to master primitive psychic energies and instincts through a higher social sense.[89] Delinquency and social problems were reinterpreted as the result of an individual's failure to control his or her psychic energies, and the social worker's role was to direct such individuals towards activities, such as hard work, or to a disciplined environment, which would sublimate the misdirected energies.[90] Divorced of his ideas on sex, which were rejected more generally in Britain, the COS were even willing to embrace the language of Freud:

> The object of the therapeutic efforts of psychoanalysis is to strengthen the EGO . . . Freud's scheme forms a useful and striking statement of the idea of constructive case work . . . far from being a rebel he offers a most powerful argument for authority.[91]

[87] J. Bowlby, *Maternal Care and Mental Health* (Geneva, 1951); J. Lewis, 'Myrdal, Klein, *Women's Two Roles*, and Post War Feminism, 1945–60', in H. Smith (ed.), *British Feminism in the Twentieth Century* (London, 1990), 180–1.

[88] M. Cosens, 'Psychology and Social Case-Work', *Charity Organization Quarterly*, 7 (1933), 63–73.

[89] The most influential psychological work was still W. McDougall's *Introduction to Social Psychology* (London, 1908), which went through numerous new editions during the interwar period.

[90] A. Raven, 'The Use of Psychology in Social Work', *Charity Organization Quarterly*, 14 (1925), 251–5.

[91] 'New Introductory Lectures on Psycho-Analysis', *Charity Organization Quarterly*, NS 8 (1934), 191; also Cosens, 'Psychology and Social Case-Work', 63–73. The discussion about the role of psychology in the COS's work began as early as 1918: Burt, 'Psychology and Social Work'. On the British reception of Freudian ideas: D. Rapp, 'The Early Discovery of Freud by the British General Public, 1912–1919, *Social History of Medicine*, 3 (1990), 217–43.

In mental deficiency work the impact of new psychological ideas was even less significant. Apart from the crude assessment of mental ability (there is little evidence that social workers undertook detailed psychological tests), visiting continued to be dominated by assessment of household cleanliness, moral behaviour, and the social efficiency of the defectives.[92]

Whatever their impact on visiting, it is clear that the new psychological ideas did provide a crucial route for the professionalization of psychiatric social work. In the long term, this would produce a new problem, for in shifting their domain of expertise from the women's' sphere of domestic and moral authority to the medicalized sphere of casework and psychiatry, social workers had to renegotiate their relation with the medical profession. In the United States, where psychiatrists did become involved in care in the community, tensions developed between psychiatrists and psychologists, but less so with psychiatric social workers who accepted a subordinate and non-threatening position in teams under the leadership of the other two professions.[93] In Britain, there was less danger of competition as psychiatrists tended to stay within the institutional sphere, and those involved in community care were muted in any criticism of institutionalization.[94] Psychiatric social workers accepted a subordinate position in teams under psychiatrists and psychologists, but, in return, they used this role to differentiate themselves as an elite from untrained volunteers. In doing so, they made the expansion of community care increasingly dependent upon provision of a combination of scarce psychiatric, psychological, and social work expertise, rather than volunteers. Ironically, therefore, professionalization acted to limit further expansion of community care.[95]

The Problematic Nature of 'Community Care'

Expansion of community was also held back by a series of unresolved tensions over the meaning of community care.[96] At the heart of the 1913

[92] See Ch. 7.

[93] Grob, *Mental Illness and American Society, 1875–1940* (Princeton, 1983), 234–87; J. Burnham, 'The Struggle Between Physicians and Paramedical Personnel in American Psychiatry, 1917–1941', *Journal of the History of Medicine*, 29 (1974), 93–106.

[94] I. Laird, 'Some Psychological Problems in "Institutionalising" Defectives', *Mental Welfare*, 19 (1938), 1–7.

[95] C. P. Blacker, *Neurosis and the Mental Health Services* (London, 1948), 92–100. See Ch. 8.

[96] E. and S. Yeo, 'On the Uses of 'Community': From Owenism to the Present', in S.

Act there was a fundamental ambiguity, in that care was intended to protect the defective from the community, yet also to protect the community from the defective. Even the first side of this equation—protection of the defective from the community—reflected a mistrust of the community from which the defectives came. In this respect, the views of CAMW member Mrs Emmett were fairly typical:

> We all agree that the best place for a mental defective is in a good home with suitable parents, but we know that many of these homes are very unsuitable: bad housing, lack of decent sleeping accommodation, bad environment, bad associates, and some of the parents have little or no conception of the importance of their responsibilities. In fact we sometimes wonder which is the worst, the certified defective or the parents?[97]

Because of such fears, community care remained a second-best alternative to institutional care, and the authorities were reluctant to allow the parents or families of defectives to serve as guardians in the community.[98] The second side of the equation—protection of the community—lay at the heart of community care. The main term chosen to describe community care—'supervision'—reflected the emphasis on policing and reforming behaviour, rather than providing aid.[99] Supervisory control was advanced as being essential for the moral and eugenic health of the community.[100] Although this could promote community care at times of national emergency, it was also a potential weakness since it exposed community care to a backlash whenever supervisory control proved to be ineffective.[101]

A second shortcoming in the interwar conception of community care was the reluctance to address the social and economic conditions of the community. Despite the fact that certification depended, to a large degree, on whether defectives could support themselves or be supported by their families, there were few attempts to offset the need

Yeo (ed.), *New Views of Cooperation* (London, 1988), 229–58; M. Bayley, *Mental Handicap and Community Care* (London, 1973), 10–11; S. Baldwin, 'From Community to Neighbourhood', in S. Baldwin and J. Hattersley (eds.), *Mental Handicap: Social Science Perspectives* (London, 1991), 145–66.

97 Mrs Emmett, *CAMW Conference Report 1926*, p. 54.

98 *BOC 1923*, pp. 44–51; *BOC 1926*, pp. 53–4.

99 'Community Schemes for the Social Control of Mental Defectives', *Mental Welfare*, 11 (1930), 71.

100 PRO MH 51/691 CACMD to Ministry of Reconstruction, 24 Oct. 1917; *CACMD Annual Report 1918–19*, p. 16.

101 *BOC 1923*, p. 60; *BOC 1924*, p. 80.

for institutional care through providing material aid.[102] Of course, this reluctance was nothing new, but it was exacerbated by two interwar trends: first, there was a shift in public health strategies, away from a focus on the environment and towards a focus on the individual;[103] and secondly, the growing influence of psychology promoted the individualization and psychologization of social and mental problems.[104] Significantly, it was in the care of mental defectives—the area of mental welfare where the impact of psychological approaches was least marked—that attempts to tackle the social roots of mental problems were better developed. The voluntary sector, particularly the CAMW, led the way in these experiments. The most impressive development was the introduction of community occupation centres, providing work for defectives and respite for their families.[105] By the start of the 1930s there were over 150 such centres, catering for nearly 4,000 defectives.[106] In the 1930s the strategy began to be applied on a regional basis, with a scheme to employ defectives in farms and domestic service introduced throughout the north east.[107] The success of such schemes demonstrates that the social efficiency of mental defectives was only relative, suggesting that a more effective way to have tackled the problem would have been through social and economic reform. Unsurprisingly, this was never seriously contemplated in the interwar policy debate. Such a reorientation only emerged in an *ad hoc* fashion in the late 1930s, when the manpower demands of the wartime emergency enabled the CAMW to negotiate a reduction in the minimum wage level and an exemption from national insurance to encourage employment of defectives.[108]

Finally, the appeal and limitations of community care also reflected contemporary attitudes towards citizenship within the community. Citizenship was a central theme in interwar social and political thought, with those on both sides of the political spectrum believing that a

[102] There is some sign of an ambivalent recognition that aid was needed (though no sign of a concerted attempt to provide this): *BOC 1933*, pp. 66–7.

[103] J. Lewis, *What Price Community Medicine?* (Brighton, 1986), 6–8.

[104] Brown, 'Methods of Social Case Workers', 379–401; E. Harrison, 'The Changing Meaning of Social Work', in A. H. Halsey (ed.), *Traditions of Social Policy* (Oxford, 1976), 96; N. Rose, *The Psychological Complex: Psychology, Politics, and Society in England, 1869–1939* (London, 1985).

[105] This would add some support to R. Lees, 'Social Work, 1925–50: The Case For a Reappraisal', *British Journal of Social Work*, 1 (1971), 371–9.

[106] Rooff, *Voluntary Societies*, 126.

[107] *CAMW Annual Report 1937–8*, p. 40.

[108] PRO MH 51/574.

successful society depended on citizens becoming actively involved in improving the conditions of their community.[109] The continuing strength of such attitudes helps to explain why citizens were still attracted to serve in voluntary organizations. Groups such as the National Council for Social Service used the rhetoric of 'active citizenship' to defend their voluntary work against the charge of paternalism, and to represent it instead as a two-way process in which the clients were helped to help themselves and were encouraged to play an active role as citizens in their own community social service.[110] However, it was difficult to extend this model to a group such as the mentally defective, for here the recipients were viewed, by definition, as being incapable of entering into the two-way contract; likewise, any possibility of an alliance of voluntary action which incorporated the parents of defectives tended to be undermined by the assumption that they were of the same incapable stock as their offspring.

The idea of citizenship was also crucial because of the way it influenced the definition of mental deficiency. Under the 1913 legislation, this definition depended on the way an individual's intelligence was manifested in terms of 'social efficiency'—ability to work and support oneself in the community, and conformity to socially and morally acceptable standards of behaviour.[111] In an era when political citizenship was being extended to all adults, the category of mental deficiency assumed an important status as a dividing line which could contain and define the boundaries of citizenship. In turn, the definition of mental deficiency was itself reshaped in response to changing ideas about good citizenship. In everyday casework the notion of good citizenship was one of the factors which came into play when deciding whether or not a dull or backward individual needed to be certified as defective and placed under care. Probably more important still was the way that fear of certification helped to enforce good behaviour amongst the child population well beyond the borders of mental deficiency. Some people wanted to enforce this influence by extending the very definition of mental deficiency:

[109] R. Plant and A. Vincent, *Philosophy, Politics, and Citizenship: The Life and Thought of the British Idealists* (Oxford, 1984); C. Attlee, *The Social Worker* (London, 1920); S. Baldwin, *This Torch of Freedom* (London, 1935), 51.

[110] 'Social Service: What Does It Mean?', *Social Service Review*, 17 (1936), 19–25; M. Brasnett, *Voluntary Social Action: A History of the NCSS 1919–1969* (London, 1969).

[111] See Ch. 7.

provision of care and protection made for certifiable defectives does not cover the larger group of social failures whose incapacity for responsible citizenship is due to poor mentality . . . in the combination of temperamental emotional instability with poor mentality the vital factor calling for the protection of the individual is his temperamental rather than his mental make-up.[112]

However, such controversial proposals made little progress. For a while the ambiguity within the Mental Deficiency Act left enough scope to reach out to the 'borderline' group, but gradually more voluntaristic and socially acceptable forms of mental welfare, such as child guidance and psychotherapy, took over as tools to regulate the boundaries of normality, and much of the passion which had fuelled interest in mental deficiency consequently ebbed away.

OUTCOMES AND ADJUSTMENTS

Political Roles

In the 1920s, 'pressure group' tactics were an integral part of the CAMW's work.[113] It advanced its ideas on policy through the press and publications, it sent national circulars to local government departments, and it organized public campaigns.[114] The CAMW's most overt political actions were directed towards extending control to the borderline groups who were outside the statutory reach of the 1913 Act: it lobbied for an alteration in the law which would enable grant-aided supervisory care to be provided for special-school-leavers, even if they were not certifiable under the 1913 Act;[115] and it recommended the creation of a new category of mental disorder to cover those who lay

[112] E. Fox, 'Modern Developments in Mental Welfare Work', *Eugenics Review*, 30 (1938), 165.

[113] The political role of interwar social work is recognized in: Lees, 'Social Work', 371–9. Although interest groups obviously had a role previously, the term 'pressure groups' only began to appear as a dominant theme, and often in a positive sense of extending the policy-making network, in the literature of the 1950s; this was perhaps stimulated by the focus upon the BMA role in the formation of the NHS: W. H. Greenleaf, *The British Political Tradition* (London, 1988), iii. 488–99; G. Wootton, *Pressure Groups in Britain, 1720–1970* (London, 1975), 1–12.

[114] PRO MH 51/374, Brock to Pelham, 29 Sept. 1932; Rooff, *Voluntary Societies*, 136–9.

[115] PRO Ed 50/112. 12 Dec. 1920; Ed 50/226, L. G. Duke to Maudslay, 24 Jan. 1936.

between mental deficiency and mental illness.[116] Although these more public campaigns were largely unsuccessful, it would be wrong automatically to attribute this to the ineffectiveness of the CAMW's pressure group tactics, especially since it faced trenchant governmental caution on such issues.

The CAMW was more successful when negotiating directly with the Government. It exploited its close personal links to the Board of Control and Board of Education, playing off one government department against another.[117] The CAMW Secretary, Evelyn Fox, was notorious for her forceful and persistent negotiating manner.[118] The effectiveness and style of her tactics were recalled by Percy Barter, Secretary of the Board of Control:

> When Dame Evelyn Fox came to see us at the Board of Control there was never any vagueness about the proceedings. She would come knowing exactly what she wanted; she would waste no time in preliminaries; and she would propound the object of her visit in challenging precision . . . She was formidable in argument not only because of her immense knowledge and experience, but because of her directness of approach and her alert mind. She could however be subtle in argument if, for example, she wanted us to adopt a somewhat unorthodox interpretation of some statutory provision. Of official caution she was often impatient, as indeed she was entitled to be; for she was a pioneer. And when [the conference over] she took her leave there was always a rather quizzical light in her eye, which seemed to say 'Thank you for agreeing to some of my points and I shall come again about the others'.[119]

Through such tactics the CAMW was successful in gradually extending the reach of the Act: for instance, it influenced the government to grant financial support to all occupation centres in 1923, and it played an important role in the decision to oblige local authorities to provide occupational training for defectives and to notify special-school-leavers for supervision under the Mental Deficiency Act of 1927.[120] Equally, the CAMW's support for a government investigation into sterilization

[116] *Roy. Com. 1926, Minutes of Evidence*, ii. 21099–529.

[117] PRO MH 51/374, Brock to Pelham, 11 Nov. 1932.

[118] P. Norman, *In the Way of Understanding* (Godalming, 1988), 61–3.

[119] *Mental Health*, 15 (suppl.) (1955), 2. For a first-hand description of comparable tactics and attitudes towards civil servants in lobbying for the British Social Hygiene Council: S. Neville-Rolfe, *Social Biology and Social Welfare* (London, 1949), 13, 42–7.

[120] *Mental Deficiency (Amendment) Act*, 1927 (17 & 18 Geo. V, c. 33).

in the 1930s was important in boosting the credibility of the campaign in the eyes of the Government.[121] Thus political action continued to be an integral part of the CAMW's vision of social service. However, the shift from public campaigning to private lobbying of the Government from an expert position was both more effective and more appropriate for a group which was itself moving towards professional status.

Survival of Voluntarism

In the 1930s, as mental welfare work was taken over by local authorities, the public began to lose interest in voluntary groups. Their finances suffered, and there are signs that their futures were threatened, with the number of local voluntary groups declining from over fifty at the start of the 1930s to under forty by the end of the decade.[122] In reaction, the voluntary sector reasserted the claim that it had an essential role to play.[123] Taking advantage of their greater flexibility to experiment and their freedom to undertake 'preventive' work outside the Act, the CAMW, larger local associations, and groups of amalgamated smaller associations developed into agencies selling specialized services to local authorities.[124] Because of the CAMW's national structure, local voluntary groups could take advantage of services which were prohibitively expensive to deliver in a single authority and were not provided by inactive central government.

Financially, the CAMW survived on the income raised by its new services. These included a national scheme to provide guardians for defectives in the community, a contract with Middlesex County Council for the running of occupation centres, national training courses for medical officers, teachers, and social workers, the loan of personnel to local authorities, and even a holiday camp scheme for defectives.[125] By the 1930s, although the CAMW was still technically part of the voluntary sector, its staff were predominantly paid professionals and its main sources of income came from local or central government, either as grants or in return for services rendered.[126] It is therefore misleading to

[121] PRO MH 58/103, Greenwood to Scott, 9 Jan. 1930; Scott to Greenwood, 23 Jan. 1930.

[122] *CAMW Annual Report 1936–7*, pp. 41–2.

[123] *CAMW Annual Report 1936–7*, pp. 13–16.

[124] *CAMW Annual Report 1936–7*, p. 47.

[125] Rooff, *Voluntary Societies*, 123–7.

[126] On the development of a salaried bureaucracy within the voluntary sector: Political and Economic Planning (PEP), *The British Social Services* (London, 1937), 174–7.

use the CAMW as a straightforward example of the continuing success of philanthropy. Instead, it functions as an example of the fluidity of the voluntary–statutory division in the interwar mixed welfare system.

From Care of Mental Defectives to Mental Hygiene

Because of the desire to reach the borderline cases who were left outside the Mental Deficiency Act, the focus of mental deficiency work rapidly shifted to the wider arena of 'mental welfare' in the early 1920s. Reflecting this change in emphasis, the title 'Central Association for the Care of Mental Defectives' was altered to 'Central Association for Mental Welfare', a change which was paralleled among local branches, while the title of the CAMW journal was changed from *Studies in Mental Inefficiency* to the more all-embracing *Mental Welfare*.[127] With the spread of interest to mental welfare, and the emergence of exciting new opportunities in child guidance clinics and psychiatric social work, mental deficiency work lost its allure. The new generation of trained social workers was eager to work instead in areas where they could better apply their psychiatric knowledge.[128] As Evelyn Fox, the Secretary of the CAMW lamented: 'It is a most disquieting feature of the present time that the centre of interest has swung over so completely to mental disorder that mental defect has become the Cinderella of Mental Hygiene.'[129] An area which had been at the forefront of developments in the first decades of the century was now in danger of becoming a neglected backwater.

The CAMW had always been interested in extending its work beyond the borderline of mental deficiency, but in attempting to make this transition it met the challenge of other voluntary groups. The first and oldest of the groups was the Mental After Care Association (MACA), founded in London in the late nineteenth century. The MACA continued to represent a very traditional style of philanthropy, confining its work to provision of food and clothes (not money, which might have been spent irresponsibly) to assist mental hospital convalescents to re-establish their independence in the community.[130]

[127] *Annual Reports of the CAMW*; *BOC 1921*, pp. 56–7.

[128] M. Ashdown and S. C. Brown, *Social Service and Mental Health: An Essay on Psychiatric Social Workers* (London, 1953), 190.

[129] E. Fox, 'Modern Developments in Work for Mental Defectives: A Historical Survey', *CAMW Conference Report 1934*, 45 (44–59).

[130] Rooff, *Voluntary Societies*, 93–8, 115–19; P. Winterton, *Mending Minds: The Truth About Our Mental Hospitals* (London, 1938).

Despite the fact that its funding increasingly came from the London County Council, the MACA objected to becoming the employed agent of the state.[131] Perhaps because it rejected statutory aid and wanted voluntarism to be firmly rooted in an understanding of the local community, it chose not to follow the CAMW in attempting to establish a national network of branches. The second group interested in mental welfare was the National Council for Mental Hygiene. This group adopted a far more exclusive and professional strategy than that of the CAMW. Rather than directing its energies at encouraging lay volunteers or establishing mental welfare services, it concentrated on publicizing the psychiatric views of its expert members in the hope of improving standards of 'mental hygiene' in everyday life.[132] The third important mental welfare group was the Child Guidance Council, which had been established under the influence of the Commonwealth Fund in 1927. The aim of the Council was far more specific than that of the CAMW: it concentrated on providing a single service—the child guidance clinic—and a cadre of trained psychiatric social workers to serve in these clinics. Finally, there were groups, such as the Homes and Schools Council, which were interested in standards of mental welfare but were not organized as providers.[133]

The overlap of these groups opened the voluntary movement to criticism for placing jealous particularism above efficiency.[134] Since the CAMW had always claimed that one of its great strengths was the ability to co-ordinate chaotic state services, it was forced to take notice of such criticism.[135] As early as 1930, the Commonwealth Fund had recognized the need to co-ordinate the array of voluntary groups. However, in the first half of the 1930s, the differences between Evelyn Fox and her CAMW, on the one hand, and the Commonwealth Fund-led Child Guidance Council, on the other, disrupted any chance of amalgamation. For a while, the Commonwealth Fund hoped that the National Council for Mental Hygiene might prove a unifying force, but this was not to happen as it was far less important than the other groups in terms of resources and manpower and too closely tied to the con-

[131] JMS 75 (1929), 348.

[132] Sir M. Craig, *Mental Hygiene*, 1/12 (1935), 2–3; R. D. Gillespie, 'Mental Hygiene as a National Problem', *Mental Hygiene*, 1/4 (1931), 1–9.

[133] Feversham Report, 1–5.

[134] Political and Economic Planning, *Report on the British Health Services* (London, 1937), 400.

[135] Norman, *In the Way of Understanding*, 63–4.

troversial Tavistock Clinic.[136] In the light of PEP's criticisms of the chaotic organization of social work, an investigation was set up in 1937 under the chairmanship of the third Earl of Feversham and containing members of each of the groups. The discussion was heated, but the Committee did produce some positive conclusions.[137] In 1939, Feversham financed the publication of a blue-book style Report, a copy of which was dispatched to every local government department and mental hospital in the country. Having considered the three main options of continued independence, federation, or amalgamation, all of the groups, apart from the MACA, were agreed in recommending the radical third option of amalgamation.[138] The MACA dissented because it wanted to maintain its independence and avoid closer co-operation with the state; such objections highlight how far the CAMW, by contrast, had moved from a nineteenth-century voluntarism.[139] Because of the onset of war, amalgamation was temporarily delayed, but in 1942 the National Association for Mental Health was founded, a departure which marked the end of the separation between mental deficiency and mental health work.[140]

Not only had the Feversham Report recommended an end to voluntarist particularism, but it also envisaged a reorientation in the nature of voluntary mental welfare work, from service provision to public education. It was acknowledged that ascertainment and supervisory work would increasingly be taken on by the statutory authorities. However, there was still a crucial role for voluntary groups in helping to co-ordinate services and guiding the general public as to the services available.[141] The Report emphasized that public education was essential if the stigma surrounding mental health services was to be broken down so that preventive mental health care could become a reality. Here, voluntary status was seen as an advantage: if the state lectured to the public about mental health it would achieve little and put many people off; person-to-person contact, through trusted voluntary workers, had a much greater chance of changing attitudes.[142] The Report provides further evidence of the adaptability and enterprise of interwar voluntarism. On the outbreak of the Second World War there was still considerable confidence that, in this area at least, a rationalized volun-

136 Norman, *In the Way of Understanding*, 63.
137 Ibid. 64. 138 *Feversham Report*, 228–31.
139 Norman, *In the Way of Understanding*, 63–4.
140 Rooff, *Voluntary Societies*, 149. See Ch. 8.
141 *Feversham Report*, 179–93. 142 Ibid. 180–8.

tary sector could continue to play a vital role working alongside and complementing a state social service.[143]

CONCLUSION

As early as the interwar period, a policy of community care had been developed to tackle the problem of mental deficiency. In the form of supervision, guardianship, and occupation centres, the policy reached over 46,000 defectives by 1939. It was encouraged by central government, and the terminology of 'community care' was even used. As such, it surely qualifies as an unrecognized chapter in the history of community care, and suggests the need to reconsider the idea that such policies only emerged in the last four decades of the century.

Despite its impressive early development, expansion was restricted by some fundamental problems. First, because of the ambiguity of the early twentieth-century concept of 'community care', with its aim to protect the community by controlling as well as caring for the defective, it always remained a second-best alternative to institutional care. Secondly, although arguments about the special qualities of voluntary and female social work were crucial in establishing community care, they also proved a difficult hurdle to overcome in negotiating for resources with the state and in attempting to claim professional status.[144] In this respect, the history of interwar community care provides another example of the important role played by women in innovating new areas of welfare, yet it also supports those who have argued that there were inherent limits in 'maternalist' strategies.

The early history of community care also adds to our understanding of the evolving role of the voluntary sector in interwar Britain. Undoubtedly, it was a time of rapid change in this area, one in which there was a moving and interactive frontier between the voluntary and statutory sectors. However, in the case of community care for defectives a strict statutory–voluntary dichotomy is misleading, since the membrane between the two was often so porous. For a more accurate model, and to give a more thorough understanding of the evolution of

[143] L. Brock, 'Towards a Unified Mental Health Service', *Mental Welfare*, 20 (1939), 93–7; M. Buck, 'Co-operation in Mental Health Administration', *Mental Welfare*, 17 (1936), 104–8.

[144] Walton, *Women in Social Work*, 261–3.

statutory–voluntary relations in the interwar welfare system, one would have to add a third element—that of professional status, which became increasingly important during the period.

Finally, not only does the community care solution provide us with a new perspective to revise our view of the problem of mental deficiency, but it also helps us to better understand the two other solutions of segregation and sterilization. Community care was not perceived as a separate solution to institutional care. The two policies developed in dialogue: the colony as community existed alongside forms of community care which extended institutional discipline into the outside world. Thus the community care solution neither subverted nor aimed to replace the segregative ideal, but rather it extended the influence of the institution into the community and modified the nature of that influence. Despite impressive early developments, those forms of community care which demanded provision of facilities, such as the occupation centre, remained on too limited a scale to have a significant impact in dealing with the overall problem. The great attraction of supervision was that it seemed to provide an inexpensive supplement to a system of institutional care which simply could not expand any faster. However, the continued success of this solution depended on the supply of cheap labour. As psychiatric social work became professionalized and labour consequently became more scarce, the attraction of community care was undermined. Moreover, with limited resources, it was always difficult for supervision to provide tight moral and sexual control over defectives in the community. Therefore, by the 1930s many of those within the mental welfare movement began to consider linking community care to a third strategy of tackling the problem of mental deficiency—the policy of sterilization.

5
The Sterilization Solution

INTRODUCTION

The third proposed solution to the problem of mental deficiency was to sterilize mental defectives. The sterilization issue has attracted considerable attention from historians. The failure of the campaign to introduce voluntary sterilization legislation has been attributed to the relative weakness of eugenic arguments by the 1930s, and to resistance from the Labour movement, the Catholic Church, and the medical profession.[1] Because there is already a substantial body of literature on this issue, I shall not repeat here a now fairly familiar narrative about the political defeat of the campaign, particularly as the party-political and governmental dimensions of this question have already been addressed (and queried in some respects) in the first two chapters of this book. Instead, I shall consider the sterilization campaign from three new angles, each related to the broader concerns of this book. First, I shall demonstrate in a narrative summary that the fluctuating fortunes of the campaign were closely related to the two other solutions to the problem of mental deficiency and to internal developments within the eugenics movement; the course of the campaign cannot be explained simply in terms of the political viability of a eugenic social policy. In a second section I shall use the sterilization campaign to explore the politics of professional society in interwar Britain. I shall argue that the campaign was so intimately related to debates over professional identity and to struggles between competing professional interests within the eugenics movement that assessments of the campaign which are based simply on the success or failure of its legislative proposals only half comprehend its intentions, its success or failure, and its significance. Finally, I shall conclude the chapter with an examination of what the sterilization campaign tells us about the process of adjusting

[1] Macnicol, 'Eugenics and the Campaign for Voluntary Sterilization in Britain Between the Wars', *Social History of Medicine*, 2 (1989), 147–70.

to democracy, and in particular what it tells us about the role of professional and pressure groups within that adjustment.

THE STERILIZATION CAMPAIGN

From Segregation to Sterilization

A serious debate over sterilization began as early as the 1890s, when American penal and mental institutions began to castrate male inmates to diminish their sexual appetites and prevent masturbation. Attempts to pass legislation to gain state backing were initially unsuccessful. Castration seemed to violate the constitutional prohibition against cruel punishments. Furthermore, doctors were wary of supporting an operation which was of no obvious medical benefit to the recipients. Two factors changed this situation. First, Harry C. Sharp of the Indiana State Reformatory developed the new operation of vasectomy, which was simple to perform and relatively safe. Secondly, sterilization began to be advocated for eugenic reasons, and therefore there was a stronger case to present it as a medical operation rather than a punishment. In 1907 Indiana passed the first state law legalizing compulsory sterilization: the operation could be applied to habitual criminals, rapists, the insane, epileptics, and idiots. By 1917, a further eleven states had some kind of sterilization law, and after 1918 fifteen more states joined them.[2]

In Britain, the superintendents of mental institutions tended to reject the idea of adopting sterilization as a tool of management.[3] The controversial policy had too many drawbacks for a profession that was desperately seeking medical respectability. The British eugenics movement was also cautious about backing sterilization. As Barker has pointed out, sterilization raised a series of moral problems for eugenicists: in particular, there was resistance to any measure which might encourage sex to be separated from reproduction, and there was a fear

[2] J. Trent, Jr., *Inventing the Feeble Mind: A History of Mental Retardation in the United States* (London, 1994), 194–7.

[3] *JMS* 58 (1902), 48–75.

that sterilized mental defectives would be sexually exploited and would spread venereal disease in the community.[4]

In the early 1920s, despite the disappointingly slow expansion of segregation, there was no great upsurge in support for sterilization from within the eugenics movement. The Eugenics Society continued to be influenced by a morally conservative constituency at the grass-roots level which restrained it from supporting birth control, let alone sterilization, as a eugenic tool. Privately, the Board of Control expressed some interest in using sterilization as early as 1923, but ruled it out as being too far ahead of public opinion.[5] The example of sterilization in the United States suggested that even if a law was passed it could face considerable problems in practice: by 1921, only ten out of twenty-five states still had a sterilization law, and elsewhere the operation was held back by the reluctance of state governments to provide funding.[6] In 1927, the Eugenics Society President, Leonard Darwin, drafted a voluntary sterilization law. Darwin's strategy was to legalize sterilization through the backdoor: in proposing to make it illegal for anyone to be sterilized on other than medical grounds, his real aim was to gain a ruling in law which effectively established that there was no existing legal impediment. Once this had been established, medical grounds could probably be stretched to include eugenic operations.[7] Although the proposal was published in the *Eugenics Review*, it was still too far ahead of eugenic, let alone public opinion, and was not taken up by the Society's supporters in the House of Commons.

In the late 1920s a series of events made the climate for sterilization proposals far more favourable. The legality of eugenic sterilization was recognized by the United States Supreme Court after Oliver Wendell Holmes passed his notorious judgment on the *Buck v. Bell* case. The ruling gave eugenic sterilization a new legitimacy, and there was a rapid increase in the number of operations performed in the United States as sterilization was combined with parole to alleviate the pressure on institutions.[8] News of this reached Britain in the work of E. S. Gosney and Paul Popenhoe.[9] But probably more important in stimulating Brit-

[4] Barker, 'How to Curb the Fertility of the Unfit, The Feeble-minded in Victorian Britain', *Oxford Review of Education*, 9 (1983), 197–211. It is significant that one of the early leading advocates of sterilization was sexual radical Havelock Ellis: 'The Sterilisation of the Unfit', *Eugenics Review*, 1 (1909), 206.

[5] PRO MH 79/292, F. Willis to A. Robinson, 11 Jan. 1931.

[6] Trent, *Inventing the Feeble Mind*, 197.

[7] L. Darwin, *The Need for Eugenic Reform* (London, 1926), 176–9, 186–205.

[8] Trent, *Inventing the Feeble Mind*, 200–17.

[9] E. S. Gosney and P. Popenhoe, *Sterilization for Human Betterment* (New York, 1929).

ish interest was the publication in 1929 of the (Wood) Report of the Interdepartmental Committee on Mental Deficiency. The Report only addressed sterilization briefly, but its evidence greatly strengthened the case for seriously considering the policy. Its claim that the scale of the mentally defective population was substantially larger than previously estimated, and was still increasing, cast considerable doubt on the adequacy of the existing segregative solution. And its claim that high-grade defectives were concentrated in the 'social problem group'—the lowest tenth of the population—suggested that mental deficiency was simply the tip of this much more important problem of biologically caused social incapacity. To solve the problem of mental deficiency it would be necessary to restrict the reproduction of all members of this social problem group. If this could be achieved it would also alleviate a broad range of social problems which were concentrated in this residual section of society.[10] Thus, by the start of the 1930s two ideas had come together to give sterilization a forceful attraction: first, that segregation was unable to cope with demand, and that sterilization could alleviate pressure on institutions by providing eugenic control as part of supervision in the community; and secondly, that sterilization would reverse the current trend of rising mental deficiency, and improve the eugenic fitness and social welfare of the population as a whole.

THE CAMPAIGN

After Darwin's aborted attempt to introduce legislation, a younger generation of intellectuals and scientists within the Society began to consider a more practical strategy. At the centre of this group was C. P. Blacker, a member of the Eugenics Society since 1927, a friend of E. O. Lewis (the chief investigator behind the Wood Report), a former pupil in zoology under Julian Huxley in Oxford, and trained in psychiatry, working at the Maudsley and Tavistock Clinics in London during the 1930s.[11] During a yachting trip in July 1929, Blacker, Huxley, and a third Eugenics Society member, C. P. Chance, devised a scheme to fund

[10] J. Macnicol, 'In Pursuit of the Underclass', *Journal of Social Policy*, 16 (1987), 293–318.

[11] D. Kevles, *In the Name of Eugenics: Genetics and the uses of Human Heredity* (London, 1975), 171.

a sterilization clinic at University College Hospital which would carry out operations on 'overburdened' mothers and high-grade mental defectives. Since it had never been absolutely clear that sterilization was illegal (the operation was, after all, regularly performed on the rich in private clinics), the group hoped that the clinic's work would be accepted and would thereby demonstrate that there was no legal impediment to its more widespread adoption.[12] However, it soon became clear that the legality of the operation was too uncertain to risk establishing a public clinic. Willis, the former Chairman of the Board of Control, privately warned Blacker that the operation would be deemed illegal unless it was for therapeutic purposes, and University College Hospital turned down the scheme, arguing that it lacked space and that contraception could be used as an alternative.[13] This did not end the issue, however, for the Society had been given a substantial bequest of £60,000 by Australian sheep-farmer, Henry Twitchin, and could now consider building its own clinic—a project which would have cost £12,000–£20,000.[14] Yet the Society held back from this scheme, deterred by the fact that such a clinic would only cope with a small number of cases, and by concern that the legal issue would deter subscribers and might lead to a costly legal battle.

Despite this set back, the Society had been provided with a new sense of direction by the sterilization issue. It shifted its strategy, forming a Committee for Legalizing Eugenic Sterilization in December 1929. This upset some existing supporters. Havelock Ellis argued that legislation would ultimately restrict individual freedom by placing power in the hands of the state.[15] And Marie Stopes condemned the Society for jeopardizing that gradual adoption of sterilization which was already taking place in birth control clinics. Linking sterilization too closely to the problem of mental deficiency would so stigmatize the operation that women who were too poor or mentally and socially incompetent (and in her view eugenically unfit) to cope with more children would be far less likely to accept it on a voluntary basis.[16] However, the limited and

[12] C. P. Blacker, 'The Sterilisation Proposals', *Eugenics Review*, 23 (1931), 239–40.

[13] BL 6.7.22, 15 Mar. 1931; EUG C 337, Tredgold to Blacker, 11 Jan. 1931.

[14] EUG C 87; L. Darwin, 'Henry Twitchin: An Account of the Society's Most Generous Benefactor', *Eugenics Review*, 22 (1930), 91; Kevles, *Name of Eugenics*, 172.

[15] EUG D 204E, Moore to Blacker, 6 Dec. 1930; H. Ellis, *BMJ* 1 (1923), 835.

[16] BL 6.7.10, M. Stopes, 21 Apr. 1935. It is difficult to know how widespread such practice was. Such clinics are described in C. Hodson, *Human Sterilization Today: A Survey of the Present Position* (London, 1933), 41–5; *The Times* (15 Dec. 1933); EUG D 50, Blacker to Brock, 17 July 1933.

patchy adoption of sterilization for the feeble-minded in the United States in the absence of Federal legislation convinced the Society's leadership that real success would depend on an Act of Parliament.[17] From the start, Blacker supported voluntary, rather than compulsory legislation. He recognized that the latter would be politically unacceptable, and he seems to have genuinely believed that people would willingly come forward for sterilization if it was made a matter of choice. Despite the sympathy of other Society members for a compulsory measure, Blacker's position held sway.

Almost immediately, the prospects for sterilization were damaged by controversy. At a meeting organized by the CAMW in 1930, the Eugenics Society's representatives were seriously embarrassed on the issue of the genetics of mental deficiency.[18] They were attacked over an article written by R. A. Fisher in the *Eugenics Review*, which estimated that sterilization (or complete segregation) would decrease the number of mental defectives in each generation by 17 per cent, so that after three generations the total number would fall by over 50 per cent.[19] Blacker ensured that the official campaign subsequently dissociated itself from such inflated figures, retreating to the view that only a 5 per cent reduction was likely in a single generation. Nevertheless, the CAMW made it clear that it would not accept the need for sterilization until the issue had been investigated by a government commission. Without the CAMW, and with the uncertainty about the eugenic impact of sterilization exposed, the Society's position was seriously weakened.

In 1931 the opportunity to introduce a sterilization bill was taken by A. G. Church, a member of the Committee for Legalizing Eugenic Sterilization. Though unsuccessful, the vote indicated that there was a already a significant and cross-party body of support.[20] However, it was also clear that no government was likely to take action unless it had the clear backing of experts in the field, and therefore the campaign shifted to lobbying for an official investigation into the issue. In June 1932, the government responded by setting up a Departmental Committee under Chairman of the Board of Control, Brock. The choice of a Departmental rather than Royal Commission was fortuitous, since it would

[17] EUG D 50, Evidence of Eugenics Society to Brock Committee, 1 Mar. 1933.

[18] CAMW *Conference*, 1930, pp. 141–65. BL 1.1.57, Blacker to Darwin, 12 Dec. 1930.

[19] R. A. Fisher, 'The Elimination of Mental Defect', *Eugenics Review*, 16 (1924), 114–15.

[20] For some time, Blacker continued to believe that the opposition of both the Labour Party and the Catholic Church might be overcome: BL 1.3.74, Blacker to Darwin, 24 Mar. 1937.

report more rapidly and could be guided by the views of Brock, who was in close contact with Blacker and was fairly sympathetic towards eugenics. On the other hand, the presence on the Commission of both R. A. Fisher and one of his main critics, A. F. Tredgold of the CAMW, suggested that it would reach a divided conclusion. When the Brock Report was published in 1934 the Eugenics Society was delighted with its conclusions. Although the Report left the genetics of mental deficiency uncertain, it concluded that defectives were too socially inadequate to serve as parents and therefore justified sterilization on social grounds, regardless of the genetic arguments. Yet the Report also confirmed that sterilization would not end the need for segregation: only 5–10 per cent of those currently in institutions might be released if they were sterilized.[21] Thus, although the Report gave much greater authority and political respectability to sterilization it also curtailed ideas about its potential scope.

On these more moderate terms, sterilization became acceptable to the CAMW and to a broad range of groups involved in social work and medical care. By contrast, many eugenicists realized that sterilization, on such a limited basis, was only of marginal importance. They therefore began to shift their energies towards new eugenic policies which would affect the whole population. This strategic shift was also a reaction to a radical change in demographic anxieties during the 1930s. A series of alarming predictions suggested that through fertility restriction the populations of Western Europe were ceasing to expand and would soon begin a rapid decline in overall numbers.[22] The eugenic implications of this new demographic research were clearly spelt out to the Eugenics Society in Alexander Carr-Saunders' Galton lecture of 1934: the emphasis of eugenic policy had to be shifted from prevention of dysgenic births to the encouragement of eugenic reproduction; of course, preventing groups such as the mentally defective from reproducing was still desirable, but it was not of central importance, and anyway it could be undertaken by contraception rather than sterilization.[23] The Eugenics Society reacted by establishing a Positive Eugenics Committee in 1934, with Carr-Saunders, Huxley, Blacker and Fisher as members. Subsequently it set up the Population Inves-

[21] *Report of the Departmental Committee on Sterilisation (Brock Report)* (PR, Cmnd. 4485, xv, 1934).

[22] e.g. E. Charles, *The Twilight of Parenthood* (London, 1934).

[23] A. M. Carr-Saunders, 'Eugenics in the Light of Population Trends', *Eugenics Review*, 37 (1935), 11–14.

tigation Committee and the Population Policies Committee (jointly with the middle-opinion pressure group Political and Economic Planning), attracting progressives such as Richard Titmuss and David Glass.[24] Thus, somewhat ironically, just as the sterilization campaign began to attract support from respected outside groups such as the CAMW, its importance to the Eugenics Society itself was diminishing; before long, the obsession with sterilization of mental defectives would be seen by the younger generation of eugenicists as a distraction from more important eugenic work.[25]

Despite the shift in eugenic priorities, the sterilization campaign had an inexorable momentum. It was further spurred on by the enthusiasm of rank and file eugenic activists, by an increasing number of sympathetic outside groups, and by the positive recommendations of the Brock Report. In July 1934, the Eugenics Society founded the Joint Committee on Voluntary Sterilization to co-ordinate the broad coalition of groups who now backed legalization. A new sterilization bill was designed, but the Society lacked the political support to introduce this in the Commons. In the hope of breaking down suspicions within the Labour Movement, it funded a National Workers' Committee for the Legalizing of Voluntary Sterilization, though this enterprise seems to have had little success.[26] The campaign also placed pressure on the Ministry of Health and Board of Control, hoping that the Government might introduce its own bill. Despite its vigorous efforts in organizing lectures, publishing books and pamphlets, and broadcasting its views on the radio and on film, the campaign was unable to convince the Government that public opinion was solidly behind it.[27] By the end of the decade, the prospects for legislation were bleak. The Society was forced to reconsider its initial limited strategy of establishing legality through practice in a privately funded clinic.[28]

[24] *Eugenics Review*, 27 (1935), 255.

[25] R. Soloway, *Demography and Degeneration: Eugenics and the Declining Birthrate in Twentieth-Century Britain* (London, 1990), 203.

[26] Macnicol, 'Eugenics and the Campaign for Voluntary Sterilization', 162–4.

[27] EUG C 27, 'What We Have Done', 7 Oct. 1931; Blacker, 'Voluntary Sterilization: The Last Sixty Years', 18–19.

[28] Macnicol, 'Eugenics and the Campaign for Voluntary Sterilization', 166.

EUGENICS, STERILIZATION, AND THE POLITICS OF PROFESSIONAL SOCIETY

As the narrative of the sterilization campaign has suggested, internal and 'professional' politics were just as important to its course of development as were the external politics of party and central government. Indeed, by embedding the campaign within the politics of professional society it will become apparent that we need to revise many of our views about its aims, fluctuating fortunes, and achievements.

Professional Society and the Problem of Mental Deficiency

The eugenics movement occupied a transitional position in the history of social relations. On the one hand, it was the product of a society which was still fundamentally divided along class lines, emerging against the background of heightened turn-of-the-century concerns about differential class fertility and the menace of the urban residuum.[29] In many respects, eugenics can be seen as a direct ideological reflection of middle-class interests: it naturalized social inequality by explaining social success and failure as a direct reflection of innate ability; and it provided justification for the replacement of a social order of aristocratic privilege, based on inheritance of unearned status, by a meritocracy, based on ability and eugenic fitness.[30] On the other hand, this meritocratic ideal cannot be explained in terms of class alone; it was intrinsic to the individualistic creed of self-help which reached well beyond the middle classes. The problem of mental deficiency became the premier object of eugenic obsession because it brought together these class and broader meritocratic anxieties. Inherently mentally incompetent, defectives would follow an inevitable downward path to social failure and a position in the residual class of society. By excising this population and preventing it from spreading, one was also providing what seemed to be a solution to the broader problem of social failure in society, and one was leaving the rest of society to run on a meritocratic basis. Posed in such terms, the problem transcended mere 'class interests' and therefore attracted the concern of a much broader constituency.

[29] H. Perkin, *The Rise of Professional Society: England since 1880* (London, 1989), 60–1.

[30] G. R. Searle, 'Eugenics and Class', in C. Webster (ed.), *Biology, Medicine, and Society, 1840–1940* (Cambridge, 1991), 217–42; D. Mackenzie, *Statistics in Britain, 1865–1930: The Social Construction of Scientific Knowledge* (Edinburgh, 1980).

The Professionalization of Eugenics

Although the conjuncture of meritocratic and middle-class interests undoubtedly helps to explain the widespread attraction of eugenics and the great interest in mental deficiency, both factors are too sweeping and crude when it comes to explaining the more subtle contours of the eugenics campaign. Instead, the eugenics movement and its institutional focus in Britain—the Eugenics Society—need to be analysed as complex and internally divided.[31] In particular, we need to take into account the dialogue between class and professional interests. The two could be mutually reinforcing, as was sometimes the case with the meritocratic ideal; but a professionalism which was based on efficiency and which positioned itself as disinterested could conflict with middle-class interests.[32]

Internal divisions also arose from disagreements about the proper relationship between science, public understanding, and the state. The formation of the Eugenics Education Society was an attempt to occupy a territory between the dry and cautious science of the laboratory and the lay public and political arenas. The Society wanted to position itself as an expert body, with the authority to educate the public and politicians about applying the science of eugenics to real social problems. However, this provoked criticism from those eugenicists, such as Karl Pearson, who were more interested in the science of genetics, and who felt that the Society's reckless use of evidence would give eugenic science a poor reputation.[33] The tension continued throughout the interwar period as scientists on the right, such as R. A. Fisher, and then scientists on the left, such as Julian Huxley and J. B. S. Haldane in the 1930s, fell out with the Society over its cavalier use of scientific arguments.[34] Such criticism could be very damaging to a group which represented itself as the source of authoritative scientific opinion. It could be equally damaging to give the impression that the Society was foisting its views into government policy without democratic consent.

[31] Macnicol, 'Eugenics and the Campaign for Voluntary Sterilization', 149–51.

[32] Perkin, *Rise of Professional Society*, 116–70. The persistence of professional aspirations for 'gentlemanly' status is particularly important since it could cause tensions with many of the other interests: M. Wiener, *English Culture and the Decline of the Industrial Spirit, 1850–1980* (Cambridge, 1981), 14–15.

[33] Kevles, *Name of Eugenics*, 104–5.

[34] R. A. Fisher, 'Eugenics Academic and Practical', *Eugenics Review*, 27 (1935), 105–7; BL 1.1.91, Blacker to Darwin, 10 Nov. 1933; L. Hogben, 'The Interaction of Heredity and Environment', *JMS* 79 (1933), 590–601; L. Penrose, *The Lancet* (29 June 1946).

Throughout its history, therefore, the Society had to carefully negotiate the balance between its expert and lay factions; between those who wanted to shift the society in a professional and scientific direction, and those who wanted it to maintain a high profile through propaganda and political action. The sterilization campaign takes on a new light when it is interpreted in these terms.

During the first two decades of its existence the Eugenics Society had been dominated by its lay members. Symbolizing the gentlemanly amateurism of the early Society, its President Leonard Darwin had no expertise in genetic science, but he brought an aura of scientific respectability through his illustrious family lineage (son of Charles Darwin and cousin of Francis Galton).[35] The Society was regarded as rather eccentric and out of touch with scientific and social reality. The opinions expressed in the *Eugenics Review* tended to be highly conservative, with eugenics often presented as a way to avoid social expenditure. The Society's main research initiative was to support E. J. Lidbetter's rather unsuccessful attempt to prove that a chain of social problems, running from mental defect to alcoholism, criminality, and even unemployment, could be traced back through hereditary lines of descent within East End of London problem families.[36] Far more typical was the Society's propagandist activity. Reflecting this focus, the majority of the members were lay people, including a substantial number of women.[37] The dominance of this morally conservative lay constituency lay behind the Society's caution on the birth control issue. The complexion of the Society began to shift in the late 1920s. Blacker's appointment as Secretary in 1931, replacing Cora Hodson, symbolizes the transition which was taking place: a psychiatric expert replacing a lay woman. The membership of the Society also reflected this 'professionalization', with female membership falling and the number of professionals, medical practitioners, and academics rising.[38] Also significant was the replacement of Darwin by Sir Bernard Mallet as President in 1927; Mallet, formerly a civil servant in the Treasury and Private Secretary to Conservative Prime Minister, Arthur Balfour, brought a new political sophistication to the Society and was keen to rid it of its eccentric image.[39] Finally, the appointment of Maurice Newfield as the

[35] M. Keynes, *Leonard Darwin, 1850–1943* (Cambridge, 1943).
[36] Macnicol, 'Pursuit of the Underclass', 305–13.
[37] I. Brown, 'Who were the Eugenicists?', *History of Education*, 17 (1988), 295–307.
[38] Ibid. 302.
[39] *Eugenics Review*, 20 (1928), 153; ibid. 24 (1932), 271; *WWW, 1929–40*, p. 894.

new editor for the *Eugenics Review* enabled Blacker to foist an increasingly reformist and progressive agenda on the still socially conservative rank and file of the Society, and to attract progressives like Richard Titmuss to contribute to the *Review*.[40]

Blacker hoped that the sterilization campaign would provide a more practical focus for the energies of the Society's lay rank and file, while the journal was left free to become the mouthpiece for a core of scientifically oriented eugenicists to present proposals for the application of eugenics to mainstream social policy.[41] The sterilization issue associated the Society with the expansion of birth control technology (as was the case in the funding of research into chemical contraception and discussion of 'eutelegenesis'—the idea of creating sperm banks from the eugenically fit). Young scientists, intellectuals, and even socialists, including Julian Huxley, David Glass, and Richard Titmuss, were drawn into involvement with the Society.[42] The proposed sterilization clinic was to become a research bureau for the collection of genetic information on defective inheritance. This scheme attracted the support of respected academics such as Julian Huxley, William Beveridge, A. M. Carr-Saunders, W. Bowley, John Ryle, and R. A. Fisher.[43] Progressives looked to voluntary sterilization as a highly effective form of birth control with a social as well as a eugenic purpose: it would help prevent the social problems caused by unwanted multiple births among the poor; and it would offset the dysgenic differential fertility between classes.

Blacker also recognized the importance of attracting support from the medical profession. He therefore always attempted to portray sterilization as an advance of medical rights, rather than a restriction of rights or a punishment. In initially proposing to legalize sterilization for a wide range of genetic carriers—not just the mentally defective, but also specific groups of the insane, the blind, and the physically defective—he hoped to reach out to as many different groups of medical professionals as possible (he attracted considerable support from surgeons and eye specialists for instance). He also argued that sterilization would prevent individuals from being forced to resort to abortion, an operation which was taking place illegally on a huge scale and was far more dangerous than sterilization, often fatally damaging otherwise fit

[40] Titmuss Papers, Wellcome Unit for the History of Medicine, Oxford, Newfield to Titmuss, 13 Dec. 1939.

[41] BL 1.1.45; BL 6.7.16; BL 6.7.30.

[42] *Eugenics Review*, 27 (1935), 121–31.

[43] BL 6.7.16, E. Moore to Blacker, 26 Oct. 1929; BL 6.7.30.

mothers or leading to the birth of defective children.[44] By contrast, the operations of vasectomy and salpingectomy were presented as safe, easy to perform, and medically innovative—wholly different to the mutilation of castration.[45] Finally, Blacker drew parallels between voluntary sterilization and the voluntary treatment for the mentally ill which had recently been introduced under the 1930 Mental Treatment Act. In both cases, voluntarism would help break down stigma and facilitate a shift from institutional to clinic-based and community care.

Despite changes in both the membership and the policies of the Eugenics Society, historians have been divided over the extent to which there was a real shift towards what has been termed 'reform eugenics'. Jones, for instance, has argued that 'Blacker's efforts to modernise the Society were largely in the realm of rhetoric and nomenclature'.[46] Undoubtedly, the sterilization question has confused the issue, particularly since there has been a tendency to equate 'negative' policies, such as sterilization, with an older, conservative, 'mainline' eugenics.[47] However, the situation was more complex than this. Sterilization was in certain respects an integral part of the shift towards a more reformist eugenics in the 1930s. Voluntary sterilization was used to bring together the two sides of the Society—the lay and professional, the conservative and the progressive. By the mid-1930s the character of the Society had shifted substantially; in this sense, the strategy had been a success. However, in the long term this shift in character undercut the sterilization strategy, as the younger generation of progressive eugenicists turned away from what seemed to be the increasingly marginal, politically insoluble, and controversial issue of mental deficiency, to promoting positive eugenics through family allowances, population investigations, and changes in the taxation system. There was also a strategic shift, away from propagandist activity, and towards establishing expert groups to advise the government, promote scientific investigations, and publish research findings. Retreating to a more cautious 'professional' politics, and realizing that its own image could deter support, the Society began to advance its interests through establishing and funding other groups to front its campaigns.

[44] Blacker, *CAMW Conference 1930*, p. 144.
[45] R. L. Dickinson, *Sterilization Without Unsexing* (New York, 1929).
[46] G. Jones, *Social Hygiene in Twentieth-Century Britain* (London, 1986), 109.
[47] Kevles, *Name of Eugenics*, 175–6.

Relations with other Professional Groups

Not only was the sterilization campaign carried out in tandem with an internal transformation—one might say a 'professionalization'—of the Society, but it also attempted to build bridges with a broad range of outside professional groups, both to advance the campaign itself, and to spread eugenic ideas among those who were in control of implementing medical and social policy. In doing so, the original primarily eugenic aims of the sterilization campaign were further complicated and compromised by a variety of professional interests. In fact, since the pattern of support corresponds so closely to professional self-interests, we have to be very cautious about assuming that those groups who supported the campaign (or those who opposed it) did so simply because of their attitudes towards eugenics.

It was a major coup to gain the support of the CAMW after the publication of the Brock Report. Up until then, the CAMW's lack of support had seriously weakened the campaign.[48] However, in incorporating the CAMW into the campaign, compromises had to be made. The moderate tone of the Brock Report seems to have calmed CAMW suspicions that sterilization was being advanced simply to avoid providing care for mental defectives. The CAMW now supported voluntary sterilization, not because of a sudden conversion to eugenics, but because it was growing increasingly clear that sterilization might prove a useful adjunct to extend community care.[49] Within the Eugenics Society there was considerable suspicion of this alliance, and a feeling that the CAMW's Evelyn Fox and A. F. Tredgold were attempting to highjack the Society's campaign.[50] However, by locating the campaign within the Joint Committee on Voluntary Sterilization, the Society itself was protected from a takeover, and it could limit the amount of resources which went into the campaign.[51]

Almost as important as the support of the CAMW was that of the

[48] See e.g. L. Scott, 'More Institutions Needed', *The Times* (23 Feb. 1929); A. F. Tredgold, *Mental Welfare*, 10 (1929), 10–11.

[49] In 1928 only eight members of the CAMW were also members of the Eugenics Society, and by 1937 this had risen to only ten: for Eugenics Society membership lists see EUG A 74, EUG A 78; for CAMW membership see *CAMW Annual Reports*.

[50] EUG C 337; EUG C 338; BL 1.3.34, Blacker to Darwin, 12 Mar. 1934, and Darwin to Blacker, 23 Mar. 1934.

[51] BL 1.2, Blacker to Darwin, 12–13 Mar. 1934. Parallels were drawn to the problem encountered in the Society's first years when the campaign over venereal disease had threatened to subsume the whole purpose of the Society: S. Neville-Rolfe, *Social Biology and Welfare* (London, 1949), 18–21.

local government organizations, the County Councils Association and the Municipal Corporations Association. Their support for sterilization was a crucial influence on the central government's decision to set up the Brock Committee.[52] Again, self-interest, in this case a desire to reduce local expenditure levels, lay behind the support. The Society itself attempted to keep clear of the economic argument, realizing that it would scupper the chances of attracting support from most of those currently involved in care and also from the political defenders of welfare provision.[53]

The campaign failed to gain the full backing of a third influential group—the Royal Medico-Psychological Association (RMPA), but again this reflected professional self-interest, as much as support or opposition to eugenics. An internal survey undertaken by the RMPA showed that the majority of the respondents were in favour of voluntary sterilization for mental defectives, and also for sufferers of mental disorder, the physically handicapped, and carriers of hereditary defects. However, the RMPA Council decided that the wording of the question had been ambiguous and that an insufficient number of replies had been received. Therefore, it circulated a second questionnaire, this time asking whether further research was needed before action could be taken. The response, again from a minority of the Association, was 175 for immediate sterilization and 190 favouring further research. Therefore the Association withheld its support from the campaign, though it did send a representative to the Voluntary Sterilization Committee.[54] The leaders of the RMPA seem to have been reluctant to lend the Association's support to such a controversial issue. If sterilization became a tool for institutional management, or if the operation was performed in institutions before releasing defectives into the community, this would almost inevitably tarnish mental hospitals with yet another stigma.[55] By contrast, the call for further research was clearly in the interests of a profession which was keen to promote its status

[52] PRO MH 79/292, Brock to Robinson, 22 Apr. 1932.

[53] The Eugenics Society's *Handbook for Speakers* advised speakers to avoid arguing '[T]hat sterilisation would lead to substantial economies in the rates or relieve local authorities from the duty of providing accommodation': *Handbook for Speakers*, point 6, p. 32.

[54] JMS 80 (1934), 759–72; *JMS* 82 (Suppl.) (1936); EUG D 50.

[55] The unacceptability of using sterilization in this way is evident in the Board of Control's hostile reaction when a workhouse was found castrating boys simply because they caused a nuisance masturbating: PRO MH 79/291. See also the criticism of the *Brock Report*, p. 42.

through more medical science within mental institutions. Blacker hoped that the research opportunities opened up by a sterilization programme would gain the support of junior medical officers who were frustrated by their limited career opportunities.[56]

In fact, strikingly few members of the RMPA were active in the eugenics movement: for instance, just seven members of the RMPA held Eugenics Society membership in 1927, rising to only ten by 1937.[57] It is, though, unclear whether this demonstrates a lack of sympathy towards eugenics, or simply the fact that superintendents had little time or opportunity to engage themselves in activities outside the closed and isolated world of the institution. The psychiatrists who were more involved with the eugenics movement were almost always those, such as Blacker, whose work took them outside the institution, and who saw sterilization as a way to extend the reach of psychiatric care into the community. This also helps to explain why support for sterilization came from the Society for Medical Officers of Health. At first the support of this group, with its strong tradition of environmentalism, might seem surprising.[58] However, in a decade of stretched resources and with much of their time taken up administering the old poor law hospitals, Medical Officers of Health were attracted by sterilization because it offered a preventive strategy to relieve institutions from the pressure of long-term care.[59]

Overall, the support from expert and professional groups for sterilization by the second half of the 1930s was fairly impressive, and it seems rather unfair to argue that the campaign failed to enlist a powerful lobby of expert support.[60] Undoubtedly, the unwillingness of the British Medical Association to back the campaign was damaging. However, a number of points should be borne in mind when interpreting this. First, although the BMA withheld official support, it never actually

[56] EUG D 50, Blacker to Brock, 29 Nov. 1935.

[57] Calculated from Eugenics Society Membership Lists: EUG A 74; EUG A 78, and RMPA membership lists in the *JMS*.

[58] For instance, it fits oddly with the picture of Medical Officers of Health presented in D. Porter, 'Enemies of the Race: Biologism, Environmentalism, and Public Health in Edwardian Britain', *Victorian Studies*, 34 (1991), 159–79.

[59] Sir W. Savage, 'Our Future: Presidential Address to the Society of Medical Officers of Health', *Public Health*, 49 (1935), 42–7. On the waning of an environmentalist philosophy: J. Lewis, 'The Public's Health: Philosophy and Practice in Britain in the Twentieth Century', in E. Fee and R. Acheson (eds.), *A History of Education in Public Health: Health that Mocks the Doctor's Rules* (Oxford, 1991), 202–16.

[60] Macnicol, 'Eugenics and the Campaign for Voluntary Sterilization', 159.

voiced outright opposition to sterilization.[61] Moreover, by the end of the decade, it acknowledged that sterilization might be useful in certain cases, and in 1939 it appointed a committee to reinvestigate the issue.[62] Secondly, in contrast to the BMA, strong support for sterilization was given by other important institutions within the medical profession: most notably, the *Lancet* continued to give strong backing to sterilization right up to the outbreak of war; support also came from the Royal College of Surgeons and Physicians, reflecting considerable interest among those doctors who wanted to secure the legality of their own surgical work.[63] Finally, it is again necessary to bear in mind that the BMA's position on the issue was partly a reflection of professional self-interests. As the official representative of the medical profession, it was always likely to take a cautious position on issues, such as this, which threatened to associate the profession with a controversial and possibly coercive policy. It also wanted to protect the profession's interests in maternal and infant welfare: for as Henry Brackenbury, Vice Chairman of the BMA Council, argued, the focus on inheritance of defects threatened to deflect public attention and resources from pre-natal care.[64]

Science and Sterilization

The view that the campaign was defeated because of criticism on scientific grounds also needs modification. Undoubtedly, in some respects it is valid. Lidbetter's project of mapping degenerate family trees among the London poor was clearly a failure.[65] The project was dogged by a series of financial and methodological problems, causing its publication to be delayed until 1933. Even then, it was forced to adopt a cautious tone, and plans to extend the project failed to materialize.[66] Under criticism, the advocates of sterilization were forced to withdraw from some earlier exaggerated claims about the importance of inheri-

[61] 'Report of the Mental Deficiency Committee of the BMA', *British Medical Journal* (25 June 1932); A. M. McCutcheon, 'The BMA Report on Mental Deficiency', *Mental Welfare*, 13 (1932), 80–3.

[62] EUG C 338, Blacker to Tredgold, 24 Oct. 1938.

[63] The wide range of professional and medical groups supporting the campaign is listed in C. P. Blacker, 'Voluntary Sterilization: The Last Sixty Years', *Eugenics Review*, 54 (1962), 9–23. Another interesting area of support was from groups involved in the care of the blind, where there was a strong belief that sterilization would help prevent hereditary blindness.

[64] H. Brackenbury, 'Maternity in its Sociological Aspects', *Social Service Review*, 18 (1937), 37–47; EUG C 338, Tredgold to Blacker, 28 Apr. 1937.

[65] Macnicol, 'Pursuit of the Underclass'.

[66] E. J. Lidbetter, *Heredity and the Social Problem Group* (London, 1933).

tance. Perhaps most importantly, dispute about the hereditary basis of mental deficiency caused delays which hindered the chances of legislation. As first the Brock Committee investigated the issue, and then there were calls for a further scientific investigation, civil servants were given a ready excuse to justify their inaction.[67]

The significance of scientific arguments against sterilization can, however, be exaggerated.[68] It is true that by the late 1930s work on the genetics of mental deficiency, particularly that of Lionel Penrose, showed that feeble-mindedness could not be attributed to a single gene and was simply one end of the spectrum of normal fluctuations of intelligence in a population.[69] But such arguments did not fatally undermine the sterilization campaign; indeed, Penrose's findings were readily published in the *Eugenics Review*.[70] Though he disproved some of the cruder Mendelian ideas about inheritance, he still suggested that the child of two low-intelligence parents was likely to inherit low intelligence and perhaps high-grade deficiency (Penrose estimated that in 12 per cent of cases the feeble-minded had mentally defective parents, and that in 70 per cent low intelligence of the parents was one of a complex of factors).[71] Likewise, though the development of a more sophisticated genetics based on mathematical modelling eroded faith in Mendelianism, it did not necessarily undermine the case for sterilization. The new methods were taken up by advocates as well as opponents of sterilization: by R. A. Fisher in Britain, and even more significantly by Ernst Rüdin in Germany.[72] In the 1930s Blacker was still able to attract serious scientists to contribute to a volume investigating the existence of a 'social problem group'.[73] Ultimately, scientists' views about sterilization depended on their moral and political inclinations, as much as any scientific theory.

[67] For instance: PRO MH 51/559, Report of Kingsley Wood in House of Commons, 17 Apr. 1937.

[68] D. Barker, 'The Biology of Stupidity: Genetics, Eugenics and Mental Deficiency in the Inter-War Years', *BJHS*, 22 (1989), 347–75; D. Paul, 'Eugenics and the Left', *Journal of the History of Ideas*, 45 (1984), 567–90.

[69] Kevles sees Penrose's work as being crucial in scientifically discrediting eugenics in the 1930s: *In the Name of Eugenics*, 146–63.

[70] L. Penrose, 'Mental Deficiency: The Subcultural Group', *Eugenics Review*, 24 (1932), 289; id., 'The Complex Determinants of Amentia', *Eugenics Review*, 26 (1934), 121. As late as 1939, Penrose's work was still being published: id., 'Eugenic Progress with Respect to Mental Deficiency', *Eugenics Review*, 31 (1939), 35.

[71] L. Penrose, *A Clinical and Genetic Study of 1280 Cases of Mental Defect* (London, 1938), 70.

[72] P. Mazumdar, *Eugenics, Human Genetics and Human Failings* (London, 1992), 198–255.

[73] C. P. Blacker (ed.), *A Social Problem Group?* (Oxford, 1937).

For instance, geneticist and socialist J. B. S. Haldane vehemently opposed what he regarded as the hereditarian elitism of eugenics and mocked the crude Mendelian view of feeble-mindedness. On the other hand, he accepted that there was a hereditary factor within intelligence, and he admitted that his attitude towards sterilization depended less on biology than the relative value he attributed to the health of the community compared to the liberty of the individual.[74]

The most fundamental problem with the view that the campaign was defeated by scientific arguments is that the case for sterilization rested as much on social as biological grounds: whether or not they were going to pass on their defect, mental defectives were deemed to be too socially inadequate to be given the responsibility of parenthood. Such a view was given a powerful endorsement by the Brock Report. It is worth noting, in this respect, that one of the most influential pieces of evidence in the Report was that given by the National Society for the Prevention of Cruelty to Children on the suffering of the children of defective parents.[75]

EUGENICS, STERILIZATION, AND THE ADJUSTMENT TO DEMOCRACY

A policy of moderation?

To a large degree, the course taken by the sterilization campaign reflects the ideological constraints which were set by the interwar British polity. One of the striking aspects of the sterilization issue is the relatively moderate and cautious position adopted by the organized eugenics movement. In an international comparative context, the absence of a sterilization law in Britain places her alongside Catholic France, rather than her more obvious compeers in Protestant Europe and North America.[76] Throughout the period, the Eugenics Society generally accepted that sterilization could not replace segregation: sterilization

[74] J. B. S. Haldane, *Heredity and Politics* (London, 1938), 80–117; Paul, 'Eugenics and the Left', 561–90; G. Werskey, *The Visible College* (London, 1978), 94–8.

[75] *Brock Report*, 5, 80–91.

[76] M. Adams (ed.), *The Wellborn Science: Eugenics in Germany, France, Brazil and Russia* (Oxford, 1990), 218–20; N. Röll-Hansen, 'The Progress of Eugenics', *BJHS* 26 (1988), 295–331; *Brock Report*, 34–6.

was a supplement for those defectives left in the community, and not a cost-cutting alternative. Until the late 1920s, the moral concerns about releasing sterilized defectives in the community undermined support for the policy among socially conservative eugenicists. After the Wood Report showed that segregation only reached a small percentage of mental defectives, sterilization became more acceptable. However, the tenor of the debate remained notably moderate: sterilization was supported as a voluntary, rather than compulsory, measure; the campaign chose a democratic legislative route; and it was soon acknowledged that sterilization would not replace segregation and would not substantially decrease the number of mental defectives in future generations.[77] There was a genuine hope that if sterilization could be legalized it would gradually be adopted voluntarily, because of economic interest and through subtle persuasion and advice from doctors, to provide a highly effective form of birth control among the lowest tenth of the population—'the social problem group'—and thereby redress what was deemed to be a dysgenic differential fertility problem. Undoubtedly, many supporters of voluntary sterilization saw it as a first stage before a compulsory measure, or hoped that, in practice, doctors would be able to coerce and pressurize members of the social problem group to be sterilized.[78] However, it was generally recognized that a compulsory measure would never be accepted by a British parliament, and even if it was it would be politically impossible to enforce. Instead, the key to success was to make the operation socially acceptable, through a dual strategy of providing it on a voluntary basis and instilling a eugenic conscience as an integral part of citizenship.

The Eugenics Society attempted to portray its proposals as uniquely moderate compared to schemes in Germany and the United States. The experience of compulsory sterilization in the United States indicated that medical authorities were reluctant, in practice, to resort to this extreme.[79] In Germany, where the state was willing to enforce such a measure, there were a series of dysgenic side-effects: it frightened the public, leading to falsification of diagnosis, concealment of genetic information, and a reluctance to come forward for treatment of mental disorders.[80] The German sterilization law was given some coverage in

[77] BL 12.2.21, Eugenics Society, *Handbook for Speakers*, 31–2.

[78] BL 1.3.1, Darwin to Blacker, 17 Jan. 1936.

[79] EUG D 50, Evidence of Eugenics Society to Brock Committee, 1 Mar. 1933.

[80] EUG D 204, Blacker to Miss Pocock, 23 Jan. 1934. On German programme: P. Weindling, 'Compulsory Sterilization in National Socialist Germany', *German History*, 5 (1987), 10–24.

the *Eugenics Review*, including a report by a 'German Official'; however, it received criticism on a number of grounds.[81] Its detailed specification of hereditary conditions demanding sterilization was viewed as unnecessarily rigid, difficult in all cases to justify scientifically, and likely to deter the sick from coming forward for essential treatment.[82] Instead, the Society looked to the Swiss voluntary sterilization law as a model.[83] By 1936, Blacker was becoming concerned that the adverse publicity attracted by the German measure might undermine the British campaign.[84] And certainly by the end of the decade, eugenics was being attacked from the left because of its association with Nazi racial science.[85] On the other hand, this attack came too late to explain the defeat of the sterilization campaign in Britain. Moreover, although the racism and anti-Semitism of the Nazi regime was common knowledge by the end of the 1930s, it is not so clear that sterilization was seen as a particularly Nazi policy; indeed, as late as 1940, the Eugenics Society continued openly to support voluntary sterilization and to disassociate its own plans from the policies of the Nazis.[86]

Blacker was also concerned that sterilization should not be viewed as an anti-welfarist attempt to pare down social expenditure. He accepted the Brock Report's estimate that only 5–10 per cent of sterilized cases would be discharged and that any savings on institutional care would be minuscule and equalled by the extra cost of maintaining the discharged defectives in the community.[87] Concerned about the suspicions of the Labour movement, he attempted to distance eugenics from its anti-progressive and anti-welfarist image, presenting it as a tool which had no inherent ideological bias and was equally essential for capitalist and socialist societies.[88] As he pointed out, socialism could be seen as having eugenic advantages over capitalism: the elimination of environmental inequality would make it easier to distinguish eugenic fitness from social

[81] 'The German Sterilization Law', by a 'German Official', *Eugenics Review*, 26 (1934), 137–40.

[82] *Eugenics Review*, 27 (1935), 89.

[83] H. Maier, 'Practical Experience of Sterilization in Switzerland', *Eugenics Review*, 26 (1934), 19–25; E. Mapother, 'Safeguards in Eugenic Sterilization', *Eugenics Review*, 26 (1934), 17.

[84] BL 1.3.21, Blacker to Darwin, 27 July 1936. Blacker later argued that the association of sterilization with Nazism was one of the factors behind the failure of the campaign: *Eugenics: Galton and After* (London, 1952), 141–5.

[85] L. Hogben, *Dangerous Thoughts* (London, 1939), 44–58.

[86] *Eugenics Review*, 31 (1940), 205–10; C. P. Blacker, *Eugenics in Retrospect and Prospect* (London, 1945).

[87] BL 1.1.90, Blacker to Darwin, 28 Aug. 1933.

[88] BL 1.3.74, Blacker to Darwin, 24 Mar. 1937.

advantage, and it would also be easier to implement programmes of interventionist social engineering; while capitalism's intense competition was resulting in dysgenic fertility restriction of the most able.[89] However, there were serious tensions underlying this attempt to realign eugenics: many eugenicists continued to fear that indiscriminate welfare policies would have a dysgenic effect through artificially protecting the unfit; that policies such as family allowances would take generations to have a significant eugenic impact; and that the Society could do little on its own to improve environmental conditions, whereas it could have a major impact if it concentrated on the narrower field of eugenic reform, particularly negative strategies such as sterilization.[90]

Sterilization and Individual Rights and Responsibilities

Sterilization raised important questions about the extent to which individuals or the state had the right to interfere with individual bodies, and to what extent this changed according to the medical and social circumstances of the individual. Initially, it was unclear whether voluntary sterilization was illegal at all. Therefore the Eugenics Society commissioned an investigation into the issue by the lawyer Cecil Binney. The main piece of statute law which might have blocked the operation was the Offences Against the Person Act of 1861. However, this law was vague about the type of mutilation which it outlawed and it did not include medical operations; as such, it was unclear whether it made sterilization illegal. A second potential legal barrier was that under common law it was illegal to injure individuals without their consent. This suggested that a voluntary operation might be legal when it involved adults who were *compos mentis*, but not if it involved children, lunatics, or idiots; in the case of the feeble-minded it was unclear whether or not voluntary consent could be established. Complicating the issue further, under common law any operation necessary for the health of an individual was lawful even without consent. Binney concluded that if the notion of 'health' could be stretched to cover eugenic health and social suffering, there would be a reasonably strong case to

[89] BL 1.4.10–11; BL 1.5; Blacker, *Eugenics in Retrospect and Prospect*, 9. This ambiguity underlies what Searle has described as the failure of eugenics to articulate class interests: 'Eugenics and Class', 217–42. For the ability of eugenics to attract support from socialist regimes: L. Graham, 'Science and Values: The Eugenics Movement in Russia and Germany in the 1920s', *American Historical Review*, 52 (1977), 1133–64.

[90] BL 1.3.18, Blacker to Darwin, 30 July 1933; BL 19.7.34.

defend sterilization as a legal operation, certainly if it was voluntary, and perhaps even in a compulsory form.[91]

In fact, sterilization operations were already being performed in Britain, and there had been no legal attack on this. For those who could pay, the operation seems to have been fairly easy to obtain from private medical practices.[92] As late as 1937, an article in a leading medical journal advised general practitioners that sterilization might 'be recommended for such patients as are suitable and can afford it'.[93] In the early 1930s, a few voluntary birth control centres began to advise clients to accept sterilization. In some instances, the operation was targeted at people with disabilities, but it was also used as a form of birth control for working-class women who wanted to avoid bearing any more children.[94] The operation was justified on a therapeutic and social basis. In reality, it also served a eugenic role of addressing differential class fertility.[95] In Leicester, the operation was actually encouraged by the local authority. The Leicestershire National Health Insurance Committee interpreted the preventive side of its work as including the eugenic welfare of future generations. It set up a Eugenics Services Sub-Committee to promote education in heredity and biology and to provide information on heredity to insurance practitioners and families, advising any individual who was likely to pass on a hereditary defect to be sterilized at a small independently financed hospital.[96] Such private, voluntary, and local authority activity would probably have extended gradually on an *ad hoc* basis. Abortion, faced by similar legal impediments against unnecessary injury, had expanded in such a manner, reaching about 100,000 cases a year by 1939; though doctors still risked prosecution, the operation could usually be defended on the social grounds that the mother was too mentally unstable to bear a child, or on the medical grounds that the child was likely to inherit a serious disease.[97] Because sterilization permanently removed an individual's right to reproduce, there was always likely to be more caution over the issue.[98] There were doctors

[91] C. Binney, *The Law as to Sterilization*, 2nd edn. (London, 1931).

[92] *The Times* (15 Dec. 1933); EUG D 50, Blacker to Brock, 17 July 1933.

[93] K. B. Aikman, 'The Treatment of Mental Defect', *Medical Press and Circular*, 193 (1936), 270; similarly, see *The Lancet* (21 July 1934), 140–1.

[94] Hodson, *Human Sterilization Today*, 41–5; PRO MH 79/291.

[95] For instance: Mrs Laski, *Voluntary Sterilization Conference 1932*, p. 24.

[96] C. J. Bond, *Voluntary Sterilization Conference 1932*, pp. 15–16.

[97] J. Keown, *Abortion, Doctors and the Law: Some Aspects of the Legal Regulation of Abortion in England from 1803 to 1982* (Cambridge, 1988), 49–79.

[98] Though sterilization, unlike abortion, did not involve what might be seen as the more fundamental issue of rights over life and death.

who advocated going ahead with the practice and countering any legal objections with evidence gained from successful practice. However, such doctors were the exceptions and tended to be mavericks.[99] The medical establishment was too concerned about the reputation and trustworthiness of the profession to countenance such risks.[100] There was little prospect of sterilization expanding through practice in the same way as abortion. As long as doctors felt insecure about their legal position, and until the state was willing to lend its financial support, sterilization would never be practised on more than a limited scale.

If sterilization was to gain political acceptance it had to be framed in such a way that it appeared compatible with prevailing beliefs about the rights of individuals and their duties to the community and state. The earliest proposals, with their language of 'duties' and 'restricted rights', had an authoritarian tone which was out of touch with the political mainstream. Leonard Darwin's aborted bill of 1928 proposed that normal individuals should lose their (presumed) right to be sterilized since this was an abnegation of their social duty to reproduce. Cora Hodson suggested that since rights should only be awarded in return for duties, mental defectives—who had not contributed anything to society—were unfit for the duty of parenthood and should automatically be excluded from the right to reproduce. It was the parents, with legal responsibility over their defective children, who had the voluntary right to chose sterilization, and who had the duty to do so in the interests of the community. It was the right of the unborn child not to have to suffer the misery of inheriting a serious defect.[101] During the 1930s, the discourse shifted away from duties and restricted rights to a more politically acceptable emphasis on extending positive rights. The campaign made great play with the 'injustice' of a situation in which the operation was effectively available to the rich in private clinics but withheld from the poor. Stressing the voluntary nature of sterilization, campaigners presented themselves as defenders of individual rights. However, in doing so, they encountered the fundamental problem that it was illogical and contradictory to argue that 'socially incompetent' mental defectives could exercise voluntary rights.

In retrospect, the Eugenics Society made a strategic blunder in

[99] For instance, support for such a strategy came from the leading advocate of social medicine: J. A. Ryle, 'Medicine and Eugenics', *Eugenics Review*, 30 (1938), 10–12.

[100] This strategy was supported by Lord Dawson: EUG C 338, Tredgold to Blacker, 22 Dec. 1938, and Tredgold to Dawson, 25 Mar. 1939.

[101] Hodson, *Human Sterilisation Today*.

adopting a legislative strategy, for it highlighted the issue of rights. It had been hoped that by concentrating on mental defectives—if the public lacked sympathy towards the rights of any group in society, this was probably it—legislation would be possible. However, this underestimated the extent to which the legislature felt duty-bound to protect the interests of mental defectives. In the long term, legalization of sterilization came through the more gradualist strategy of introducing the operation in practice so that the legislature only had to approve a *fait accompli* in the 1960s.[102] The operation was directed at the very class of women who had been the concern of the eugenicists in the 1930s, though this was now justified on social rather than eugenic grounds.[103] Significantly, sterilization of those who were *non compos mentis*, remained a highly contentious issue.[104]

CONCLUSION

This account of the sterilization campaign has diverged from existing interpretations in a number of respects. Sterilization has been presented as both more and less than a solution to the problem of mental deficiency. It was more, in the sense that its aims went beyond mental deficiency to cover other defective groups and ultimately embraced the whole of the 'social problem group'; it only narrowed to concentrate exclusively on mental deficiency for political reasons. It was less, in the sense that it was never a solution on its own: it had to be combined with the two other solutions of segregation and community care. In forming alliances with the advocates of these other solutions, plans for sterilization were heavily circumscribed and modified. The sterilization solution was also shaped by an internal transition within the eugenics movement, from a lay and propagandist pressure group, to a professional bureau for the circulation of scientific advice and social engineering

102 BL 9.2.145.

103 S. Trombley, *The Right to Reproduce: A History of Coercive Sterilisation* (London, 1988), 159–74, 210–13.

104 C. Dyer, 'Sterilisation of Mentally Handicapped Women', *British Medical Journal*, 2 (1987), 825; E. Kingdom, 'Consent, Coercion, and Consortium: The Sexual Politics of Sterilisation', *Journal of Law and Society*, 12 (1985), 19–29. Similarly, in the United States the use of sterilization for institutional patients has become legally difficult: P. B. Reilly, *The Surgical Solution: A History of Involuntary Sterilization in the United States* (Baltimore, 1991), 148–65.

plans. And finally, the sterilization solution was constrained and shaped by the dominant discourses surrounding the issues of morality and rights. As such, support for or against sterilization reflected a complex of attitudes and should not be viewed, as it tends to be, simply as a gauge for the impact of eugenics.

In this chapter, I have also attempted to use the sterilization issue to cast further light on the place of professional power in Britain's adjustment to democracy. The divisions over the issue of sterilization show how misleading it can be to generalize about the interests of the professional classes, or to produce monolithic models of eugenics or social hygiene.[105] Any general professional disposition towards eugenics on meritocratic and class grounds was complicated by the fact that a policy such as sterilization also raised issues about rights and morality that cut across the professional classes as a whole. And more importantly still, support or opposition towards sterilization often reflected the self-interest of each profession—not only their interest in maintaining or expanding a monopoly of professional power, but also views about the best strategic role for legitimating professional power. Thus, the sterilization issue highlighted some fundamental divisions within professional society over the way in which professionals should relate their expertise to the democratic polity: should they confine their role to scientific research, or act on probability and be political publicists for social policy based on scientific hypothesis; and should they be strictly confined by the letter of the law, or be able to interpret and reformulate social policy through their expertise? The answer to both of these questions was that the professional classes tended to take the more cautious route. In this respect, the sterilization issue provides further evidence that British social policy did not generally fall into the hands of a new echelon of professionals and experts during the interwar years; this was prevented, both because of the constraints of the democracy, and because of the internal dynamics within professions themselves.

[105] For the idea that eugenics embodied the interests of welfare professionals: Searle, 'Eugenics and Class', in Webster (ed.), *Biology, Medicine and Society*, 217–42. The use of the term 'social hygiene' to encompass a broad range of ideas about social policy during the period has been developed by Greta Jones: *Social Hygiene*, 5–21. For problems with this: D and R. Porter, 'What was Social Medicine? An Historiographical Essay', *Journal of Historical Sociology*, 1 (1988), 90–106.

6
The Geography of Mental Deficiency

INTRODUCTION

Although the history of social policy now provides us with a very full picture of change over time, we still know little about the spatial dimension of policy and social problems.[1] In this chapter, I shall attempt to map out the 'geography of mental deficiency', dissecting this into three interrelated elements: first, perceptions of its geographical location and distribution; secondly, the relationship between reported levels and related social, economic, and medical factors—the 'ecology' of mental deficiency; and thirdly, the geography of local authority provision.[2]

Although local case-studies of social policies are not uncommon, there have been few comparative studies of local provision across the nation as a whole; the local tends to be viewed in isolation, rather than used as a way to question the assumptions which underlie the notion of 'national policy'. Thus, even though local government continued to be far more important than central government as a provider of social services in the interwar period, we know relatively little about the way local conditions shaped policy.[3] This chapter will attempt to tease out the relationship between the problem of mental deficiency and its local context and will thereby suggest avenues for future investigation in this and other areas of social policy.

In terms of thinking about social policy and the adjustment to

[1] Historical geographers have recently alerted historians of social policy to some of these shortcomings. See e.g.: F. Driver, 'Bodies in Space: Foucault's Account of Disciplinary Power', in C. Jones and R. Porter (eds.), *Reassessing Foucault: Power, Medicine and the Body* (London, 1994), 113–31.

[2] For a detailed case-study adopting this approach see: M. Thomson, 'Social Policy and the Management of the Problem of Mental Deficiency in Interwar London', *London Journal*, 18 (1993), 129–42.

[3] For a valuable exploration of the importance of local politics in NHS mental health policy: J. Pickstone, 'Psychiatry in District General Hospitals: History, Contingency and Local Innovation in the Early Years of the National Health Service', in J. Pickstone (ed.), *Medical Innovations in Historical Perspective* (London, 1992).

democracy, examination of the local politics of mental deficiency is essential, for local government was at the heart of Britain's democratic culture.[4] The Mental Deficiency Act provides an example of an important shift in power in early twentieth-century social policy: an instance of central government imposing a new policy on the country as a whole, asserting a new degree of administrative control, and demanding uniform standards of provision. As such, the Act provides a case-study to test whether national policy ignored local variations in needs and means, and, if so, to explore how effectively local government agencies adjusted to new policy demands and to the new political culture.

NATIONALIZATION OF THE PROBLEM

Perceptions

In the nineteenth century the main innovations in policy provision for mental defectives had emerged at the local level: the first institutions for idiots were private and charitable initiatives; local schools for defective children expanded more rapidly under permissive legislation than after they were made compulsory in 1914; and after-care services were introduced locally by voluntary agencies. As a result, provision of services was extremely patchy, surprisingly well developed in some urban areas, but non-existent in many others and in most of the countryside. This reflected a society in which social policy was still primarily dictated by local perceptions of need. Mental deficiency, like poverty, tended to be thought of as a localized problem which was largely confined to a slum-dwelling residuum. Under the 1913 Act, however, mental deficiency was reformulated as a national problem which demanded a national policy. Provision of a basic level of institutional and community care services was made the obligation of every county and county borough authority in England and Wales. No account was taken of variations in the need for services or the means for provision.

Crucial to this reconceptualization was the first 'national' survey of mental deficiency undertaken by the Royal Commission of 1904–8. Although the survey consisted of a series of localized studies which

[4] C. H. Wilson (ed.), *Essays on Local Government* (Oxford, 1948), 12–21.

revealed considerable variation, the selected areas were deemed to represent the nation as a whole and their average level of mental deficiency was presented as a single national rate.[5] The Report tended to massage local variations, attributing them to exceptional local factors or to the limitations of the survey. The local investigators, most of whom had very little experience in this new field, tended to be guided by the levels which had already been set by the chief medical investigators A. F. Tredgold and W. A. Potts.[6] If their findings had been wildly different it would simply have fuelled doubts about the meaningfulness of this category 'mental deficiency', and it would have done little to advance the medical profession's claims to a diagnostic expertise. Both leading investigators had a professional interest in portraying mental deficiency as a national problem. As a Medical Officer of Health, Potts was keen to demonstrate that mental deficiency was not confined to the urban residuum; he wanted to advance his profession's role in mental deficiency work, but at the same time he wanted to extend this work beyond the slums to the better-off areas.[7] Tredgold was soon to emerge as the country's leading expert in mental deficiency, with his *Mental Deficiency* of 1908 establishing itself as the accepted medical textbook on the subject throughout the interwar period. Through his private practice and through personal experience of rural life, he too was convinced that mental deficiency was not confined to the slum population.[8] Therefore, although the local surveys suggested a series of correlations between mental deficiency and local economic, social, and demographic factors, the Report as a whole emphasized the existence of the problem throughout the nation. In fact, the one variation which did gain considerable attention—the higher levels in rural than urban areas—did so exactly because it emphasized the national extent of the problem.

The portrayal of mental deficiency as a national problem was crucial if legislation was to be made acceptable to all areas of the country; expenditure on legislation which was exclusively directed at the urban residuum, but which would be paid for by middle-class ratepayers in suburbs and rural areas, was always likely to be defeated. The 1908 Report showed that the problems of the towns could not be separated from those of the country. The two were umbilically linked through the

[5] On the role of surveys in constructing social problems during this period: D. Armstrong, *The Political Anatomy of the Body* (Cambridge, 1983), 42–53.

[6] *Roy. Com. Feeb.*, vi. 15.

[7] Ibid. 115–16.

[8] A. F. Tredgold, *Mental Deficiency*, 6th edn. (London, 1937; first pub. 1908), 2–15. Obituaries: BMJ, 2 (1952), 726–7; *The Lancet*, 2 (1952), 642–3; *The Times* (18 Sept. 1952).

movement of population: rural levels of mental deficiency were inflated by the migration of the ablest to the towns and by the emigration of the least able away from towns to a less demanding life in the countryside. By showing that the rate of mental deficiency was higher in seemingly healthy rural areas than it was in urban areas, the Report strengthened the view that mental deficiency was a hereditary rather than an environmental problem and, as such, threatened the population as a whole.[9] Although it was confirmed that mental deficiency was worst amongst the residuum, this was now conceptualized as a national rather than a localized residuum: a biological sub-class within the population, rather than the population of a specific slum area.[10]

In 1924 the government initiated a second major survey of the number of mental defectives in England and Wales. Published in 1929, the (Wood) Report of the Inter-Departmental Committee on Mental Deficiency again selected a series of local areas to represent an idealized national population.[11] The leader of the survey, E. O. Lewis, had impressive credentials: he had been a school teacher, a university lecturer in Psychology, an Assistant Medical Officer for the LCC, and was now a Medical Inspector for the Board of Control.[12] The survey was far more sophisticated than its predecessor, using a series of mental-testing techniques, employing trained investigators, and differentiating between grades of mental deficiency.[13] It detected significant regional variations, including higher overall levels in rural areas and in the 'geographic foci' of the 'social problem group'—the latter term coined by this report.[14] In explaining this distribution, the Report adopted a far more sophisticated hereditary model than its predecessor of 1908. The fairly equal distribution of low-grade mental deficiency across the country was attributed to the random inheritance of recessive genes which struck all classes and areas of the country alike. By contrast, the disproportionately high rate of feeble-mindedness in

[9] *Roy. Com. Feeb.*, viii. 234–6.

[10] Thus, just three years later, we see feeble-mindedness being used to reconfigure the East End in national and biological terms: H. Boyle, 'Methods of Dealing with Feeble-Minded Cases Bordering on Lunacy', *Report of the National Association for Promoting the Welfare of the Feeble-Minded Annual Conference* (1911), 19.

[11] Significantly, the names of the areas were hidden at the time due to fear that it would inhibit local authorities from co-operating if this simply showed up their neglect: *Report of the Interdepartmental Committee on Mental Deficiency, 1925–1929* (*Wood Report*) (1929), pt. IV, p. 4. Later the areas were named: E. O. Lewis, 'Mental Deficiency and Social Medicine', *JMS* 94 (1948), 258.

[12] PRO MH/107; *The Lancet* (21 Aug. 1965), 395.

[13] *Wood Report*, pt. IV, pp. 1–2, 21–53.

[14] Ibid. 201–4, app. tables 21–3.

socially deprived areas was attributed to the inheritance of a poor pool of genes from the weak-minded individuals who were concentrated at the bottom of the social scale.

The Wood Report was not simply premised on hereditarian attitudes. Lewis was also interested in why people of a similar low intelligence would be certified as defective in one area but not in another.[15] Lionel Penrose later claimed that this interest was at the centre of Lewis's work and reflected his experience of growing up within a tightly knit Rhondda Valley community where problems such as delinquency were unknown.[16] Lewis recognized that the social component, or as he termed it 'socialization', was crucial in determining whether an individual could cope within the community without certification. For instance, the unusually high rate of mental deficiency in a Midlands mining area was attributed to the rapid influx of population in recent generations which had left it without that strong sense of community which was so important in providing support, guidance, and education for the borderline young.[17] Conversely, when abnormally high rates were reported among the more stable populations of rural and cotton manufacturing areas, it was set down to the greater visibility of social problems in such tightknit communities.[18] High rates in other rural areas were explained in terms of a breakdown of community and the marginalization of defectives from the rural workforce, as agricultural work became increasingly technical and responsible, and as large paternalistic estates were replaced by small landholding farmers who were more exploitative of their workers.[19]

Despite this sensitivity to the local dimension of the problem, the Report's main impact was to revive the idea that mental deficiency was a national threat. Public attention concentrated on the Report's contentious conclusion that national levels of mental deficiency had risen significantly since 1908, from 4.6 to 10.49 defectives per 1,000 of the population.[20] The Report claimed that this rise represented a serious

[15] As Armstong has pointed out, the survey was a 'two edged sword', constructing individuals into social problems, but also individualizing and therefore problematizing these social problems: *Political Anatomy*, 51–3.

[16] Penrose Archive, University College London, 147/3, RMPA meeting, 17 Feb. 1966, p. 6.

[17] *Wood Report*, pt. IV, pp. 7–8.

[18] Ibid. 5–6, 70–4.

[19] Ibid. 9–10, 70–4.

[20] Ibid. 175, app. table 7.

escalation and was not simply a reflection of improved survey methods or changed standards: if the latter had been the case, the proportion of low-grade defectives—who were already easily recognizable in 1908—should have fallen in the Wood Report, rather than remaining constant.[21] Worse still, if the level of mental deficiency was rising, it seemed inevitable that this would be mirrored by a similar rise in those on the 'borderline'—those who were not certifiable, but were likely to become criminals, paupers, or unemployed as a result of their mental weakness. Since such individuals were regarded as naturally irresponsible and sexually promiscuous, there seemed to be a great danger that the problem would continue to worsen as they outbred and swamped the respectable classes. Through such logic, the problem of mental deficiency was presented as the key to many of the broader social and eugenic problems of the nation.

It is significant that the areas surveyed were not even named. Anonymity shifted attention away from the local context and enabled the reported levels to serve as national averages. It also helped to shift attention away from the politically controversial issue of national inequalities of wealth and employment; for instance, in surveying mining areas during the General Strike, the Report was touching on a highly sensitive area. The main geographic claim to attract attention was that overall level of mental deficiency was higher in rural than urban areas.[22] This finding sparked off a vigorous public debate over the relative merits of the countryside and city.[23] However, this reflected anxiety about rural and national decline rather more than it reflected serious interest in the local context of mental deficiency.[24] Thus, like the 1908 Report, and despite providing evidence of a geographically varied map of need, the main impact of the Wood Report was to promote mental deficiency as a national rather than local problem.[25]

[21] Ibid., pt. I, pp. 75–82.

[22] This had already been detected by the Royal Commission of 1908 and had been confirmed in a minor survey of 1924: G. K. Bowes, 'Mental Defect and Degeneracy in a Rural Area', *The Lancet* (16 Aug. 1924), 313–18.

[23] Report of Cyril Burt's lecure on high rates of backwardness, *The Times* (26 Oct. 1933); A. Wilson, 'A Defence of Rural Values', *The Times* (28 Oct. 1933); letters on debate, *The Times* (1, 3, 10 Nov. 1933).

[24] R. Williams, *The Country and the City* (London, 1975), 9; M. Wiener, *English Culture and the Decline of the Industrial Spirit 1850–1980* (Cambridge, 1981), 48.

[25] *Wood Report*, pt. IV, pp. 201–3, app. tables 21–2.

THE ECOLOGY OF THE PROBLEM

With their flawed methodologies and clear biases we should certainly not accept the 1908 or Wood surveys at face value. However, they do provide the best available evidence, and their suggestion that the rate of mental deficiency was subject to local variation is worth further investigation. Since mental deficiency, by definition, had a social competence component, we should not be surprised that reported levels might be higher in those areas where it was hardest for defectives and their families to cope with life in the community and where backward or socially problematic individuals were most likely to attract the attention of the authorities.[26] Equally, it would be foolhardy to dismiss out of hand the possibility that rates of mental deficiency might vary in relation to material factors, such as poverty, nutrition, housing, and the standard of medical care.

In the interwar period there was considerable variation in the labour market between different parts of the country. For example, unemployment was more of a factor in depressed single-industry areas such as Wales, where the rate was 36.5 per cent among insured workers in 1932, than in London and the south east, with a figure of just 13.7 per cent.[27] Such substantial variation must have had some effect on levels of reported mental deficiency. Among the first to be marginalized from the labour force, high-grade defectives became visible through unemployment as the local authorities attempted to provide them with an alternative source of support, occupation, and control.[28] Similarly, unemployment among family members undermined intrafamilial support and brought dependent defectives to the attention of public authorities. On the other hand, since defectives provided cheap labour, economic depression could also improve their chances of employment.[29] Employers hoped to save on national insurance contributions

[26] *BOC* 1924, pp. 64–5. See Appendix I. A similar bias occurred in the labelling of mental illness: K. Davis, 'Mental Hygiene and the Class Structure', *Psychiatry*, 1 (1938), 55–66. The need to look at the relationship between mental disorder and socio-economic conditions is raised in D. Ingleby, *Critical Psychiatry: The Politics of Mental Health* (London, 1981), 52.

[27] J. Stevenson, *British Society, 1914–45* (London, 1984), 266–95; R. Pope (ed.), *Atlas of British Social and Economic History Since c.1700* (London, 1989), 150–69.

[28] Compare the way the young were squeezed out of the Labour market in some areas: W. R. Garside, 'Juvenile Unemployment and Public Policy between the Wars', *Economic History Review*, 2nd ser., 30 (1979), 322–9.

[29] Compare the rejoinder to Garside, 'Juvenile Unemployment', by D. K. Benjamin and L. A. Kochin, *Economic History Review*, 2nd ser., 32 (1979), 523–33, which argues that interwar circumstances were not detrimental to juvenile employment.

if they employed defectives as part of a guardianship arrangement or on licence from institutions.[30] Licensed female defectives were particularly attractive as a source of low-paid domestic staff for middle-class families. The fact that the greatest success in employing special-school-leavers in Birmingham came at the height of the Depression again suggests that defectives were being used to undercut the labour market.[31] Thus, although unemployment and the problem of mental deficiency were intimately related, the search for a direct relationship is unlikely to be successful.

Interwar social surveys persistently located high rates of mental deficiency in the poorest areas of cities, sometimes seeing the slums as a contributory material cause, but more often arguing that the inability of mental defectives dragged them to the bottom of society where they bred with their like to produce offspring who perpetuated the slum conditions.[32] These high reported levels were more a reflection of perceptions than anything else. Residence in a slum area increased the visibility of mental defectives.[33] Standards of assessment were influenced by prejudices about 'slum-dwellers'; epitomized by one visitor's description of 'the slum mind, [as] a fungus kind of mental deformity bred from slum conditions'.[34] An added concern was that the close proximity of the sexes in overcrowded housing encouraged sexual abuse of defenceless defective children and left normal children unprotected against possible abuse from defective siblings; a concern which was reflected in the high rate of feeble-minded cases (269 out of 1027) among cases of 'moral corruption' reported by NSPCC in the early 1930s.[35] Despite finding a high correlation between mental deficiency and slum areas, the Wood Report dismissed poor environment as a direct cause of mental deficiency. Even when slum dwellers had been rehoused to new, commodious, suburban dwellings they still showed very high rates of mental deficiency in subsequent generations and they

[30] PRO MH 51/574.

[31] PRO MH 51/726, Birmingham Voluntary Association Grants 1923–1944.

[32] H. L. Smith (ed.), *The New Survey of London Life and Labour*, vi (London, 1934), 359; P. Garside, 'Unhealthy Areas: Town Planning, Eugenics and the Slums, 1890–1945', *Planning Perspectives*, 3 (1988), 24–46.

[33] J. White, *Campbell Bunk: The Worst Street in North London* (London, 1986), 45.

[34] J. Conquest, *The Naked Truth . . . About the Slums* (London, 1935), 106; M. Gaskell (ed.), *The Slums* (Leicester, 1990).

[35] PRO MH 58/102, Brock to Robinson, 30 Nov. 1932 (this concern extended to rural areas where inbreeding was seen as a possible cause of high levels of mental deficiency). On the problem of incest: L. Gordon, *Heroes of Their Own Lives: The Politics and History of Family Violence* (London, 1989), 360–1.

rapidly recreated the slum conditions in the new location—environment was the outcome, rather than the cause of mental inability.[36] It was far less common for mental defect to be attributed to the vast inequalities in the standard of environment and housing which existed in interwar Britain.[37] One piece of research which did point in this direction was an Edinburgh study linking high rates of mental defect and ill health to overcrowded living conditions in tenements.[38] However, there was a very fine line between the conclusions drawn from environmental and hereditary epidemiologies: by stressing the high rate of mental deficiency in slum areas, both reinforced the tendency to view mental deficiency as a condition which was largely confined to the lowest social classes; and they were equally likely to be pessimistic about ameliorating the situation, and equally draconian in their remedies. For instance, the Edinburgh study recommended containing this source of disease and disorder within ghettos, policed and segregated from the outside community.

Perhaps the most obvious area to look for the impact of material conditions on the geography of mental deficiency is that of living standards and nutrition. The degree of poverty and malnutrition in the interwar period has been contested by historians, though the existence of pronounced inequalities between regions is accepted.[39] Whether poverty was severe enough to have affected infant and maternal mortality remains unclear.[40] However, the serious problems of poverty and consequent malnutrition experienced by some mothers and children during the period almost certainly affected morbidity levels.[41] It is unlikely that nutritional deficiency was severe enough to have caused brain damage to infants at birth, but it could well have

[36] *Wood Report*, pt. IV, p. 74. See also Garside, 'Unhealthy Areas', 24–46.

[37] On regional disparities in housing conditions, see C. Brown, 'Urbanization and Living Conditions', in R. Pope (ed.), *Atlas of British Social And Economic History* (London, 1989), pp. 176–9.

[38] *Annual Report of the Public Health Department of Edinburgh* (1934), 98; ibid. (1930), 96–7.

[39] C. Webster, 'Healthy or Hungry Thirties?', *History Workshop Journal*, 13 (1982), 110–29.

[40] J. M. Winter, 'Unemployment, Nutrition, and Infant Mortality in Britain, 1920–50', in id. (ed.), *The Working Class and Modern British History* (Cambridge, 1983), 232–56. For criticisms of Winter, and for the argument that poverty continued to be a cause of infant and maternal mortality: L. Bryder, 'The First World War: Healthy or Hungry?', *History Workshop Journal*, 24 (1987), 141–57; C. Webster, 'Health, Welfare and Unemployment during the Depression', *Past and Present*, 46 (1985), 204–29.

[41] I. Loudon, 'On Maternal and Infant Mortality 1900–60', *Social History of Medicine*, 4 (1991), 48–51.

hindered the mental development of young children and affected school performance to the extent that it significantly increased the chances of being certified as educationally defective. The conclusions of more recent nutritional work would indeed suggest that the inadequate and poorly balanced diet of the poorest working-class mothers and infants could have significantly raised rates of mild mental defect and retardation.[42] In the interwar period, however, the possibility of such a link attracted little attention.[43] In part this reflected the predominance of hereditary over environmental explanations. It also probably reflected the political sensitivity which surrounded this issue. If mental deficiency was linked to nutrition, it tended to be an unbalanced diet, rather than poverty, which was blamed.[44]

Wide variations in the standard of infant and maternal welfare services may also have shaped the geography of mental deficiency. The period saw a rapid improvement of urban provision which was not matched in rural areas. Partly as a consequence, urban infant mortality fell below rural rates for the first time in the 1920s.[45]. The parallels with the reported higher levels of rural mental deficiency are striking, suggesting that backward rural services may have been a contributory factor. Standards of maternal and child welfare may also help to explain why there was an unusually high proportion of low- to high-grade defectives in prosperous urban areas. The superior care may have improved the survival rate of these handicapped children;[46] or in some cases interventionist neonatal care (which led to surprisingly high rates of maternal mortality) may have damaged children at birth.[47]

[42] M. and A. Wynn, *Prevention of Handicap and the Health of Women* (London, 1979), 30–66, 80.

[43] An exception was Dr Newton, *Public Health*, 33 (1920), 165–70.

[44] C. Burt, *The Backward Child*, 2nd edn. (London 1946); first pub. 1937), 122–3, felt that the low occurrence of mental defect among poor Jews was an indication that the balance of the diet was more important than poverty. Interest increased after demonstration of the affect of starvation during WWII: C. Asher 'The Relationship of Birth Weight to Mental Development', *Archives of Disease in Childhood*, 23 (1948), 142; B. Landtman, 'On the Relationship Between Maternal Conditions During Pregnancy and Congenital Malformations', *Archives of Disease in Childhood*, 23 (1948), 237–48.

[45] Loudon, 'On Maternal and Infant Mortality', 54–5. On falling rates and geographic variation, see Brown, 'Urbanization and Living Conditions', 180–2.

[46] *Wood Report*, pt. IV, p. 58: the predominantly middle-class extra-metropolitan area had an infant mortality rate of just 57/1,000 under 1 year old, compared to 99/1,000 and 84/1,000 in the other urban areas. This corresponded to an incidence of low-grade defect among children of 0.85/1,000, compared to 0.73/1,000 and 0.68/1,000.

[47] I. Loudon, 'Deaths in Childbirth from the Eighteenth Century to 1935', *Medical History*, 30 (1986), 1–41. Wynn, *Prevention of Handicap and the Health of Women*.

However, it is most likely that the high proportion was largely an illusion, reflecting the fact that high-grade defectives usually went unrecognized if they came from middle-class homes. Whatever the case, the Wood Report's discovery of a high rate of low-grade defectives in prosperous areas casts doubts on those who blamed mental deficiency on lack of medical intervention at birth or on the physical incapacity and neglect of mothers.[48]

Finally, the map of mental deficiency was related to patterns of demographic change.[49] Detection of high rates in rapidly expanding urban areas was partly a reflection of the disproportionate number of school children—the most visible section of the population—in such towns. In addition, the disruption of familial and community networks of care probably did make the problem more visible and pressing.[50] It is also worth considering the argument that high rural rates were sometimes related to the emigration of the young and able, leaving a disproportionate number of the least able and dependent behind.[51] More generally, geographic variation in the demographic transition towards smaller families, more households, and a rise in the proportion of dependent elderly and young meant that mental deficiency was more likely to be pushed out of the familial context to become a social problem in some areas of the country rather than others.[52]

In summary, largely because of the social competence definition, but also because of social and economic inequalities, there was a geographically varied map of perceived and real need for mental deficiency services. On the whole, the areas with the greatest social and economic problems were likely to be those where the problem of mental deficiency was the most acute. However, the interaction between socio-economic factors was complex, and therefore we should not expect a consistent correlation between social conditions and rates of mental deficiency. Moreover, since levels of unemployment, infant welfare, housing, and poverty fluctuated within individual administrative areas,

48 H. Brackenbury, 'Maternity in its Sociological Aspects', *The Social Service Review*, 18 (1937), 37–47; K. Vaughan, 'The Shape of the Pelvic Brim as a Determining Factor in Childbirth', *BMJ* 2 (1931), 939–41; S. W. Savage, 'Intelligence and Infant Mortality in Problem Families', *BMJ* (1946), 86–7, reported higher chances of retardation and mortality in children of defective mothers due to maternal neglect.

49 For an overview of regional variations: S. Jackson and G. Timmins, 'Demographic Changes, 1701–1981', in Pope (ed.), *Atlas of British Social and Economic History*, 134–48.

50 *Wood Report*, pt. IV, pp. 3–8.

51 Ibid. 7–8.

52 An interwar survey which did look at mental deficiency in these terms, though only as a side issue, was: B. S. Rowntree, *Poverty—A Study of Town Life* (London, 1941), 44–7.

any significant correlation would depend upon a detailed breakdown of a single area. In Britain, unlike in the United States, there were few serious attempts to construct detailed maps of mental hygiene.[53] The reliability of those studies which did take place is questionable. The leading British work was Cyril Burt's research into the distribution of the retarded, defective, and delinquent in London, which did indicate some significant correlations with poverty, although this tended to be attributed to the behaviour of the poor as much as their material conditions.[54] A further problem with accepting figures for the distribution of mental deficiency as 'real' was that they depended upon the ability and will of local authorities to detect and provide services for defectives.

GEOGRAPHICAL PROBLEMS OF ADMINISTRATION

Nationalization of the problem of mental deficiency ignored considerable variation in the administrative problems facing local authorities.[55] The 1913 Act required every county and county borough authority in England and Wales to provide institutional care, to organize community supervision and guardianship, and to ascertain the local defective population. This work was to be organized by the existing asylums

[53] For instance, R. Faris and W. Dunham, *Mental Disorders in Urban Areas* (Chicago, 1939). W. L. Neustatter felt that the factor of race in the United States made socio-economic factors more important than in Britain: 'Social and Economic Factors in Psychiatry', *Eugenics Review*, 34 (1942), 55–61. The limitations of the British social survey in contrast to interwar sociological surveys in the United States are discussed in: M. Bulmer, 'The Development of Sociology and of Empirical Research in Great Britain', in id. (ed.), *Essays in the History of British Sociological Research* (Cambridge, 1985), 3–36; M. Bulmer, K. Bales, and K. Kish Sklar (eds.), *The Social Survey in Historical Perspective, 1880–1940* (Cambridge, 1991).

[54] C. Burt: *The Young Delinquent*, 4th edn. (London, 1945; first pub. 1925); id. (*The Backward Child; The Subnormal Mind*, 3rd edn. (London, 1955; first pub. 1935). See especially his mapping of the problems on to Booth's poverty map of London (problematically somewhat out of date by then), and his breakdown of correlations between retardation and social factors: *The Backward Child*, 97, 100–1, 104–33. For the view that Burt's work was deliberately falsified to serve his ambitions and biases: L. S. Hearnshaw, *Cyril Burt, Psychologist* (London, 1979). This biography has now itself been accused of bias, and Burt placed in a more favourable light: R. B. Joynson, *The Burt Affair* (London, 1989).

[55] W. Robson, *Development of Local Government* (London, 1930).

committee or by a new mental deficiency committee of the local council. In addition, under the 1914 Education Act, local authorities were obliged to provide special schools for defective children. However, these services were to be administered by the Local Education Authorities (LEAs), rather than the local mental deficiency or asylums committees. The majority of LEAs came under the county and county borough authorities, but a few were under the separate administrative head of the urban district councils. The continuing role of the poor law in providing institutional care for many defectives complicated the picture further, since this care was organized by a third tier of administrative organization—parishes and joint poor law unions.[56]

Not only was the existence of multiple and overlapping local authorities confusing, but it was also sometimes ill-suited to needs. The distribution of population within the county was a considerable problem.[57] In county boroughs, with their large and concentrated populations, provision of specialist mental deficiency services could be reasonably straightforward, but it was far harder in rural and sparsely populated county authorities. Rural children usually had to leave home to attend residential schools, which were expensive, unpopular among parents, and, as a consequence, rarely successful. The sparse population also made it difficult to organize supervisory services; it could take a welfare visitor all day to visit perhaps three cases.[58] On the other hand, the geographical problems did encourage certain types of community care, such as home visiting by teachers and experts, and it was felt that the countryside was a more appropriate environment for defectives to be placed with guardians.[59]

Rural areas were less geographically disadvantaged in terms of institutional care. There was a tradition of locating mental hospitals in the countryside, away from what seemed to be the physical and moral corruption of the city.[60] With their dispersed design, it was easier to locate colonies in rural areas, where the land was relatively cheap and readily available, and where there were less likely to be objections from

56 H. Finer, *English Local Government* (London, 1945), 40–179.

57 E. Fox, 'Modern Developments in Mental Welfare Work', *Eugenics Review*, 30 (1938), 165–78.

58 For a detailed case study: E. Pinsent, *The Mental Health Services in Oxford City, Oxfordshire, and Berkshire (Oxford Survey of Social Services)* (Oxford, 1937).

59 L. Beach, 'The CAMW Guardianship Scheme', *Mental Welfare*, 13 (1932) 29–38.

60 See Ch. 3. For regional variation in institutional provision, compare Felix Driver, 'The Historical Geography of the Workhouse System', *Journal of Historical Geography*, 15 (1989), 269–89.

local residents.[61] By 1938, those counties which still lacked any arrangement for institutional provision were mainly isolated and sparsely populated areas where the geographic problems were the most acute.[62]

The main disadvantage faced by county boroughs was the lack of convenient sites to build colonies. Instead they often established joint schemes with adjoining counties, or sent the town's defectives away to outside institutions. On the other hand, the greater density of population made it easier to ascertain defectives and to provide supervision, special school, and community care services. They also benefited from greater administrative unity, with education and mental deficiency services both under the council, although the existence of poor law unions within boroughs still caused some boundary complications. Most advantaged of all were metropolitan areas, such as London or Birmingham, where the sheer scale of the population justified the provision of a series of specialist services.[63]

THE PARTY POLITICS OF LOCAL ADMINISTRATION

Since the 1913 and 1914 Acts took little account of variation in needs and means between local authorities, levels of provision were largely determined by the local politics of mental deficiency. The interwar period has been seen as a golden age for local government, with local social services expanding to probably their greatest ever extent.[64] Legislation such as the Mental Deficiency Act placed new duties on local authorities. Through its power to pass orders and provide grants in aid, the Board of Control could pressurize local councils to provide services, but ultimately the extent to which the councils acted on this remained

[61] A. Scull, 'A Convenient Place To Get Rid Of Inconvenient People', in A. King (ed.), *Buildings and Society: Essays on the Social Development of the Built Environment* (London, 1980), 39–60.

[62] *BOC 1938*, p. 51: counties: Anglesey, Caernarvon, Hereford, Huntingdon, Isle of Wight, Merioneth, Monmouth, Soke of Peterborough, and East Sussex; by contrast, there were only three county boroughs, all resort towns: Eastbourne, Hastings, Newport.

[63] Thomson, 'Social Policy and Mental Deficiency in Interwar London', 129–42.

[64] A. T. Peacock and J. Wiseman, *The Growth of Public Expenditure in the United Kingdom* (London, 1961), 96–120; M. Loughlin, M. Gelfand, and K. Young, *Half a Century of Municipal Decline, 1935–1985* (London, 1985); R. Lee, 'Uneven Zenith: Towards a Geography of the High Period of Municipal Medicine in England and Wales', *Journal of Historical Geography*, 14 (1988), 260–80.

in their own hands.[65] The political geography of welfare which resulted can be examined by constructing comparative patterns of local expenditure.[66] The analysis will be based on local government financial returns for the year ending March 1933, by which time local authorities had two decades to set up an administration.[67] These will be compared with the Board of Control figures for 1933 on levels of ascertainment, reported defectives, and defectives in institutions. Unfortunately, similar detailed community care statistics were not published until 1937. Nevertheless, these figures have been included for comparative purposes, and it is unlikely that they differ drastically from those of 1933.[68]

County Borough Councils

Expenditure on mental deficiency services can be ranked according to its contribution or 'rate' in pence out of the overall rate levied.[69] The county boroughs with the highest rate were Walsall, West Bromwich, Birmingham, Ipswich, Bristol, East Ham, Smethwick, Tynemouth, West Ham, and York. Six of these authorities were also among the ten leading per capita providers of institutional care. Of the other authorities with high rates for mental deficiency, Bristol and Ipswich had very good records as service providers among the top ten authorities for defectives reported, ascertained, in institutions, and under community care, and a similar, but less notable, pattern existed among the other leading spenders. Mid-sized prosperous southern municipalities, such as Oxford, Plymouth, Reading, and Exeter, had low rates but compensated for this through their greater resources to provide high levels of defectives reported, ascertained, in institutions, and under

[65] On the other hand, such national legislation was gradually shifting power from local to central government as the 'centralization' effect led to increasing dependency on central government grants: Peacock and Wiseman, *Growth of Public Expenditure*. For the view that this threatened local democracy: W. Thornhill, *Central Government and the Local Authorities: A Study of Methods of Control* (Sheffield, 1954).

[66] Such an approach remains surprisingly undeveloped in histories of social provision. Recently, the value of such an approach has been recognized: Lee, 'Uneven Zenith', 260; also S. V. Ward, *The Geography of Interwar Britain: The State and Uneven Development* (London, 1988), which compares provision of housing, highways, and parks; M. Powell, 'The Geography of English Hospital Provision in the 1930s', *Journal of Historical Geography*, 18 (1992), 307–16.

[67] *Ministry of Health Local Taxation Returns. Local Government Financial Statistics for 1932/3* (HMSO).

[68] *BOC 1933*, pp. 57–9; *BOC 1936*, pp. 42–4.

[69] *Local Government Financial Statistics 1932/3*, table III, pp. 24–5, 90–1.

community care. Resort towns were among those with the worst services, reflecting limited demand in populations which were predominantly made up of the retired elderly and visiting holidaymakers.[70]

County Councils

To compare the expenditure of largely rural county councils with that of urban county boroughs, one has to look at the combined rate devoted to lunacy and mental deficiency services, since the county statistics did not differentiate between the two. The average county rate was 3.2d., compared to 3.7d. in the county boroughs. The lower county figure can partly be attributed to a lower average annual rateable value of £5. 5s. per head, compared to £6. 5s. in the county boroughs. This slightly lower county expenditure may be reflected in levels of service provision. However, the difference between county and county borough provision was not dramatic. It would seem reasonable to assume that poor services might result from the difficulties of providing services in sparsely populated areas. However, this is not supported by figures on the most and least densely populated counties, which show that the average level of provision in the densest six was slightly lower than that of the six most sparsely populated counties in the country. In fact, the main geographical trend visible from the statistics is a lower level of provision per head in those counties with the highest populations. This suggests that the resources of a single administration were overstretched by the demands of providing for such large and scattered populations; this factor may also help account for the poorer overall performance compared to county boroughs, especially in the area of community care.

Party, Ideology, and Welfare

With the escalation of local government expenditure and rates since the late nineteenth century, party politics had become an increasingly important element in local elections.[71] However, the ideological content of this party politics has been given little consideration by historians

[70] J. Walton, *The English Seaside Resort: A Social History, 1750–1914* (Leicester, 1983), 128–55.

[71] N. McCord, 'Ratepayers and Social Policy', in P. Thane, *The Origins of British Social Policy* (London, 1978), pp. 21–35; J. Harris, 'The Transition to High Politics', in M. Bentley and J. Stevenson, *High and Low Politics in Modern Britain* (Oxford, 1983), 58–79.

and tends to be dismissed as secondary to the dominant issue of the rates. Because there has been so little study of the local politics of interwar welfare, we are left with a series of crude caricatures of party positions. It is generally assumed that the Conservatives opposed the extension of virtually all services, and that Labour were just as ineffectual, 'labourist', and devoid of ideas locally as they were in national politics. The interwar alliance in many areas of the country between the Liberal and Conservatives is explained as an anti-socialist and anti-spending opposition to Labour, rather than analysed in terms of competing ideas over specific policies.[72] The view that party differences were relatively unimportant in practice, is supported by E. D. Simon's account of Manchester politics in the interwar period. Simon suggested that although party was by this time an important factor in the election of candidates it did not determine policy within the council: chairmen of committees were initially elected on a party basis, but would then stay for life whichever party was subsequently in control. Party rarely intruded into the administration of social services, apart from differences between 'economists' and 'progressives' which emerged when issues of extension or new provision of services were at stake.[73] However, since Simon had a vested interest in presenting the Liberals as interceding above faction, he probably underplayed political differences. Even if his account is accepted, it would be rash to assume that the Manchester pattern applied elsewhere.

Although study of local Conservative and Liberal Party welfare policy is still virtually non-existent, a number of recent studies have begun to advance a more sophisticated picture of the Labour Party.[74] It has been suggested that Labour's interests extended beyond protection of traditional working-class values and interests to a programme of 'municipal socialism' that embraced demands for higher wages, municipal jobs, and more generous levels of relief.[75] Moreover, the political

[72] C. Cook, 'Liberals Labour and Local Elections', in G. Peele and C. Cook (eds.), *The Politics of Reappraisal, 1918–1939*, 166–88; B. K. Lucas and P. G. Richards, *A History of Local Government in the Twentieth Century* (London, 1978), 111–19; G. W. Jones, *Borough Politics: A Study of the Wolverhampton Town Council, 1888–1964* (London, 1966), 53–74.

[73] E. D. Simon, *A City Council From Within* (London, 1926), 31, 130–50.

[74] London is an exception: e.g. H. Jones, 'Conservatives and the LCC after 1934', in A. Saint (ed.), *Politics and the People of London: The London County Council, 1889–1965* (London, 1989), 237–50, questions whether Labour was that more radical than the Conservatives in London between 1934–9.

[75] J. S. Rowett, 'The Labour Party and Local Government: Theory and Practice in the Interwar Years', unpub. D.Phil. thesis (Oxford, 1979), 171.

concerns of Labour varied across the country according to local circumstances, from the problems of a boom town in Coventry, to the problems of Catholics in John Wheatley's Glasgow, to the interests of working women in an area such as Preston. Many non-labourist issues, such as issues of welfare, which have generally been dismissed as non-political, deserve reconsideration in this light.[76] But for a full picture we also need to look at the local Conservative and Liberal Party attitudes to 'non-political' issues. This examination of mental deficiency policy in relation to local political control will begin to explore the prospects and problems in tackling such questions.

At a general level, the hypothesis that party control was important in welfare provision is supported by the county borough financial returns for 1933.[77] These show that, in proportion to rateable value, rates were on average higher in areas where Labour was in control during some part of the period (and was probably strong enough to have an influence on policy even when it lacked a majority).[78] On the other hand, the average rate levied for mental deficiency services was slightly lower than average in these areas. Similarly, if levels of provision are compared, Labour authorities fell below the overall average. This suggests that Labour gave mental deficiency services a comparatively low priority compared to the Conservatives.

There are two possible explanations for the Conservative commitment to mental deficiency services. First, from a class and moral perspective, one might expect them to have been particularly concerned about the social and eugenic threat posed by mental defectives. Secondly, their normal resistance to expenditure may have been overcome by the prospect that mental deficiency colonies would save ratepayers

[76] N. Kirk, 'Traditional Working Class Culture and the Rise of Labour', *Social History*, 16 (1991), 203–16.

[77] *Ministry of Health Local Taxation Returns. Local Government Financial Statistics for 1932/3*. For further information on this financial analysis and for a detailed breakdown of local authority provision and spending: M. Thomson, 'The Problem of Mental Deficiency in England and Wales *c.*1913–1946', D. Phil. thesis (Oxford, 1992), 220–3.

[78] Before 1946 there are no easily accessible published data as to party control of local government. The information has been reconstructed from the Appendices of Rowett, 'Labour Party and Local Government', 369, 381, and C. Cook, *The Age of Alignment: Electoral Politics in Britain, 1920–1929* (London, 1975), 27–87. Political control fluctuates in many of these areas during the period, but the seats would have been marginal, with the controlling Council forced to consider Labour views. Therefore, the liberty has been taken of including all areas with some history of Labour control during the period as 'Labour areas'. This has the advantage of simplicity and provides a higher number for comparison.

the even greater costs of prisons, asylums, and unemployment, and the long-term costs of bringing future generations of defectives into the world. The position of the Labour Party was more ambiguous. In social policy it generally stood up for the interests of the weakest members of society: for instance, it expressed some support for special education services, and it showed a keen interest in policies for the blind.[79] Moreover, Labour women were often among those supporting the extension of mental welfare, especially when it involved helping women and children.[80] On the other hand, the language of eugenics and anti-working class bias which surrounded so much debate about mental deficiency policy probably alienated many Labour councillors who objected to social policies which placed the working-class family under the coercive power of experts.[81] Such suspicions were expressed by Mr T. Hill of the Newcastle Board of Guardians:

> We as Labour members have our own opinions . . . the great majority of the people who are certified to be mentally deficient are not really mentally deficient. I believe if certain preventive measures took place in the earliest periods of their lives, they would not at a later period of life require certification.[82];

Similarly, a Labour member of South Shields Board of Guardians deplored the way that the 'psychology of the worker' was neglected in discussions dominated by experts, and argued that the parents of children were the best placed to determine the need for care. It would

[79] *Labour Party Report of 31st Annual Conference* (1932), 270–1. P. Thane, 'The Labour Party and State Welfare', in K. D Brown, *The First Labour Party* (Beckenham, 1985), 191–2.

[80] *Report of the 19th National Conference of Labour Women, 1938*, pp. 47, 110–11; C. Steedman, *Childhood, Culture and Class in Britain: Margaret McMillan, 1860–1931* (London, 1990), 53–7, 156–72. Labour women were often more vigorous supporters of social services than men in the Party: Thane, 'Labour Party and State Welfare', in Brown (ed.), *First Labour Party*, 183–216; P. Thane, 'Visions of Gender in the Making of the British Welfare State: The Case of Women in the British Labour Party and Social Policy, 1906–1945', in G. Bock and P. Thane (eds.), *Maternity and Gender Politics: Women and the Rise of the European Welfare States, 1880–1950* (London, 1991), 93–118. The relative strength of Labour women may be an important factor in explaining variations in policy between local Labour councils: J. Mark-Lawson, M. Savage, and A. Warde, 'Gender and Local Politics: Struggles over Welfare Policies, 1918–39', in L. Murgatroyd et al. (eds.), *Localities, Class, and Gender* (London, 1985), 195–231.

[81] H. Pelling, 'The Working Class and the Welfare State', in id., *Popular Politics and Society in Late-Victorian Britain* (London, 1968), 1–18; P. Thane, 'The Working Class and State Welfare', *Historical Journal*, 27 (1984), 877–900.

[82] *CACMD Conference Report 1920*, p. 98.

be far better, he asserted, if the authorities concentrated their efforts upon environmental improvement.[83]

Despite the suspicions of individual Labour MPs there is little indication of overwhelming opposition towards institutional care at the ground level. It was probably more a case of mental deficiency institutions being a low priority at a time when the effects of the Depression and unemployment were hitting the whole of the population. However, there are signs that Labour was more sympathetic than other parties towards providing material aid to families as part of community care and as an alternative to placing defectives in institutional care. For instance, Mr Andrews of Abertillery complained of the parsimony and means testing which held back the practice of providing material aid as part of community care guardianship orders:

> I come from a very hard-working—when work is there—industrial area, and we would resent any suggestion of exploiting a mental defective for personal gain. I am positive that no income on the Scale could show any personal gain to a person who took a mental defective in from a boarding point of view.[84]

Under such local pressure, local authorities began to appoint parents to serve as guardians so that they could receive direct financial aid.[85] Unfortunately, the Board of Control's community care statistics do not enable us to tell whether Labour authorities were at the forefront of providing this type of material aid; the figure for community care was mainly a reflection of the numbers under supervisory control, and in fact the average level in Labour areas was slightly lower than in others. However, the marked disparity between the level of community care and institutional care in a series of Labour areas does suggest that there may have been a preference for the former over the latter.[86]

An obvious way to examine the importance of party politics in shaping local policy is to compare attitudes on the controversial issue of sterilization. Greta Jones has argued that the local opposition to

[83] *CAMW Conference Report 1924*, pp. 122–3.

[84] Councillor Andrews of Abertillery, *CAMW Annual Conference Report For 1932*, p. 58.

[85] The Board of Control disapproved of this practice as it did not provide the control which was desired under the Act: *BOC 1933*, p. 67. The rise in guardianship in the later 1930s from 1,591 in 1930 to 4,723 in 1939 is, however, probably an indication of the increased use of parents; see Ch. 7.

[86] For example, Smethwick stood at 2.31/1.000 compared to 0.74/1,000, Swansea at 1.44/1,000 compared to 0.21/1,000, West Ham at 1.87/1,000 compared to 0.72/1,000, Sheffield at 1.55/1,000 compared to 0.79/1,000, and Hull at 1.58/1,000 compared to 0.63/1,000. It is also significant that five Labour authorities were in the top eleven providers of community care.

sterilization was Labour-led, while support was primarily southern and Conservative.[87] In fact, taking evidence from the whole of the 1930s, the pattern of support was rather more complicated. Geographically, support was not confined to the south; in particular, there was also support from East Anglian, Welsh, Midlands, and some northern areas in the later 1930s. Support from areas such as Caernarvon and Herefordshire, which still had no mental deficiency institution of their own, and from counties such as Rutland, which suffered from the geographical disadvantages of providing for a sparse population, suggests that sterilization was a reaction to administrative problems as much as a response to ideological predispositions.[88] The pattern of support also reflected the spread of lobbying campaigns between adjacent areas: in 1933–4, resolutions of support came from Dorset CC, Exeter City, Devon CC, Poole Guardians, Poole Borough, and Plymouth; while in 1937–8 a resolution reviving the issue was sparked off by Hampshire CC and followed up by Southampton CC.[89] The fact that the politically cautious Association of Municipal Corporations and County Councils Association were both willing to play prominent roles in the campaign demonstrates how widespread and cross-party support actually was.[90] Part of the problem in our current understanding of the issue is the assumption that support for sterilization necessarily reflected moral conservatism. In fact, on economic grounds, many of the same authorities who supported sterilization also supported an extension and liberalization of community care—policies which were somewhat at odds with a morally conservative agenda.[91] Other authorities supported sterilization as part of a broader programme of liberalizing access to birth control.[92] Thus, in explicating the pattern of local authority support for sterilization, party was important but it was mediated by a complex set of factors which included the standard of local provision,

[87] G. Jones, *Social Hygiene in Twentieth-Century Britain* (London, 1986), 101.

[88] EUG D 203, 2 June 1938. For sterilization: Hampshire CC, Brecon, Caernarvon, Cumberland, Denbigh, Devonshire, East and West Suffolk, Gloucestershire, Herefordshire, Holland Lincs., Western Lincs., Leicestershire, Montgomery, Norfolk, Pembrokeshire, Rutland, Surrey, West Riding Yorkshire, Warwickshire. Against sterilization: Bedfordshire, Buckinghamshire, Cambridgeshire, East Sussex, London, Oxfordshire, Shropshire, East Riding Yorkshire.

[89] EUG D 203, 2 June 1938 on Hampshire CC resolution; PRO MH/100–3.

[90] *Municipal Review* (Mar. 1931), 109–10; (Dec. 1931), 496; (Mar. 1932), 123; Lucas and Richards, *History of Local Government*, 180–95.

[91] *Municipal Review* (May 1933), 203–4.

[92] *Municipal Review* (June 1930), 204.

the presence of local activists, the proximity of other campaigning authorities, and attitudes on related issues such as birth control.

A detailed examination of the LCC's vote against sterilization in 1934 can be used to demonstrate the subtle contours of the local politics of mental deficiency. As part of the Brock Committee investigation into sterilization, local authorities had been requested to submit information on the inheritance of mental defect. London's contribution was considered to be one of the most convincing pieces of evidence in favour of sterilization.[93] With the publication of the Brock Report, local authorities were again asked to express their opinions. In light of the LCC's earlier survey, it is hardly surprising that the Mental Hospitals Committee initially backed the Report's recommendations. However, in the election of March 1934, Labour won control from the Municipal Reformers (Conservatives) and this led to a change of personnel on the Mental Hospitals Committee, with Labour members as the new Chairman and Vice Chairman. The new Commitee was divided on the issue, splitting twenty-two votes to nineteen, and then twenty-six to eleven in favour of sterilization, with the Labour Chairman adding a note that he was not in agreement.[94] Recognizing the political sensitivity of the issue, responsibility was passed on, first to the General Purposes Committee, and then to the full Council itself.[95] After a fairly heated debate, the Council voted to express no opinion on the Brock Report (it should be noted that this was not a vote against sterilization as such). The majority of those voting to withhold support were Labour members, outnumbering the Municipal Reformers by forty votes to twenty. On the other hand, this was partly a reflection of an overall Labour majority on the Council of sixty-nine to fifty-five. Of the votes expressing support for the Brock Report, there was an equal split, with twenty votes from both the main parties.[96] Considered in these terms, this was neither a resounding defeat for sterilization, nor a clear-cut party vote.

Attitudes towards sterilization depended upon personal as well as party factors. Doctors and women tended to vote for passing a

[93] *LCC Annual Report 1933*, vi. 34–5.

[94] GLRO LCC MIN/598 Minutes of the Mental Hospitals Committee, 15 May 1934.

[95] GLRO LCC MIN/6291 Minutes of the General Purposes Committee, 18 June 1934.

[96] GLRO LCC Minutes of Proceeding of Council, Report of the General Purposes Committee to Council, 18 June 1934. Party affiliation from A. E. Caffrey, 'The Transformation of Working-Class Politics and Political Culture in London, *c.*1900–1929' (forthcoming D.Phil. thesis, Oxford).

resolution in favour of the Brock Report, while the majority of clergy voted against this. Support for sterilization was also associated with support for birth control: Mr Silkin of Labour stated that the speeches in favour of sterilization would apply equally well to birth control, and Mrs Bolton of Labour attacked the injustice of limiting forms of family restriction to the well-off. Religious opinion was also an important influence on the vote. Leader of the London Labour Party, Herbert Morrison, argued that any sign of support for sterilization would offend the religious sensibilities of a substantial proportion of London voters, particularly the large Catholic constituency. Indeed, the Catholic News warned the Labour Party that it had lost considerable Catholic support because of support for sterilization from some Labour councillors.[97] In summary, as in the national politics of mental deficiency, factors such as gender and religion complicated party positions.

THE NON-PARTY POLITICS OF MENTAL DEFICIENCY

The division of responsibility between and within local authorities formed a political level which was probably even more influential than party politics in shaping local provision. The Mental Deficiency Act stipulated that a special committee should be formed to administer the Act at the local level. Just as the Board's creation at the centre had raised opposition, so did this specialization of administration in the localities. The work of the mental deficiency committees went beyond that of the existing asylum visiting committees: in addition to inspecting conditions in institutions, they were to have responsibility for ascertainment of defectives, provision of institutional care, and supervision in the community.[98] Such work demanded full-time staff and generous funding if it was to be adequately performed. However, mental deficiency committees had a low status within councils, with little influence over the important finance and organizational committees, and therefore they struggled to attract resources away from higher profile services.[99]

[97] *The Times* (4 July 1934); *The Daily Express* (4 July 1934); *Catholic News* (14 July 1934).

[98] F. Robinson, 'Local Administration of the Mental Deficiency Act 1913', *Public Health*, 29 (1916), 78–93.

[99] *BOC 1930*, pp. 13–14; B. S. Miller, 'Mental Deficiency and the Local Government Bill', *Mental Welfare*, 10 (1929), 29–40.

To understand the position of mental deficiency committees within councils, it is also necessary to take into account the role of women. Patricia Hollis has demonstrated the importance of integrating gender into an understanding of the politics of local government.[100] This is especially relevant for an area such as mental deficiency, since the 1913 Act stipulated that at least one member of the local mental deficiency committees had to be a woman, and invariably the number of women appointed was far higher. In the late nineteenth century it was women who pioneered local services for the feeble-minded. They were drawn into this area because it was seen as a natural site for their 'nurturing qualities', and because men paid it little attention, accepting it as part of a 'women's sphere' of welfare.[101] This type of 'maternalist' strategy enabled women to develop a series of new areas of municipal welfare and to enter public life in the second half of the nineteenth century.[102] However, as party politics became more important in local government, elected party women were generally placed on to committees, such as education or welfare, where it was harder to make an impact than previously on the more independent school boards and boards of guardians, while the decision-making committees, such as finance, remained in male hands.[103] Hollis argues that the role of these new local government women became far more bureaucratic, and that voluntary work became less valued as local authorities began to employ welfare workers.[104] This viewpoint probably underestimates the continuing importance of the voluntary roles played by women. Studies of maternal welfare indicate that voluntary work and female activism can still help to explain regional variations in services during the interwar period.[105] To a certain extent the same was true for mental deficiency.

[100] P. Hollis, *Ladies Elect: Women in English Local Government, 1865–1914* (Oxford, 1987). [101] Ibid. 466. [102] Ibid. 410, 464–5, 62.

[103] Ibid. 421. Recent historiography on gender and the welfare state has perhaps rather exaggerated the impact of maternalism on the development of British welfare. For an overview and criticism of this literature: J. Lewis, 'Gender, the Family, and Women's Agency in the Building of "Welfare States": the British Case', *Social History*, 19 (1994), 38–55. [104] Hollis, *Ladies Elect*, 422–3.

[105] R. Dingwall, 'Collectivism, Regionalism, and Feminism: Health Visiting and British Social Policy', *Journal of Social Policy*, 6 (1977), 291–315; L. Peretz, 'Maternity Services for England and Wales: Local Authority Maternity Care in the Inter-War period in Oxfordshire and Tottenham', in J. Garcia, R. Kilpatrick, and M. Richards, *The Politics of Maternity Care: Services for Childbearing Women in Twentieth-Century Britain* (Oxford, 1990), 30–46; L. Marks, 'Irish and Jewish Women's Experiences of Childbirth and Infant Care in East London 1870–1939: The Responses of Host Society and Immigrant Communities to Medical Welfare Needs', unpub. D.Phil. thesis (Oxford, 1990).

Here, the limitations of the local administrative apparatus were often overcome by employing female-dominated voluntary groups to provide services, and by co-opting women to serve voluntarily on local government committees. Voluntary group activity is undoubtedly an important factor in explaining local variations in provision. Unfortunately, the Board of Control statistics for levels of community care provision only began in 1937, too late to highlight the impact of pioneering voluntary work, and by which time many voluntary groups had been absorbed by the local authorities.[106] Significantly, there had been active voluntary groups in nearly all of the county boroughs with the best mental deficiency services. The high number of defectives 'reported' (and probably under some kind of informal 'voluntary supervision'), rather than officially 'ascertained', demonstrates the considerable extent of voluntary work which was still being undertaken outside the official remit of the Act.

A second potential strategy to overcome the geographical and political problems of administration was to place mental deficiency work within local authority public health departments so as to utilize the existing staff of health visitors. In 1913, MOHs had demanded a central role in the administration of mental deficiency. The creation of new specialized local authority health committees like that for mental deficiency appeared to herald a break-up of public health.[107] MOHs argued that mental defectives were the vectors for a wide range of public health problems and therefore needed to be monitored by public health departments so that disease patterns could be traced back to their families and homes.[108] Similarly, School Medical Officers of Health defended their role in detecting defective children against the threat of psychologists, arguing that they were uniquely suited to this task because of their experience and knowledge of the normal child.[109] A dislike of the hereditarian basis of the Act may also have antagonized some MOHs. Certainly this appears to have been part of the reason for the opposition of

[106] A list of areas with voluntary associations can be collated from *CAMW Annual Report* lists. However, by 1937 many of these groups had been or were in the process of take-over by local authorities.

[107] D. Watkins, 'The English Revolution in Social Medicine', unpub. Ph.D. thesis (London, 1984); D. Porter and E. Fee, 'Public Health, Preventive Medicine and Professionalisation' in E. Fee and R. Acheson, *Education in Public Health* (Oxford, 1991), 38–40; D. Porter, 'Stratification and its Discontents: Professionalisation and Conflict in the British Public Health Service', in Fee and Acheson, *Education in Public Health*, 83–113.

[108] *Public Health*, 40 (1927), 286–90.

[109] *Public Health*, 37 (1924), 301–2.

leading MOHs Eichholz and Newsholme towards the proposals of the Royal Commission on the Feeble-Minded.[110] The voluntary groups, who had been involved in the alarmist campaign of 1908–13, and who now threatened to usurp the role of public health, were unsurprisingly viewed with initial suspicion by MOHs.[111] However, this was gradually overcome through the experience of co-operation. MOHs began to recognize that special knowledge was needed to deal with the problem of mental deficiency, and they began to attend training courses run by the CAMW.[112] Co-operation between the public health administration and voluntary groups produced more effective services, and was particularly important if a successful system of ascertainment of defective school-leavers was to be established.[113]

The local politics of education was also crucial in shaping the problem of mental deficiency. In some areas of the country, special schools were never successfully established. They were opposed as unnecessary, especially in areas where the population was too small or dispersed to justify a special school. Others objected on the grounds of expenditure. The fact that it cost more to educate children in special schools than in elementary schools seemed irrational and unfair when it was only the latter who would ever grow up to return society's investment.[114] Cheaper alternatives were sometimes adopted, such as sending low-grade defectives to occupation centres, which increased from 57 in 1924 to 191 in 1939. There was an increasing interest in setting up special classes within elementary schools. This option had the support of many parents and teachers who wanted to avoid the stigma of special school education.[115] School Medical Officers also began to support integration, particularly since it maintained and extended their role in mental

[110] A. Eichholz, *Roy. Com. Feeb.*, i. 205–11.

[111] *Public Health*, 37 (1924), 302.

[112] M. Rooff, *Voluntary Societies and Social Policy* (London, 1957), 129–30.

[113] In Birmingham, statutory–voluntary co-operation dating back to the late-nineteenth century had established one of the best after-care services in the country: PRO MH 51/726.

[114] At the Board of Education it was calculated that if all defective children were dealt with in special schools it would cost £17m./year as compared to £1.3m. actually being spent in 1925: PRO Ed 24/1199, A. H. Wood to Sec., 23 Oct. 1925. Such concern precipitated the *Wood Report*, partly as a delaying tactic to put off any action: J. Welshman, 'The School Medical Services in England and Wales, 1907–1939', unpub. D.Phil. thesis (Oxford, 1988), 274–90.

[115] It was especially difficult to provide special schools in rural areas due to the scattered population and parental dislike of residential schools: PRO MH 51/706 Somerset CC to L. Brock. N. Rose, *The Psychological Complex: Psychology, Politics, and Society in England, 1869–1939* (London, 1985), 132–8.

testing. By the 1920s, they were beginning to acknowledge that the attempt to train the 'ineducable' had not proven a success. Instead, they called for a reorientation towards the dull and backward, who were neglected in the current system and who might really benefit from special education.[116]

With special education in such flux, there was a considerable degree of variation between local services. As a result, the very definition of mental deficiency could vary between areas, shaped by the local complexion of professional and administrative forces.[117] In London, mental testing was developed very early on, stimulated by the presence of voluntary groups and psychological experts within the public health service.[118] This led to a well-developed special-school system which had already reached over 8,000 places by 1915.[119] The head of this system, Dr James Kerr, protected special schools and opposed placing defective children under the 1913 Act.[120] The relative abundance of special school places made it possible to accept children of a higher measure of intelligence than in other areas of the country: only 8 per cent of London special-school-leavers were deemed unemployable, compared to 20 per cent in Birmingham.[121] As the balance of professional forces in London shifted, there was a further administrative reshaping of the problem. With the development of child-guidance services in the capital, problem children were less likely to be labelled as mentally defective and were often reclassified as suffering from emotional problems.[122] A more important trend was the decline in general school rolls, with the school age population falling from 1.03 per cent of the capital's population in 1923 to 0.72 per cent in 1934. In

[116] R. Hughes, 'Educational Provision for Mentally Defective Children', *The Medical Officer*, 34 (1925), 225; *The Health of the School Child for 1925: Annual Report of the Chief Medical Officer of the Board of Education*, 85–91; *The Health of the School Child for 1933*, p. 144; PRO MH 51/374.

[117] G. Sutherland, *Ability, Merit, Measurement: Mental Testing and English Education, 1880–1940* (Oxford, 1984), 60–3.

[118] S. Sharp and G. Sutherland, 'The Fust Official Psychologist in the Wuurld', *History of Science*, 18 (1980), 181–208.

[119] *LCC Statistical Abstract 1915–20*, 26 (1921), 152–3.

[120] GLRO PH/GEN/3/11.

[121] *CAMW Conference Report 1926*, pp. 46–7; *LCC Report of the County Medical Officer and School Medical Officer for 1919* (London, 1920), 377–8.

[122] *LCC Annual Report 1936*, 60–4; *LCC Annual Report of the MOH 1959*, app. B, 'Psychological Services for Children in London from Burt to Underwood'.

such circumstances, elementary school teachers became less enthusiastic about sending children away to mental deficiency homes or special schools.[123] Because of London's huge population, it was feasible instead to create schools which were large enough to integrate the educationally backward through special classes and streaming.[124] Thus local administrative factors, which earlier had pushed London into leadership of special-school development, were directing London away from this policy by the 1930s. As a consequence, special-school rolls fell from over 8,000 in 1916, to only 2,611 in 1938.[125]

Levels of provision were also affected by the role of the local poor law authorities. Since local authorities were so slow to build specialized mental deficiency institutions, many defectives were placed in the poor law accommodation which was made available by the move away from indoor relief.[126] In some areas, local voluntary workers provided a visiting service to these institutions, and in doing so they often created a unique bridge between poor law and council services.[127] Elsewhere, co-operation was hampered by the resentment of poor law medical officers at the intrusion of outside experts into their regime. In many areas there was little co-ordination; defectives were held temporarily in poor law institutions while in need of support, and then discharged into the community without notifying the local mental deficiency administration.[128] The greatest concern centred on the poor law being used to support feeble-minded women as they brought another generation of defectives into the world.[129]

The system of care in interwar London provides an example of the advantages of close co-operation between poor law and mental deficiency services, but it also reveals some of the tensions between the two. Before 1913, the Metropolitan Asylums Board (MAB) had already

[123] *LCC Annual Report 1936*, vi. 42; GLRO EO/SS/1/6 Inter-Departmental Conference on the Ascertainment and Certification of MD Children, 28 May 1937; EO/SS/1/4, p. 2.

[124] GLRO EO/SS/1/5.

[125] *LCC Statistical Abstract (1915–20)*, 26 (1921), 251; C. Gibbon and W. H. Bell, *A History of the London County Council, 1889–1939* (London, 1939), 275.

[126] Board of Control, *The Provision of Mental Hospital Accommodation* (London, 1925), 30–48; A. Crowther, *The Workhouse System, 1834–1929: The History of an English Social Institution* (London, 1981), 90–1.

[127] M. Jeffries, 'Mental Defectives Under the Poor Law', *Studies in Mental Inefficiency*, 4 (1923), 35–8. 'The Mentally Defective and the Poor Law', *Mental Welfare*, 5 (1924), 87–90.

[128] *BOC 1919*, pp. 79–83. The Board of Control was concerned at the lack of training among poor law officers to recognize mental deficiency.

[129] *BOC 1922*, pp. 41–3.

developed a series of specialized poor law institutions for the mentally defective.[130] The MAB resented the challenge of the LCC and defended its service as more specialized, better managed, and more democratic, since it was in closer touch with the population of poor law unions than the centralized LCC.[131] The MAB was allowed to maintain its own system for mental defectives, continuing to deal with pre-1913 cases under the far simpler regulations of the poor law. The unique MAB system provided specialized institutions for different types of defective according to age and degree of mental defect: colonies were provided for the aged at Tooting Bec, the very young at the Fountain, the low-grade at Leavesden and Caterham, and the able-bodied and high-grade at Darenth, with patients transferred between colonies as their status altered. This system made management easier than in the mixed institutions envisaged under the 1913 Act, but it was only possible in an administrative area as large as London.[132] The work of the MAB until its demise in 1930 greatly alleviated the LCC's task, enabling the Council to concentrate its own resources on providing a high level of community care services and special education. The LCC only needed to build one colony of its own and was able to experiment with a more communal regime since it took only high-grade defectives.[133] With the most seriously handicapped already dealt with by the MAB, the LCC could afford to send morally suspect defectives away from the perceived corruption of the city to private homes in the countryside, often run by charities or religious denominations.[134] This mixed system of poor law, local authority, and private institutions worked effectively because the LCC was able to develop a sophisticated community care and public health system to co-ordinate, integrate, and organize transfers within the complex network of care.[135]

In the country as a whole, lack of co-ordination in local government

[130] Ayers, *England's First State Hospitals and the Metropolitan Asylums Board*, 169–78; Thomson, 'Social Policy and Mental Deficiency in Interwar London'.

[131] Revd Sprankling, 'The Work of the Metropolitan Asylums Board for the Mentally Defective', *CAMW Conference Report 1920*, pp. 83–93; *The Times* (24 Feb. 1908); *Daily Chronicle* (24 Feb. 1908); *Local Government Journal* (20 Mar. 1908); GLRO LCC/MIN/753, Sub. Com. on the MD Bill, 22 Apr. 1913. On the issues of centralization: W. Robson, *The Government and Misgovernment of London* (London, 1934).

[132] See Ch. 3; Ayers, *England's First State Hospitals*, 139–54; *MAB Annual Report 1929–30*, pp. 24–5.

[133] 'The Manor Institution, Epsom: Some Comments on Its First Two Years', *Studies in Mental Inefficiency*, 5 (1924), 25–32.

[134] GLRO LCC/MIN/724 MD Act Sub. Com., 7 May 1918.

[135] See Ch. 7.

services was addressed by administrative rationalization in the 1930s.[136] Because this process was uneven, it introduced a further geographic variable in local provision. Rationalization remedied some of the chaos in mental deficiency administration, but it also undermined the independence of this service. The first threat came from changes to the system of local government funding. Until 1929 local services had been supported by a matching grant from central government, but the Local Government Act of that year introduced a new system in which local authorities distributed funds internally from an overall block grant. This left low-priority services, such as mental deficiency, vulnerable to redistribution of their share of the grant.[137] The block grant encouraged local authorities to look at their services as a whole, and many of them recognized the economic and administrative advantages of rationalizing and combining services. This process was encouraged by the closure of the poor law system and the take-over of its services by other arms of local government. The integration into public health departments of poor law institutions, many of which housed mental defectives, strengthened the case for taking control of all mental deficiency services. As Dr Ash, MOH of Derbyshire, argued:

> any relation between mental deficiency, lunacy, and public health is of the closest nature and that any attempt at separation will be generally inadvisable. There are those who advocate specialisation to such an extent as to dismember the body medical into its component parts. They must not be surprised if the result is a collection of inert lifeless members of what should be a living whole. The result would be a disintegration into component parts each of which will be on a par with the bone setter and herbalist.[138]

By the end of the 1930s the independence of mental deficiency administration was also being challenged by calls to amalgamate mental deficiency with mental illness services under a single mental medical officer of health.[139]

In summary, mental deficiency administration, which had been one of the earliest experiments in medical specialization in local government,

[136] 'The Organisation and Correlation of Mental Health Services in Local Areas', *Mental Hygiene* 2 (1936), 47–73; M. Buck, 'Co-operation in Mental Health Administration', *Mental Welfare*, 17 (1936), 104–8.

[137] Miller, 'Mental Deficiency and the Local Government Bill', 29–40.

[138] Debate on 'Mental Deficiency and Lunacy in Relation to Public Health', *Public Health* (August 1930), Dr Ash, 345, 347–8.

[139] L. Brock, 'Towards a Unified Mental Health Service', *Mental Welfare*, 20 (1939), 93–7.

was being integrated into the rapidly expanding departments of public health by the 1930s. This rationalization had advantages in terms of the efficiency of local government services as a whole. Moreover, the shift of power away from voluntary groups and co-opted mental deficiency committees placed services more firmly under the control of the elected local government. However, those involved in mental deficiency work also recognized that the shift entailed a dilution of the activism and participatory democracy which had been so essential to the early development and innovation of these services, and that it left mental deficiency services in danger of neglect, as local government concentrated its resources on higher priority issues.

CONCLUSION

In reality, there was no single 'problem of mental deficiency'; rather it was a problem which varied according to the geography of social and economic conditions, the administrative difficulties of establishing a local administration, and the local resources to provide services. However, like many other social questions, the problem was nationalized by surveys and national political debate. It was only through making the problem a national one that central government could be forced to make implementation compulsory and urgent. In the process, geographical variations in needs and means were generally ignored, despite the fact that the implementation of policy remained largely in the hands of local government. The dilemma of how the increasingly centralized government system was to deal with severe geographical variations in means and needs reflected a more general failure of interwar British social policy. Those who might otherwise have emphasized the socio-economic roots of the problem were cautious to do so, since this ran the danger of pathologizing the poor as a 'social problem group'.

Even though every local authority in the country was expected to provide an adequate level of institutional and community care for mental defectives, considerable variation persisted. This was related to local geographical problems of service provision, social and economic conditions, financial resources, and the availability of volunteers and professionals to supplement local government effort. There were trends in party positions, on what is generally seen as a 'non-political' issue, but these were complicated by additional interests arising from religion, gender, and professional background. These interests mediated the impact of party control on the levels of provision, obscuring any clear pattern. Provision was also affected by the internal non-party politics of local government. As a comparatively low status area within councils, mental deficiency provision was often shaped indirectly through the decisions and priorities of more powerful committees and services. As often as not mental deficiency services were hit by the drive to economize: in areas with severe social problems, mental deficiency services were likely to be sacrificed because of the demands of higher-profile issues such as unemployment; in others, they came under constant pressure of being swallowed up by more powerful services in a process of economic and organizational rationalization; and it was the struggle to cut costs, as much as attitudes towards eugenics, which lay behind the widespread local government support for sterilization. An effective mental deficiency service depended, not only on the internal politics of the council, but also on a series of political negotiations with geographically adjoining authorities and statutory agencies, and with local non-governmental voluntary and professional groups. Indeed, the presence of active voluntarists and professionals was often the key to successful local provision.

The 'mixed economy of welfare' gave local authorities the flexibility to experiment with innovative forms of service delivery.[140] The combination of local government agencies and voluntary groups sometimes resulted in a surprisingly comprehensive network of care. However, in many areas of the country the opposite was the case—the mixed economy resulting in chaos, confusion, and inefficiency. The process of rationalization in the 1930s began to erode these marked discrepancies, pushing local authorities up or down to a standard, but still fairly minimal, level of provision—a level which paid less attention than ever to variations in local needs or means. In shifting power from the participatory democracy of local volunteers, co-opted committees, and professionals, to integration within the more bureaucratic but electorally accountable local government public health departments, there was a greater prospect of getting to grips with the socio-economic problems which made mental deficiency into a problem. However, there was also the danger of relative neglect

[140] For use of this term see Lewis, 'Gender, the Family, and Women's Agency', 39–41.

alongside issues which were considered more important by the local electorate. The beneficial and negative portents for the drive towards rationalization and regionalization under the NHS were therefore already in place.

7
The Micro-Politics of Mental Deficiency

INTRODUCTION

Although the early twentieth-century was the first period of mass democracy in Britain, many of the social policies of this period have been portrayed as tools of 'social control', which enabled the state to intervene in and control the private lives of citizens in an unprecedented manner.[1] Most historians have come to view 'social control' as a clumsy concept which ignores the complexity of motivations and ideas behind social policy.[2] In emphasizing the imposition of hegemonic power from above, social control explanations too often neglect the process of negotiation between providers and recipients and the modifications of policy which resulted.[3] Because of these shortcomings it seems preferable to avoid using this term. However, it is impossible to ignore the theme altogether, for contemporaries consciously designed policy to regulate and control the (social) behaviour of mental defectives in the community; quite openly, and, with no sense of its subsequent malevolent connotations, they even used the phrase 'social control' as a technical term to describe what they saw as a necessary and inevitable dimension of policy. Through examining policy at its point of delivery, I shall explore in this chapter the extent and nature of any control which

[1] For instance, M. Langan and B. Schwarz, *Crises in the British State, 1880–1930* (London, 1985). Social control has become a dominant explanation for the motivation behind welfare provision in studies such as A. P. Donajgrodzki (ed.), *Social Control in Nineteenth-Century Britain* (London, 1977).

[2] F. M. L. Thompson, 'Social Control in Nineteenth-Century Britain', *Economic History Review*, 2nd ser., 34 (1981), 189–208; G. S. Jones, 'Working-Class Culture and Working-Class Politics in London, 1870–1900: Notes on the Remaking of the Working Class', *Journal of Social History*, 7 (1974); A. Reid, *Social Classes and Social Relations in Britain, 1850–1914* (London, 1992), 43–9.

[3] L. Gordon, *Heroes of Their Own Lives: The Politics and History of Family Violence* (London, 1989); and P. Mandler (ed.), *The Uses of Charity: The Poor on Relief in the Nineteenth-Century Metropolis* (Philadelphia, 1990), especially: C. Lis and H. Soly, '"Total Institutions" and the Survival Strategies of the Labouring Poor in Antwerp', 38–67; and L. Lees, 'The Survival of the Unfit: Welfare Policies and Family Maintenance in Nineteenth Century London', 68–91.

was imposed on mental defectives and their families, but I shall also examine the way that control was informed by a complex set of ideas and motivations, was integrally linked with care, and was implemented through a process of negotiation between providers and recipients which reflected shared values between the two. As such, I hope to point towards a history of what social control meant to contemporaries themselves.

Once again, a central theme will be the relation between the politics of mental deficiency and the process of adjusting to democracy. In the nineteenth century there had been two characteristic modes of welfare provision: first, that of the poor law, where welfare was minimal, carried the stigma of 'pauperism' and the penalty of loss of political rights, and was therefore usually turned to only as a last resort; and secondly, philanthropy, where welfare depended on benevolence, rather than right, and confirmed the gulf in social status between provider and recipient. I shall examine here the extent to which the implementation of the Mental Deficiency Act perpetuated these modes of welfare or was associated with a more democratic and scientific style of provision, in which social policy was administered according to rational principles rather than moral attitudes, welfare began to be seen as a social right, and recipients had more power to chose and modify services as 'consumers'.[4] In the first half of the chapter, I shall consider whether implementation of the Act was characterized by a new, more scientific, rational, objective, and democratic style of administration; or whether, despite these new tools of administration (and perhaps assisted by them), policy continued to be manipulated according to the moral and social biases of the welfare providers. In the second half of the chapter, I shall examine the extent to which the process of adjusting to democracy affected the relationship between welfare providers and recipients: whether it increased the influence of mental defectives, their families, and communities in modifying policies. In analysing this micro-political interaction of the family, community, and state, I shall demonstrate that the political fault lines of gender, class, and professional expertise stretched down from the highest to the lowest political level. But perhaps most importantly, I shall show that for a full under-

[4] A significant body of literature on early twentieth-century welfare now suggests that nineteenth-century attitudes persisted. An important element of this was the working-class ambivalence towards accepting welfare or the intervention of the state into their lives: H. Pelling, 'The Working Class and the Welfare State', in id., *Popular Politics and Society in Late-Victorian Britain* (London, 1968); P. Thane, 'The Labour Party and State Welfare', in K. D. Brown (ed.) *The First Labour Party* (Beckenham, 1985).

standing of social policy we cannot confine our analysis to high-political intentions; we also need to analyse the gulf between intention and reality—how policy worked in practice.

I shall utilize a series of fifty-five detailed LCC case-history files.[5] The cases all begin in the early years of the Act, but are made more representative since many extend over a forty-year period until the closure of the files in 1959. Detailed notes on the method of selecting cases for preservation do not remain but it appears to have been random, the cases selected according to surname at regular intervals of the alphabet. Because of the small size of the sample, any quantitative results drawn from the material will be viewed tentatively, and wherever possible they will be backed up by statistics from the local and central authorities. What is more valuable about the case histories is the qualitative evidence which they provide. Indeed, the case files show that quantitative 'snapshots' can be misleading, for the status of cases was highly fluid, with considerable movement between institutional and community forms of care.[6] The qualitative data provides important and rare insights into issues such as the attitudes of social workers, defectives, and families, data which can easily be blunted by quantification. Therefore the main approach of the chapter will be to reconstruct the life stories of defectives and their families from the fragmentary records of case files; an approach which will entail describing in some detail the voices of the historical actors.[7]

THE TARGETING OF THE ADMINISTRATION

Defining Mental Deficiency

Protection of the liberties of the individual had been at the heart of nineteenth-century lunacy legislation and continued to be a prominent

[5] Reference will be made to file number, but not usually to individual documents, since histories are reconstructed from whole files. The files come from PH/MENT/2/10–12. They will be cited according to case number.

[6] The disadvantage of institutional studies is that they tend to hide the interaction of institutional care and community and family care: M. Ignatieff, 'Total Institutions and the Working Classes', *History Workshop Journal*, 15 (1983), 169–70.

[7] Despite a much larger sample of cases, Gordon acknowledges the quantitative problems and concentrates upon a qualitative approach: *Heroes of their Own Lives*, 18.

theme in the drafting of the Mental Deficiency Act. However, in practice, the complexity of the rules and regulations of the Act, and the ambiguity over the definition of mental deficiency, placed considerable influence in the hands of those who interpreted and implemented policy at the ground level. This was particularly the case in the process of certification.

Detection of mental defectives was at the forefront of early developments in mental testing.[8] There has been a tendency to caricature the early mental tests as blatant instruments of bias, primarily designed to justify social inequality. Cyril Burt's apparently deliberate falsification of evidence on inheritance of intelligence among identical twins is regarded as a damning symbol of the spurious nature and ideological intent of the mental test.[9] Undoubtedly, the vogue for testing did reflect broader support for a new meritocratic social order, and in this sense it was ideological; though in presenting an alternative to an elitism based on inherited privilege, it was a progressive vision in certain respects.[10] Accurate mental assessment was clearly distorted by the personal biases of examiners, environmental inequalities, and a driving professional ambition to prove the efficacy of the tests. However, to caricature and dismiss the tests as wholly spurious is unfair and does little to advance our understanding of them. In fact, contemporary psychologists, such as Burt, acknowledged and attempted to remedy many of the problems which have been mentioned. They designed complex systems of mental assessment, which combined educational, intelligence, social, and physical tests. Neither physical stigmata nor an individual's mental ratio (the term preferred in Britain to that of 'intelligence quotient' or IQ) were viewed as convincing evidence on their own; in fact, the value of both tended to be viewed with a fair degree of suspicion.

The relationship between mental testing and mental deficiency cannot be understood without an appreciation of the complex model of intelligence which prevailed in the interwar years. Certifying Officer to the Lancashire Mental Deficiency Authority, Henry Herd, in his 1930 manual on diagnosing mental deficiency, described mental ability as a composite of three mental qualities—temperament, intelligence, and conation.[11] Just as

[8] G. Sutherland, *Ability, Merit and Measurement: Mental Testing and English Education, 1880–1940* (Oxford, 1984).

[9] R. B. Joynson, *The Burt Affair* (London, 1989).

[10] A. Wooldridge, *Measuring the Mind: Education and Psychology in Britain, c.1860–1990* (Cambridge, 1994).

[11] H. Herd, *The Diagnosis of Mental Deficiency* (London, 1930). The description of certifcation which follows is based on Herd's book but is representative of mainstream ideas.

mental deficiency could result from a defect of intelligence, it could also result from a temperamental or conative defect. This made the assessment of mental defect a highly complex process. Intelligence, itself, was a compound of the qualities of sensation, perception, memory, conception, judgement, and reasoning, each of which needed to be tested.[12] Temperamental defect was assessed through behaviour: in low-grade defectives it manifested itself in a virtual absence of instinctive drives; in high-grade defectives it was revealed by inability to control instinctive drives and susceptibility to the influence of others.[13] In Herd's view, the most important mental process of all was 'conation', in particular the quality of 'attention'—a quality which leading psychologists Charles Spearman and Cyril Burt regarded as the 'general factor' which was common to all other mental abilities. Conation was particularly important in testing for mental deficiency because it was the basis of all social and practical skills. The multifactorial view of intelligence meant that it was just as important to assess attention span and temperament as it was to assess mental ratio.[14] So, although guidelines could be set down as to the normal mental ratio range of each grade of mental defect, there would be exceptions when the defect lay in temperamental or conative development; individuals with 'normal' mental ratios might still, therefore, be certifiable as mentally defective.[15]

Although tests for mental deficiency were designed to assess the wide variety of elements which constituted mental ability, there was invariably an emphasis on those which led to social competence. The most popular test in Britain was Burt's modification of American psychologist Lewis Terman's age-graded mental test. This itself was a modification of the pioneering mental tests developed by Alfred Binet in France during the first decade of the century (the process of translation reflecting a degree of sensitivity to problems of cultural difference).[16] Since knowledge-based questions were recognized as discriminating according to social and educational background, the tests aimed to keep them to a minimum. Instead, they favoured types of knowledge

[12] H. Herd, *The Diagnosis of Mental Deficiency* (London, 1930). The description of certifcation which follows is based on Herd's book but is representative of mainstream ideas, 24–35, 70–3.

[13] Ibid. 17–24, 65–70.

[14] Ibid. 35–40, 74–5.

[15] Ibid. 170–4. The normal mental ratio range was: 25 and below for idiots; 25–45 for imbeciles; 50–75 for the feeble-minded. The approximate borderline was the intellectual development of a 12-year-old.

[16] S. Sharp and G. Sutherland, 'The Fust Official Psychologist in the Wuurld', *History of Science*, 18 (1980); Herd, *Diagnosis*, 107–32.

which were regarded as essential for social competence, such as counting money and understanding everyday objects. Care was taken in the phrasing of questions to ensure that correct answers could not arise from memory alone. The aim was to test reasoning—the ability most crucial for social competence. To further diminish the influence of social background, intelligence tests were complemented by non-knowledge-based performance tests. An elaborate new technology of portable tools, games, and machines was designed by psychologists and put into operation to test these skills. Fairly typical and common was the Porteus Maze Test, which measured ability to trace a route through a maze according to a series of increasingly complex instructions. Others tested ability to imitate operations with cubes, to fit wooden blocks into holes, and to judge the relative size and weight of objects.[17] Technical assessment of temperamental qualities was more difficult. The Psycho-Galvanic Reflex Test attempted to measure emotional reactions through registering changes in the body's electrical conductivity in response to moral and ethical questions. However, the efficacy of this test was disputed, and the psycho-galvanic apparatus was too technical and cumbersome for adoption on a wide scale.[18] Instead, temperamental assessment rested heavily on the evidence of behaviour from schools and other such agencies. Finally, scholastic assessment, duly modified to take into account the relative educational opportunities of the child, was used both as a measure of intelligence and as an important indicator of a child's ability to succeed in a social situation.[19] In summary, despite the elaborate paraphernalia of the testing apparatus, mental diagnosis of defectives usually rested on a highly subjective and practical judgement: whether they were socially competent or not.

Contemporary models of mental ability also justified assessment of moral behaviour within mental diagnosis.[20] Under the 1913 Act the distinct category of moral imbecility had been established for persons showing 'mental defectiveness coupled with strongly vicious or criminal propensities'. However, the interpretation of this definition caused much controversy. It was unclear whether moral imbeciles were simply people who lacked an inherent moral sense (if such a thing existed, which was heatedly contested), or whether moral inadequacy had to be coupled with, and merely reflected, an underlying intellectual defect—perhaps even imbecility, as the term itself implied. The Mental Defi-

[17] Herd, *Diagnosis*, 135–51. [18] Ibid. 151. [19] Ibid. 154–64.
[20] Ibid. 193–207.

ciency Amendment Act of 1927 attempted to make the meaning clearer, replacing the classification of 'moral imbecile' with that of 'moral defective'. However, it remained unclear whether moral failure alone was sufficient evidence to certify an individual as mentally defective. Because of this confusion the moral imbecile and moral defective labels were the rarest used classifications under the Act.[21] Far more important was the integral part played by moral assessment in certification of all types of mental deficiency. Since individual control over moral behaviour was believed to be a direct reflection of temperamental and therefore mental development, persistent moral misdemeanour was regarded as a likely sign of mental defect. In Herd's view it was wrong to make a sharp distinction between mental defect and moral failure: the latter was invariably evidence of the former; temperamental defects were just as 'mental' as defects of intelligence.[22] On the basis of such reasoning, it was considered perfectly legitimate, even essential, to include assessment of moral behaviour in the process of mental diagnosis.

It is difficult to establish the extent to which the guidelines on testing laid down in manuals such as Herd's were followed in practice. Since there was little or no special training in testing in the earliest years of the Act, it seems likely that standards of assessment would have varied widely, and that common-sense notions of what constituted mental defect and social competence would have guided action more than expert knowledge. The evidence of surviving exam reports suggests that doctors openly compounded assessment of intellectual capacity with evaluation of behaviour, social status, and moral character. Behavioural traits, such as 'uncleanliness' and 'dirty habits', were frequently cited as evidence of mental deficiency, in part because this indicated possible intellectual problems, but equally because such behaviour was considered to be a sign of social incompetence and moral inadequacy. By the 1930s local authority medical officers had much more experience and were also now receiving training on courses organized by the CAMW. However, as Herd's manual indicates, assessment based on social and moral attitudes found new legitimation as it was codified into mental tests and embedded into a view of intelligence which prioritized social and moral competence.

Social, moral, and eugenic criteria were also integral to the process of assessing the home conditions of defectives. The official enquiry forms

[21] S. Watson, 'The Moral Imbecile: A Study of the Relations Between Penal Practice and Psychiatric Knowledge of the Habitual Offender', unpub. Ph.D. thesis (Lancaster, 1988).

[22] Herd, *Diagnosis*, 206.

embedded such biases into the machinery of the Act, with questions on whether there was a history of insanity, mental deficiency, epilepsy, hysteria, neurasthenia, asthma, deaf mutism, habitual criminality, intemperance, drug habits, syphilis, or tuberculosis within the family. Such questions, along with others on size of family, age of the mother at birth, and consanguinity of the marriage, automatically targeted certification and institutionalization at defectives in the 'social problem group'.

Finally, although mental diagnosis was shaped by the values of the examiners, it was also modified by a process of negotiation with the examined. F. C. Shrubsall, leading medical expert on mental deficiency under the LCC, described the mental examination as a duel of wits. Since children and their parents were often hostile to the process of examination, he recommended that tests be disguised in general conversation, designed with an eye to convincing the parents themselves to acknowledge the defective condition of their children. Tactics were even more important when examining adolescents or adults. Shrubsall recommended starting with a physical exam, which 'produces a greater degree of confidence and cooperation, since it fits better with popular ideas of medical practitioners'.[23] For individuals who would not cooperate in a mental test, the recommended strategy was to induce them to recount their life stories. This would help to win their confidence, and it would reveal evidence of any social incompetence or moral weakness. Even if these traits were concealed, the skill of the doctor would enable him to see through inaccuracies and inconsistencies. The life story was also important in helping to trace the individual's school records—records which were generally vital in order to fulfil the statutory requirement of proving the existence of a mental defect 'from birth or early age'. Tests which emphasized knowledge and education were deemed inappropriate for adults, who had probably forgotten much of their schooling. Instead, mental tests which laid greater stress on logic and performance of manual tasks (and thereby of social efficiency) were preferred. These 'duels of wit' suggest that the dominance of the social competence definition did not simply reflect the attitudes of those who designed mental tests, it also reflected popular attitudes towards proving mental incompetence.[24]

[23] F. Shrubsall and A. C. Williams, *Mental Deficiency Practice* (London, 1932), 70.

[24] On the importance of lay attitudes in defining mental deficiency: D. Wright, '"Childlike in his Innocence": Lay Attitudes to "Idiots" and "Imbeciles" in Victorian England', in D. Wright and A. Digby (eds.), *From Idiocy to Mental Deficiency: Historical Perspectives on People with Learning Difficulties* (London, 1996), 118–33.

In summary, because of contemporary understanding of the nature of intelligence, and through the machinery for detecting mental defect, the Mental Deficiency Act automatically targeted individuals who were the subject of moral, social, or eugenic concern. In practice, therefore, the Act provided far more than a system of specialized care for the mentally disabled and retarded: it also provided an administrative machinery whose influence had the potential to regulate the moral, social, and eugenic boundaries of good citizenship.

Distribution of Defectives by Age, Sex, and Grade

The obvious way to trace the impact of social, moral, and eugenic biases in assessment would be through a statistical breakdown of certified mental defectives according to age, sex, and social background. Unfortunately, national statistics for mental deficiency were not broken down in this way until the 1950s. The only interwar, sex-differentiated, national figures were those in the census, but this was a combined total for the population in mental deficiency and mental illness institutions. At first sight these figures appear to confirm the idea that mental illness was a particularly 'female malady', but as soon as they are disaggregated this crude stereotype begins to disintegrate.[25] For instance, the 1931 census showed that although women outnumbered men by 54 per cent to 46 per cent, men were in a majority between adolescence and age 39.[26] The first clear statistical breakdown for mental defectives alone was that provided by the Registrar General's Office from 1949 onwards. In contrast to figures for mental illness hospitals, these statistics showed that there were more men than women in mental deficiency institutions. The male majority was greatest in the age group 16–25, suggesting a possible link to delinquent and criminal behaviour which was a particular problem among men of this age group. Only above the age of 45 did female defectives outnumber males.[27] The predominance of women in the older age groups was partly a reflection of their longer life expectancy. It also reflected the fact that female defectives were being

[25] E. Showalter, *The Female Malady: Women, Madness, and English Culture*, 1830–1980 (London, 1987) 52–3. For a critique of Showalter's view: J. Busfield, 'The Female Malady? Men, Women and Madness in Nineteenth Century Britain', *Sociology*, 28 (1994), 259–77.

[26] *Census of England and Wales 1931, General Report* (London, 1931), 62–3, tables xxix, xxx.

[27] *Registrar General's Statistical Review of England and Wales 1950–51* (London, 1955), 130, 138–9.

kept in institutions for longer than men: their median residence was 9 years and 7 months, compared to 7 years and 9 months for males; and they outnumbered males by 273 to 210 among those resident in an institution for over ten years.[28]

The Registrar General's data on the social-class background of institutionalized mental defectives, from the highest group I, to the lowest V, suggests a series of correlations between class, gender, and mental deficiency.[29] These statistics enable comparison between idiot and imbecile cases—where assessment of intellectual and social difficulties was more clear-cut—and feeble-minded cases—where social and moral judgements were likely to be more significant in the decision to use institutional care. Institutionalization of male defectives was proportionately most likely among families in the lowest social groups, especially in feeble-minded cases. Among the lower grades of mental defect, the affect of low social status was less marked. In fact, a high proportion of idiots and imbeciles came from professional families. This suggests that by the end of the 1940s professional families were willing to use institutions (though probably private ones) if they had seriously handicapped family members. The statistics for female mental defectives showed similar general trends, except that extremes were less marked at either end of the social scale.

In the sample of fifty-five London cases there were thirty-one females and twenty-four males. A mental certificate recording the grade of mental deficiency survives for forty-seven of the cases: two moral imbeciles, twenty feeble-minded cases, twenty-four imbeciles, and one idiot. The sample contains a disproportionate number of feeble-minded females—fifteen compared to just five males. The two moral imbeciles (probably also high-grade in intelligence) were also females. By contrast, male imbecile (lower intelligence) cases outnumbered females by fourteen to ten. The unequal distribution between grade of deficiency and gender was paralleled by gender differences in the pattern of age at certification: under the age of 9 there was an approximately equal proportion of boys and girls; between the ages of 10 and 17, boys outnumbered girls by thirteen to eight; while in the 18 to 25 age group the balance was reversed, with women now outnumbering men by three

[28] *Registrar General's Statistical Review of England and Wales 1950–51* (London, 1955), 188.

[29] The model of classes I–V was adopted in the 1911 census and is arranged as follows: I. Professional occupations; II. Intermediate occupations: III. Skilled occupations; IV. Partly skilled occupations; V. Unskilled occupations.

to thirteen. Although the numbers involved in this sample are far too small to draw any general conclusions, they suggest significant differences between the standard case history of male and female defectives. This can be explored further through a detailed examination of the case files.

Defective Girls and Sexual Control

In the London sample, eleven of the thirteen female cases certified at age 18 or older were linked to sexual concerns: seven of these were feeble-minded, three were imbeciles, and one was a moral imbecile.[30] In at least five of the cases, the evidence suggests that mental condition was not so serious as to prevent the women from working to support themselves in the outside community. In the other cases, where some degree of mental handicap was more obviously present, sexual concern was still an important factor in the decision to send the women to institutions.

Elizabeth D. was notified from special school and placed under the supervision of the local authorities in 1915, aged 15. The local authority expressed concern over her case, reporting that she was easily led and 'would go with anyone'. She had wandered off with a man recently and only been retrieved by a suspicious onlooker. She was said to run wild in the streets unless bolted in by her parents, in which case she would stand at her window, shout at passers-by, and throw her clothes and furniture on to the street. In 1916, Southwark Diocesan Association for Rescue and Preventive Work complained that Elizabeth had been involved in immoral behaviour with other children during the holiday. This convinced the local authorities that she needed institutional care. She was not discharged until 1959; the report of immoral behaviour in 1916 remained on her file, undermining her family's requests for discharge and leave.[31] A similar pattern is found in the case of Winifred E., born in 1907, and notified by the Local Education Authority as an imbecile in 1917. Winifred's initial supervision lasted six years, but ended after reports that she was 'exposing herself' and was showing 'decided leanings to the undesirable side of life'. In these two cases, both imbeciles, the existence of a mental disability is not being ques-

[30] Adelaide B., Francis C., Ada D., Emily E., Winifred E., Laura G., Florence L., Annie M., Ellen S., Winifred T., Winifred W.

[31] GLRO PH/MENT/2/10, Case 1412.

tioned, but what is being suggested is that concern to control the sexual behaviour of the women was crucial in the replacement of supervision by long-term institutional care.

Two further case histories will show that control of sexuality could also lead to long-term institutionalization of women whose mental defect was only minor, if it existed at all. Born in 1894, Ada D. only came to the attention of the authorities at the age of 20.[32] She had passed through school without being certified, and since then she had been able enough to gain paid employment. She only came to the notice of the authorities when she became pregnant and was forced to enter the workhouse, having no other form of support. The Westminster Infirmary doctor certified her as feeble-minded, reporting that she was 'dull and tells me she has had several situations and refused to work but does not know why. Appears morally defective.'[33] Shrubsall, the writer on mental deficiency and medical officer in charge of mental deficiency in London, enforced the first doctor's view by his own certificate which specifically noted that she could not add sums of money in shillings and pence, nor could she give differences between a dog and a cat, and she only had a reading age of about 9.[34] Despite the fact that Ada had demonstrated a substantial degree of social competence in finding employment as a shop assistant in Regent Street, at a lodging house, and as a labeller in a factory, she was sent to an institution for the next thirty years. On her release, she was again able to find gainful employment.[35]

Kathleen W., born in 1903, was notified at the age of fourteen after being excluded from special school as 'ineducable'.[36] Her parents were upset about this decision, and asked for the case to be reconsidered. This led to a second examination which certified Kathleen as a moral imbecile, unfit to be with other children and uncontrollable through punishment. Her head teacher reported that she had no concentration, and had been caught stealing, talking indecently, and eating food off the street.[37] The enquiry officer accused her of inventing 'stories [in offensive language] as to what boys have been doing to her on the Common, also of the behaviour of her father and step mother'.[38] The report concluded: 'Altogether the child seems a very serious source of moral danger to all

[32] GLRO PH/MENT/2/10, Case 1418.
[33] Ibid., J. Coulson, 17 Aug. 1916.
[34] GLRO PH/MENT/2/10 F. C. Shrubsall, 10 Aug. 1916.
[35] Ibid., Enquiry Form, 22 Dec. 1915.
[36] GLRO PH/MENT/2/12, Case 2112.
[37] Ibid., 23 July 1917.
[38] Ibid., 20 July 1917.

the children of the neighbourhood and a danger to herself also unless under adequate control.'[39] Since control had to be found for Kathleen during the forthcoming school holiday, institutionalization was deemed to be urgent. She, too, was not discharged until the policy review of 1959.

As these cases demonstrate, the Mental Deficiency Act could be used to control the sexuality of young women. The use of the Act to certify female paupers bearing illegitimate children has attracted considerable attention.[40] This practice helps to explain why a higher number of women than men were regularly ascertained via the poor law system.[41] However, the cases in the London sample suggest that the Act's influence in regulating sexual behaviour of women was far broader than this. These cases, all during the first years of the Act, need to be placed in the context of a major shift in the anxieties which surrounded female sexuality: a discourse which had constructed women as the weaker, purer sex, who had to be protected against moral vice, began to be overshadowed by a parallel discourse which configured the individual female as a potential danger to the community, a vector for the spread of immorality, venereal disease, and degeneration.[42] In line with such anxieties, the feeble-minded woman, who had originally been an object for pity and protection, was now increasingly thought of as a danger. Although venereal disease was never deemed to be evidence of mental defect, the continuation of behaviour which placed one in danger of contracting or spreading the disease could encourage such a diagnosis.[43] The Board of Control encouraged the idea there was a link with prostitution in its 1914 circular, directing local authorities to pay special attention to any defective who was 'a prostitute or lodges or resides with prostitutes or otherwise lives in circumstances calculated to cause, encourage or favour her seduction or prostitution'.[44]

The view that there was a direct relationship between mental deficiency and prostitution or venereal disease diminished over the interwar

[39] GLRO PH/MENT/2/12, Case 2112.

[40] The link is sometimes exaggerated, with suggestions that the act of having an illegitimate child was enough on its own and the main route of entry into mental deficiency institutions. For instance: S. Humphries, *A Secret World of Sex: Forbidden Fruit: The British Experience, 1900–1950* (London, 1988), 63–5.

[41] For instance, the poor law was the source of ascertainment for 707 women compared to just 376 men in 1922: *BOC 1922*.

[42] F. Mort, *Dangerous Sexualities: Medico-Moral Politics in England Since 1830* (London, 1987), 153–209; L. Bland, '"Guardians of the Race" or "Vampires upon the Nation's Health"? Female Sexuality and its Regulation in Early Twentieth-Century Britain', in E. Whitelegg et al. (eds.), *The Changing Experience of Women* (Oxford, 1982), 373–89.

[43] GLRO PH/MENT/1/5. [44] *BOC 1914*, p. 51.

period. However, in a modified guise, the feeble-minded woman continued to provide a focus for broader anxieties about female sexuality. For instance, although Cyril Burt now estimated that just 1.8 per cent of prostitutes were hereditary defectives, he claimed that a disproportionately high percentage of prostitutes (12.4 per cent) were dull and backward, and that these individuals were the tip of the iceberg of a broader problem of lack of control over sexual instincts and 'amateurism' at the bottom of the social scale.[45] Supposedly unable to control or protect themselves, feeble-minded, dull, backward, and borderline women came to symbolize the moral and eugenic dangers which appeared to threaten post-war Britain; because of their status they were also the most obvious target for control.

In fact, male defectives were more likely than women to be convicted of sex offences: of 306 criminally charged male defectives in 1931, seventy-three were sex offenders, including one rapist, thirty-two cases of indecent assault on girls and children, nineteen of indecent exposure, four of carnal knowledge with a child, sixteen of unnatural offences with another boy or animal, and one case of incest.[46] However, unless they were convicted of a crime, men were far less likely to be dealt with as defective simply because of their sexual behaviour. There was a typical sexual double standard at work here: defective women were regarded as defenceless, sexually desirable, and open to sexual exploitation, yet also sexually dangerous and uncontrollable; while male defectives were regarded as unmanly, ineffectual, and unable to attract female partners—only a danger if they used force. In reality, women were also the primary target for control because this was seen as the surest way to prevent the breeding of a future generation of mental defectives.

The Delinquent Male

Just as control of sexuality helps to explain the distribution of female cases in the London sample, so control of disruptive and delinquent behaviour helps to account for the fact that all the male cases under the age of 16 were imbeciles, whereas all those between 16 and 18 were feeble-minded. Up to the age of 16 the behaviour of feeble-minded boys

[45] C. Burt, 'The Contribution of Psychology to Social Hygiene, *Health and Empire*, 1 (1926), 25–33; id., 'The Causes of Sex Delinquency in Girls', *Health and Empire*, 1 (1926), 251–71. The work of Burt has been focused upon here as he was the official psychologist for the LCC at this time. See Sharp and Sutherland, 'Fust Official Psychologist', 181–208.

[46] *BOC 1931*, pp. 86–9.

could be controlled within the educational system or in reformatory or industrial schools. Thereafter, they moved outside the jurisdiction of these juvenile systems. The Mental Deficiency Act provided a way to extend control over some of these individuals without sending them into the adult penal system. Five of the six feeble-minded males were ascertained after coming before the judicial authorities for committing a crime; the remaining case had committed no crime, but was described as 'out of control'. By contrast, in the female sample just two cases arose through the penal authorities (and one of these was arrested on suspicion of prostitution, another example of concern over sexual control). National figures also suggest that the Act was being used to control disruptive young men. In 1992, for instance, twice as many boys as girls were notified as defective under the Children Act, and notifications of boys from schools was also substantially higher.[47] Even more striking are the national figures for ascertainments involving criminal offences between 1913 and 1927, in which male defectives outnumbered females by 2,984 to 770.[48] An analysis of several case histories will cast further light on the links between the mental deficiency and penal systems.

Born in 1901, Frederick B. was reported as being slow at school, though not to the extent that he was certified as a defective. However, in 1917 he was charged with a felony at Clerkenwell Court, and rather than being sent to a penal institution he was certified and sent to a mental deficiency institution.[49] In effect, the Mental Deficiency Act was used as a form of preventive control while Frederick's father was away on military service. In 1919, with his father having returned from the forces, and his uncle—a policeman—also living with the family, the authorities granted Frederick a discharge after only two years at the institution, and despite his poor record of behaviour, which included an attempted escape from the institution. This leniency is striking when it is compared to the cases of female defectives. The case history reads more like a penal than a medical one, with first the attempted escape, then the two year 'sentence', and then the 'probation' at home under the eyes of his police officer uncle and the local voluntary group.

A similar pattern is found in the case of William S. who was born in 1901.[50] Following delinquent behaviour, he was sent to Sandwell Hall industrial school where he was recognized as being a mental defective.

47 *BOC 1922*. 48 *BOC 1934*, p. 39.
49 GLRO PH/MENT/2/10, Case 1992.
50 GLRO PH/MENT/2/12, Case 1826.

On reaching the age of 16, he was sent to Marylebone Police Court to be charged for a criminal offence, but instead of being convicted he was certified and sent to a mental deficiency institution. In 1918, the official visitors to the institution suggested that he should be given a licence for temporary leave, with a view to discharge. The trial leave with his parents was extended, and three months later he was discharged. Five years later William was arrested by the Metropolitan Police. The police medical officer discovered William's previous history as a certified defective, and he was therefore sent back to an institution under the Mental Deficiency Act. After only one week he escaped, and he was free for six months before his recapture. Within three months, he had again escaped. By the time the authorities could locate him, his original one year order under the Mental Deficiency Act had expired, and he was therefore dealt with by sentence to prison instead. From there, he was ascertained as a defective by the prison medical officers and was sent to Rampton State Institution for dangerous and criminal defectives. Subsequently, he was discharged and sent, full circle, back to a local authority mental deficiency institution. This time, however, he was not discharged until 1946. If William had served his sentence in prison he would have been released years earlier. The diagnosis at industrial school had pushed him away from the penal system, but it is difficult to tell whether there was any benefit from being sent into the mental deficiency system; Frederick's own reactions suggest that he viewed the course taken to be still one of punishment, rather than care.

From these cases it is apparent that the mental deficiency system was being used alongside the penal system to control the behaviour of disruptive young men.[51] Social anxieties about this group of the population had a long pedigree, but in the first decades of the century a series of factors produced a peak of anxiety and associated delinquency with the problem of mental deficiency.[52] Theories of child development, epitomized by Stanley Hall's work on 'periodicity', problematized adolescence in medical terms: the adolescent was viewed as unstable, sapped by physical growth, not yet in control of his or her emotions and instincts, and therefore prone to delinquency if put under too much

[51] To a lesser extent this was also the case with young women: Pam Cox, 'Girls, Deficiency, and Delinquency', in Wright and Digby (eds.), *From Idiocy to Mental Deficiency* (London, 1996), 184–206.

[52] J. Gillis, *Youth and History: Traditions and Change in European Age Relations, 1770–Present* (London, 1974), ch. 4.

pressure or not kept under strict enough control.[53] The feeble-minded, with their poorly developed control of active instinctive urges, were of particular concern, for they were fixed in a permanent state of adolescent development.

Contemporary alarm over male juvenile delinquency was also associated with the 'boy labour problem'. Changes in the labour market appeared to be leaving young men stranded in 'dead-end jobs', or worse still, unemployed. Without the discipline and training of regular employment, they were exposed to the temptations of delinquency and crime.[54] Since feeble-minded adolescents lay on the very margin of the labour force, it is easy to see why contemporaries implicated them with the problems of boy labour and delinquency. During the First World War, the scarcity of labour made it easier for high-grade mental defectives to gain employment; however, this simply provoked a new anxiety that their earning power in conjunction with the wartime lack or absence of parental control would give defectives a dangerous degree of independence. Besides, their wartime employment was seen as a temporary phenomenon, to last only until the returning troops took back their old jobs.[55]

During the interwar years there was a shift towards viewing the liminal 'borderline' area, rather than mental deficiency as a whole, as the main source of delinquency. For instance, in Cyril Burt's work, certifiable mental deficiency was detected in only 7 per cent of juvenile delinquents, but 70 per cent were found to be dull and backward.[56] With delinquency interpreted as a developmental problem—merely an extreme example of 'childhood naughtiness'—the borderline of mental deficiency was an obvious site for concern, for although these individuals seemed normal they were inherently incapable of controlling their instinctual impulses and urges (their apparent normality made them doubly dangerous); lower grade defectives were less of concern, since it was quite apparent that they were simply not bright enough to become delinquents.[57] Although the dull and backward were marked out as a

[53] S. Hall, *Adolescence—Its Psychology* (New York, 1904). Hall's work was taken up in Britain by J. W. Slaughter, *The Adolescent* (London, 1911).

[54] E. O. Sandler, 'Working Class Adolescents in Birmingham: A Study in Social Reform, 1990–1914', unpub. D.Phil. thesis (Oxford, 1987); H. Hendrick, *Images of Youth: Age, Class, and Male Youth Problem*, 1880–1920 (Oxford, 1990); E. J. Urwick (ed.), *Studies of Boy Life in Our Cities* (London, 1914).

[55] Frederick B., Case 1992.

[56] C. Burt, 'Delinquency and Mental Defect', *British Journal of Medical Psychology*, 3 (1923), 169–78.

[57] C. Burt, *The Young Delinquent*, 4th edn. (London, 1945; first pub. 1925), p. viii.

problem group, they were paradoxically also coming to be viewed with increasing sympathy as victims who needed better guidance and control, rather than as maliciously and intentionally bad. Their delinquency was explained as a response to academic failure, bullying, and teasing at school, and as an alternative way to gain the acceptance of their peers. Because of their backwardness, they were highly impressionable and easily led into delinquency by brighter ringleaders. It was these bright but temperamentally unstable ringleaders—children who were less often caught than their slow-witted accomplices—who were considered to be the real danger group.[58] Thus the most important impact of the mental deficiency diagnosis in controlling delinquency was not so much that it provided a direct form of control, but that the threat of certification extended out to control behaviour in this amorphous 'borderline' area and beyond. As such, mental deficiency did contribute to a 'psychologization' of delinquency at its high-grade margin, but in many ways its more important and pervasive function was to provide the threat of certification as an ultimate sanction to control behaviour within a much broader section of the child and adolescent population.

The relationship between the mental deficiency and penal systems needs to be placed in the context of a broader specialization and 'medicalization' of penality.[59] This transition began with specialization and reform of the juvenile system: there was a shift from punishment and custody towards preventive policies, such as probation and education, and strict demarcation between the child and adult penal systems. Since mental defectives, like children, could not be held fully responsible for their actions, the 1913 Act was to replace punishment with specialized care and long-term prevention. Not only would this be more just and effective, but it also offered a way to alleviate pressure on the penal system. The juvenile system itself came under pressure as the wartime delinquency scare stretched its resources,[60] with the number of boys sent to Home Office Schools rising from 4,632 in 1913, to 5,738 by 1916.[61] Unfortunately, there is no easy way to establish accurately the

[58] G. Keir, 'A History of Child Guidance', *British Journal of Eucational Psychology*, 12 (1952), 5–29.

[59] D. Garland, *Punishment and Welfare: A History of Penal Strategies* (Aldershot, 1985).

[60] G. K. Behlmer, *Child Abuse and Moral Reform in England, 1870–1908* (Stanford, Calif., 1982), 193–207.

[61] V. Bailey, *Delinquency and Citizenship: Reclaiming the Young Offender, 1914–1948* (Oxford, 1987), 317.

degree to which the mental deficiency system provided a direct alternative to penal sentences. The introduction of psychologists to undertake mental examination of offenders was slow, and in 1926 the number of charged children recorded as being mental defectives was only 0.16 per cent of new cases.[62] However, this figure is probably an underestimate, since it excluded defectives who were ascertained before being charged or after being sent to penal institutions.[63] The role of the mental deficiency system as a supervisory and custodial extension of the penal system probably became less important over the interwar years, as alarm over delinquency diminished, as psychological concern shifted from the behaviour of defectives to that of dull, backward, and neurotic children, and as these borderline cases were absorbed by the new child guidance system.[64]

In summary, the Mental Deficiency Act should be seen not only as an example of specialization of medical welfare, but also as part of the differentiation and specialization of penality in the early twentieth century.[65] This demonstrates the importance of studying the interrelationship between different social problems and different systems of welfare, and, in particular, the process of 'transcarceration' between penal, medical, and educational areas of the welfare complex.[66] On the other hand, it is important not to exaggerate the direct overlap between mental deficiency and delinquency. Due to a conjuncture of institutional pressures, social strains, and thinking about the nature of delinquency, there was a significant overlap in the first decade of the 1913 Act; thereafter, direct overlap was less important than the way certification functioned as a threat to control behaviour beyond its boundaries—a symbol of what could result from common 'childhood naughtiness'.

[62] Bailey, *Delinquency and Citizenship*, 312–13, table 1b.

[63] *BOC 1933*, pp. 83–4.

[64] In 1936, the seven London child guidance clinics, dealing with over 1,000 cases, reported backwardness to be the third highest reason for sending children to the clinic, not far behind nervous disabilities and behavioural difficulties: *LCC Annual Report 1936*, iii, pt II (1937), 60–4. This is supported by material for Birmingham: C. L. C. Burns, 'Birmingham Child Guidance Clinic: A Review of Two Years' Work', *Mental Welfare*, 16 (1935), 3–6. Again, however, Birmingham was not typical.

[65] Garland, *Punishment and Welfare*, 223–4.

[66] J. Lowman, R. Menzies, and T. S. Palys, 'Introduction', in id., *Transcarceration: Essays in the Sociology of Social Control* (Cambridge, 1987), 9–11.

THE ROLE OF THE FAMILY

The Attitude of the Authorities Towards the Family

In deciding how to deal with a mental defective, assessment of the home and family was just as important as the defective's own intellectual ability, social competence, and moral behaviour. 'Neglect' at home provided the most common reason for placing mental defectives in care, and although this followed the definition set out under the 1908 Children's Act, it left considerable room for interpretation:

a parent or other person legally liable to maintain a child or young person shall be deemed to have neglected him in a way likely to cause injury to his health, if he fails to provide adequate food, clothing, medical aid or lodging for the child or young person, or if, being unable otherwise to provide adequate food, clothing, medical aid or lodging, he fails to take steps to procure the same to be provided under the Acts relating to the relief of the poor.[67]

The concentration on neglect automatically targeted the Act at the poorest and least 'respectable' section of the population. If neglect was serious, the defective would probably be removed to institutional care. If it was less serious, the home would be placed under regular supervision. Visits to the family home had the dual function of helping the defective, and reforming the family itself; as one social worker described:

At first, the personal condition of the defective is often uncleanly—hair uncared for, nails dirty, and appearance generally unkempt. But by tactful management, the Home Teacher is generally able by degrees, to give her pupil an appreciation of nicely kept hair and hands, and this new appreciation of cleanliness will permeate the whole family.[68]

As child care was made into a professionalized science, it became even harder for families to satisfy expectations; for instance Shrubsall advised mental welfare workers that:

the type of control exercised in the home, the consistency and severity of any punishments or the nature of rewards and encouragements should be noted carefully . . . Recreational facilities may be inadequate. In those of older years the occupation may be detrimental to health or morals, unadapted to the

[67] *Children Act, 1908* (8 Edw. VII, c. 67, s. 12 (1)); on Children Act, Behlmer, *Child Abuse and Moral Reform*, 193–207.

[68] M. Anderson, 'The Home Teaching of Defectives', *Mental Welfare*, 17 (1936), 36.

individual, or economically inadequate and so liable to encourage disturbances of behaviour. For satisfactory home care the parents or guardians need to have some understanding of the mental condition and needs of the defective and to do their best to provide care, encouragement, occupation, training, and recreation. If, as is often the case, they doubt the fact of deficiency and are content to treat the subject as a normal individual in all ways; still worse if they are always full of complaints in his presence or upbraid him for contributing inadequately to the family exchequer, the control is likely to be faulty and the influence unfavourable.[69]

'Respectable' families, who failed to achieve such new ideals of good care, were not in danger of having their defective children forcibly removed as a result of 'neglect'; they were, however, opened up to new forms of scrutiny, advice, and regulation.

A high proportion of the London cases involved accusations of neglectful conditions at home, and only five families were described as providing 'good' homes.[70] Although the reports typically objected to filthy living conditions in the home, there was often a confusion between material and moral uncleanliness. By the 1930s, social workers had become assiduous in recording the social and economic circumstances of cases.[71] Case records included details of the rent paid by the family, the size and conditions of the home, and the employment and salary of the parents; an assessment of the defective's abilities for employment and a record of any manual training received; and record of the duration, pay, and reasons for termination of any employment obtained by the defective. However, economic analysis continued to be permeated by moralism: poverty was blamed on laziness and social inefficiency, and behavioural problems were invariably attributed to mental defect rather than economic circumstances.[72] Thus, the impetus of mental welfare work did not change course: reform of defectives and their families, rather than the provision of material assistance, continued to be the primary aim.[73]

69 Shrubsall and Williams, *Mental Deficiency Practice*, 251. For the way strict and unsuitable standards of care were imposed on mothers see E. Peretz, 'The Professionalization of Childcare', *Oral History Journal*, 12 (1989), 22–7.

70 Herbert C., Winifred E., Aaron G., Albert L., Cherry M.

71 H. L. Smith (ed.), *The New Survey of London Life and Labour*, vi (London, 1934), 364–79, annex 1: seventeen cases of nine boys and eight girls. For an insight into the background of this survey and the selection of cases: New Survey of London Archive, British Library of Political and Economic Science (NSOL), 5/5 Shrubsall to Mr Verdeer, 5 July 1933; and for more case histories, NSOL 5/7.

72 Smith, *New Survey of London Life*, vii, annex 1, Case 9.

73 Ibid., Cases 6, 11, 13, 14.

The Weakening of Family Responsibility?

In its 1930 annual report, the Board of Control argued that the continual expansion of institutional care for the mentally disordered was being driven by a decline in familial responsibility for dependent members.[74] There has been considerable historical debate over this issue.[75] Andrew Scull has argued that in the second half of the nineteenth century the strains of capitalism forced working-class families to use lunatic asylums, which he provocatively describes as 'dumping grounds' for the unwanted.[76] Although some of his suggestions about the 'social production of insanity' have been substantiated by subsequent work on the demographic and economic patterns of admission, his description of the asylum as a 'dumping ground' has been queried by evidence which explores the familial context of commital, and by evidence which shows that the average length of stay was far shorter than he claimed it to be; the asylum appears to have been a last resort and often a temporary expedient for those who could no longer be supported in the community, but who were not necessarily 'unwanted'.[77] The idea that institutionalization was fuelled by pressure from below has also been questioned by work on the late nineteenth-century poor law which suggests that through a reduction in the numbers on and level of outdoor relief there was a state-enforced continuation, rather than an erosion, of family responsibility.[78]

National statistics cite relatives as the source of ascertainment for only 252 cases in 1922 and 171 cases in 1932.[79] However, in most cases we can

[74] *BOC 1930*, p. 8.

[75] J. Benson, *The Working Class in Britain, 1850–1939* (London, 1989), 95–116.

[76] Scull, *Museums of Madness: The Social Organisation of Insanity in Nineteenth-Century England* (London, 1979), 240, 259. This powerful argument remains largely unrevised in the new version of the book: *The Most Solitary of Afflictions: Madness and Society in Britain, 1700–1900* (New Haven, 1993), 334–74.

[77] For instance, J. Walton, 'Casting Out and Bringing Back in Victorian England: Pauper Lunatics, 1840–70', in W. Bynum et al., *Anatomy of Madness* (London, 1985–8, ii. 132–46; J. Walton, 'Lunacy in the Industrial Revolution: A Study of Asylum Admissions in Lancashire, 1848–50', *Social History*, 13 (1979); D. Wright, 'The National Asylum for Idiots', Earlswood, 1847–1886', unpub. D.Phil. thesis (Oxford, 1993). A series of major projects linking asylum records with census data and the records of workhouses and prisons will soon tell us far more about the social roots of 19th-cent. insanity and seem likely to reveal a complex pattern of movement between different institutional sites.

[78] On the elderly, for example: D. Thomson, 'The Decline of Social Welfare: Falling Support for the Elderly since Early Victorian Times', *Ageing and Society*, 4 (1984), 451–82. The changing balance between institutional, outdoor relief, and family care of the elderly is discussed in: M. Pelling and R. Smith (eds.), 'Introduction', in id., *Life, Death, and the Elderly: Historical Perspectives* (London, 1991), 13–38.

[79] *BOC 1922*; *BOC 1932*.

assume that some degree of parental co-operation was obtained, for the legislation required this unless there were exceptional circumstances.[80] On the other hand, assent, or even active request for institutional care, is not necessarily evidence of a lack of responsibility; on the contrary, it may arise from this very characteristic. In some cases, it was the parents who accused the authorities of neglect in their failure to provide institutional facilities for care. For instance, the mother of Frederick W. complained:

> It was my wish that he should be allowed to attend day school, but by the Council's ruling it seems that he is to be kept almost uneducated until he is of an age when he will find it hard to learn anything. He is at present at home, restless and tiresome because he has nothing to occupy his mind. I think it is a down right shame that the Council, who are responsible for the child's education, should have put this matter off as long as this and then to have come to a decision whereby nothing is done save a visitor calling now and again giving advice which is not practicable.[81]

When families requested institutional care it was usually because they could no longer cope. Willingness to resort to state aid varied markedly at different stages of the life cycle: the ageing, retirement, or death of parents, the arrival of new siblings, and the ageing of defectives, all tended to make care more difficult and request for institutional aid more likely. Reluctance to accept the return home of a defective was also related to practical problems: for instance, the move to a smaller house or new area, or the deterioration of the family's economic situation through unemployment.[82] In the long term, the general trend away from extended family networks towards smaller nuclear families was probably making it harder to maintain care during crisis points in the life cycle. However, the London case histories show that extended familial and neighbourhood care persisted in some working-class communities. References to defectives playing with groups of children suggest that they may have been cared for within child-minding 'survival networks', akin to those Ellen Ross has described in interwar London.[83] And there

[80] Herd, *Diagnosis of Mental Deficiency*, 211–15.

[81] GLRO PH/MENT/2/12 Case 1987, letter of Mrs Wooding to Council, 19 Apr. 1917.

[82] For an example in which the circumstances of a case changed as the parents became pensioners: case 1412 Elizabeth D.

[83] E. Ross, 'Survival Networks: Women's Neighborhood Sharing in London before World War II', *History Workshop Journal*, 15 (1983), 4–27. e.g. cases 450 Elsie L., 290 Ellen S. As these cases grew older, however, the parents became concerned at their treatment of younger children.

is strong evidence to suggest that kinship networks continued to provide an important caring role within the Jewish community.[84]

To assess familial responsibility we need to go beyond structural changes to consider the emotional environment within the family. Historians have been divided over whether affection or economic instrumentality bonded the early twentieth-century working-class family.[85] In the case of the mentally defective, one has the added problem of gauging familial sentiment towards disability. A contemporary survey showed that such parents hardly ever attributed mental defect to inheritance, instead explaining it as the result of a knock on the head, an accident, or fright and worry in pregnancy; something to be viewed with pity perhaps, certainly not something deserving stigma or shame.[86] The constant reference to maternal care in case reports gives the impression of a fairly close and affectionate relationship between most mothers and their defective children. Reference to fathers is less common, but this can be explained by the fact that they were usually working outside the home at the time of visits, and that they were of less interest to the visitors, who assumed that care was the responsibility of mothers. There is little evidence to support the crude stereotype of the brutish uncaring father.[87] In cases where a father was at home, he too might be centrally involved in care, as in the cases of Winifred E., whose invalid father cared for her, and Herbert C., who worked alongside his foster father in his music shop.[88] As a cause of institutionalization, the absence of one or both parents was far more common than an uncaring father or mother: in eight of the fifty-five London cases, one or both parents were absent, while in fourteen cases economic strain split up the family and led to the defective coming under the poor law. In only two of the fifty-five cases was there any suggestion of parental cruelty or neglect simply because of a child's mental condition. The mother of Cherry M. was reported as 'very

[84] GLRO PH/MENT/2/11, Case 389. On indigenous care within Jewish community: Marks, 'Irish and Jewish Women's Experience of Childbirth and Infant Care in East London, 1870–1939: The Responses of Host Society and Immigrant Communities to Medical Welfare Needs', unpub. Ph.D. thesis (Oxford, 1990).

[85] M. Anderson, *Family Structure in Nineteenth-Century Lancashire* (Cambridge, 1971); E. Shorter, *The Making of the Modern Family* (London, 1976).

[86] *Mental Welfare*, 12 (1931), 88. *LCC Annual Report 1930*, iii, pt. I, pp. 44–5.

[87] N. Lowe, 'The Legal Status of Fatherhood Past and Present', in M. O'Brien and L. McKee (eds.), *The Father Figure* (London, 1982), 26–42; T. Lummis, 'The Historical Dimensions of Fatherhood: A Case Study 1890–1914', in O'Brien and Mckee (eds.), *The Father Figure*, 43–56. A 'child guidance and fathercraft' group was set up in Kensington: *JMS* 73 (1927), 443–6.

[88] GLRO PH/MENT2/10, Cases 1943 and 572.

indifferent at first because the child was stated to be an imbecile'.[89] More seriously, the LAMW reported that the mother of Winifred R. (an idiot, and the eighth of fourteen pregnancies) was 'beginning to find defective a burden in that she costs so much to feed, and states that she is trying to get her work as a clown',[90] though the poverty of the family, rather than cruelty, is surely the more likely explanation for this. The NSPCC reported that a high proportion of its cases involved defectives, but argued that this was due to the difficulties of caring for such children and the inadequacies of the parents, rather than parental cruelty.[91] The association of mental deficiency with a child-like state made the development of sexual interests a shock to parents, while the physical growth of the children could make them difficult to manage if they became agitated. The far higher familial notification of female compared to male defectives suggests that parents were particularly concerned about the sexual control and protection of their mentally defective daughters.[92]

Economic insecurity, rather than lack of affection, was the main reason for families to turn to institutional care. In this period of high unemployment and housing shortages there was considerable economic pressure pushing poor families to find statutory care for non-contributing dependants, such as defectives, especially if care at home prevented parents from going out to work.[93] LAMW records from 1930 indicate that parental dissatisfaction with a child's attitude and ability to work could encourage notification.[94] However, there were also some economic disincentives to the removal of a defective family member. The local authorities demanded a maintenance contribution in non-pauper cases, usually set at 2*s*. 6*d*. per week, which was less than the cost of

[89] GLRO PH/MENT/2/11, Case 1407, MD Enquiry Form 17 Oct. 1915.

[90] GLRO PH/MENT/2/12, Case 1414.

[91] PRO MH58/102, NSPCC evidence to Brock Committee on voluntary sterilization; *Report of the Departmental Committee on Sterilisation (Brock Report)* (PP, Cmnd. 4485, xv, 1934), 80–91. On the other hand, the NSPCC was an active supporter of using the Mental Deficiency Act to remove defective children, or to remove normal children from defective parents: *NSPCC Yearbook 1913/14*, pp. 46–7.

[92] For instance, in 1922, families were the source of notification for 51 males compared to 201 females: *BOC 1922*. In 1932, the figure was 25 males compared to 146 females: *BOC 1932*.

[93] M. A. Crowther, 'Family Responsibility and State Responsibility in Britain before the Welfare State', *Historical Journal*, 25 (1982), 131–45. On the importance of maternal income as a supplement to 'family wage', see E. Roberts, 'Women's Strategies, 1890–1940', in J. Lewis (ed.), *Labour and Love: Women's Experience of Home and Family, 1880–1940* (Oxford, 1986), 223–48.

[94] Smith, *New Survey of London Life*, vi. 364–78.

support at home, but met such vehement opposition that it seems likely to have cancelled out any economic incentive. In fact, defectives were not necessarily a financial burden upon their families: many contributed to the family economy with a wage, helped in the family business, or assisted their mothers in housework.[95]

In summary, there is little evidence to support the idea that institutionalization was driven by a decline in family responsibility. In most cases, families were affectionate towards mentally defective members, and attempted to keep them at home if at all possible. However, because of lack of resources, absence of parents at work, the decline of extended networks of care, and the difficulties of coping with the behaviour of mental defectives, continuation of home care was often very difficult. Therefore, familial co-operation was obtained in most cases of institutionalization, but this was an expression, as much as an abdication, of responsibility. As was the case with many other areas of welfare provision in this era of adjustment to democracy, increase in demand for services was driven by economic need, rather than by a new sense of social rights, a 'dependency culture', or the erosion of family responsibility.

The Familial Role in the Construction of Policy

In calling on the state for aid, or simply co-operating with the state when aid was offered, families played a role in mediating the implementation of mental deficiency policy. If supervision was to be effective, welfare visitors had to gain the confidence of the family:

> visitors should know how to visit, i.e., learn not to appear to interfere but yet to achieve what they have come to do. A visitor should be able to give the best practical advice to the parents as regards the care of the defective, and where possible arrange for the simple home training of the defective.[96]

Institutionalization and collection of maintenance contributions were also far easier if parental co-operation could be obtained.[97]

Detailed analysis of the case of Albert G. reveals how policy in practice could be modified through negotiation between families and

[95] In one case, a mentally defective son provided essential assistance for his blind mother: GLRO PH/MENT/2/10, Case 572.

[96] E. Fox, *The Work of the Local Associations for the Care of the Mentally Defective* (London, 1920), 8. For a discussion on the limitations of supervision: *CAMW Conference 1926*: Miss Darnell, 48–9; Mrs Emmett, 54–5; Mr Wormald, 59–60.

[97] GLRO PH/MENT/2/10, Elizabeth D., Case 1412, is an example of non-payment by the parents leading to a court case.

the local authorities. Albert G. was born in 1905 and at the age of 10 was notified as an imbecile and placed under supervisory care.[98] The LAMW visited the family quarterly. Since its reports were consistently favourable, the visits were made less frequently from 1924 onwards. However, in 1930, the LAMW reported a deterioration in the financial position of Albert's family, with the father out of full-time work, another brother forced to return home, and out-relief for Albert finished after the close of the poor law system.[99] Even when the father returned to work in 1933, he was earning so little that the family's financial position remained precarious.[100] Under such circumstances, Albert was in danger of being adjudged 'in neglect' and in need of institutional care; instead, the authorities appointed Albert's father as his official guardian, enabling family care to continue with the assistance of the guardianship fee. Because of the history of favourable supervisory reports, the authorities had ignored the fact that family and friends were generally debarred from acting as guardians.

Albert's family was also successful in modifying local authority policy over his attendance at Shoreditch craft centre.[101] In January 1935, Albert's supervising visitor complained that he was not attending the centre and needed the regular occupation provided there.[102] The fact that the boy was now under guardianship gave the authorities extra power to exert control over conduct within the private sphere of the home. His mother complained that attendance at the centre was too much of a strain, particularly the train journey and walk at the other end. The authorities reacted by providing a guide to accompany Albert from the station to the centre, even though they felt that the parental objections were unreasonable.[103] The next argument with the local authorities came after Albert slipped and hurt himself on a football match trip organized by the centre, and the family decided that it was safer to keep him at home. After consultation with the Board of Control, the LCC agreed to provide a door-to-door guide to placate the parents. The authorities were annoyed when Albert's non-attendance continued, and became highly critical of the parents.[104] The standard of

[98] GLRO PH/MENT/2/11, Case 398.
[99] Ibid., LAMW 25 July 1930; 25 Aug. 1931.
[100] Ibid., Supervision Section, 28 Dec. 1933.
[101] GLRO PH/MENT/2/11, Case 398.
[102] Ibid., Board of Control to Mental Hospitals Department, 25 Jan. 1935.
[103] Under pressure from London parents over the cost of visiting, the authorities changed their policy and began to pay parents' transport fares: *LCC Annual Report 1930*, iii, pt. I, p. 47.
[104] GLRO PH/MENT/2/11, Case 398, Supervision Section, 6 Dec. 1935.

care, which had previously been given good reports, was now attacked. The mother was chastised for allowing her son out alone, leaving him unoccupied at home, and because he was 'thin and nervous'.

The mother's side of the story reveals a gulf in understanding between the authorities and the family. She claimed that her son had been going to the park on his own for the past twenty years and had never been warned against this. He was always very well behaved and never talked to strangers, except for a couple of pensioners. Moreover, he was well known in the park and watched over by the park keeper. Under such conditions, Albert could hardly be described as 'out of control'.[105] She argued that sending her son to the occupation centre placed him under undue strain. To be ready for the guide by 9 a.m. in the morning meant getting up at 6 a.m. since it was Albert's practice to dress and wash himself.[106] Criticism that Albert did nothing in the house except for standing around and looking at himself in the mirror failed to appreciate that this, along with the trips to the park, constituted his own 'occupational regime'. Attendance at the occupation centre totally disrupted this proven and congenial routine and upset the boy so much that he became nervous and enuretic. Albert's mother argued that in other circumstances she would welcome the relief offered by the occupation centre, but attendance produced more problems than it solved. She added that in the past she had always been treated with great respect by the 'lady visitors' and medical officers, and was obviously quite upset by the new interventionist stance.[107] The authorities finally conceded over attendance at the occupation centre, even though the mother was damned as being dangerously 'over protective'—an ironic term when one considers the parallel use of 'neglect'.

Albert's case highlights some of the tensions and misunderstandings which divided providers and recipients of care, but it also reveals a significant degree of flexibility and a willingness to moderate policy in response to the complaints of the family. This concern about parental opinion was a factor which also entered into the LCC's design of occupation centres. In 1923 it reported that:

The parents appear to interest themselves in the work of these centres, and although the attendances of the children are not large, much useful work appears to have been accomplished already. Several mothers have given testi-

[105] Ibid., Mother, 12 May 1936.
[106] Ibid., Supervision Section, 16 May 1936.
[107] GLRO PH/MENT/2/11, Supervision Section, 12 May 1936.

mony to the improved deportment and usefulness of the children when at home which has followed their attendance at the centre.[108]

As this suggests, the LCC was keen for policies to satisfy the clients; instilling discipline into defectives was a response to parental desire, as much as local authority 'control'. Many families were in desperate need of some relief from their constant role of home care. A representative of the Mother's Defence League articulated the support of its members: 'I want to support the centres because they inspire the most enormous confidence in the mothers.'[109] However, this probably represented the viewpoint of middle-class mothers. As long as occupation centres were so few in number, working-class families were forced to accept what provision they could find and had little power to shape policy as 'consumers'; parents were coerced into acceptance through fear that refusal would damage their chances of keeping their children at home, or of serving as guardians.

It was hardest of all for parents to have any impact on the nature of care which was given in institutions. The realities of conditions were largely hidden from the outside world, and if parents did want to express criticism they had to proceed through a bewildering maze of bureaucracy. Letters of complaint and requests for withdrawal, leave, or licence could be sent to the institution, the local authorities, or the Board of Control, yet action upon this was hampered by administrative confusion over responsibility, by mistrust of the parents, and by the restrictive rules and regulations of the administration.[110]

In some cases, the family had agreed, perhaps under duress and deference to the expert, to initial care for their child, but had not understood that care would be long-term. Typically, in 1923, the father of the 23-year-old Elizabeth D. complained: 'I am asking if I can have her home as when she first went away she was only supposed to be away for 12 months, and then it was until she was 16.'[111] Families often accepted institutional care on the understanding that their children, having been excluded from school as ineducable, would be provided with residential education. On visiting their children and seeing the reality of the conditions, parents, such as the mother of Charlotte T. expressed disappointment: 'I don't think she is being taught anything in

108 *LCC Annual Report 1923*, iii, 52.
109 N. Boyle, *CAMW Annual Conference Report 1922*, 22 July.
110 GLRO MAB 238A, 7 Jan. 1930.
111 GLRO PH/MENT/2/10, Case 1412, letter, 25 July 1923.

particular as regard earning her living as we understood she would be. Could you be so kind as to find out for me if she could come home at the end of her two years.'[112] At other times, parents expressed anger and complained, as is evident in this sad message from one mother to her daughter:

I suppose you are longing to come home as we are longing for you. If you come home there is no need for you to work you were always very well and happy when you were at Halstead we don't understand why they want to keep you in that place for I suppose they are keeping you for their own use . . . It doesn't seem to us as though they are treating you properly if they don't let you home before very long. Father is going up and make them understand that you were in fit state when you went to that place. If they have any children of their own and had any love for them as your mother and Father has for you it would make them understand a bit better . . .[113]

The problem was that parents usually lacked the necessary understanding of the Act, the confidence, or the financial means to channel their resistance and complaints through the law. Detailed legal records of one London case do survive, in which a philanthropic employer and sympathetic solicitor intervened to help a defective.[114] However, this case cannot be taken as typical; on the contrary, its exceptional nature highlights the weak legal position of the average individual in coming to terms with the rapid growth of state medico-legal power during this period.[115]

CONCLUSION

In this chapter, I have set out to demonstrate that social policy cannot be understood through the intentions of central government alone. It is necessary to take into account the ways in which policy was modified in the process of implementation at the ground level; public and private spheres were not divorced, they interacted, particularly in the complex

[112] GLRO PH/MENT/2/12, Case 208, letter from mother, 12 Aug. 1916.

[113] GLRO PH/MENT/2/12, Case 1627, letter undated, *c.*1919.

[114] GLRO MAB 2325.

[115] The issue of legal aid appears to have attracted little attention from historians. For a contemporary account of the poverty of legal aid and the inequalities within the legal system: *Justice in England* (London, 1938) 'by a Barrister'. Compare the analogous problem of workmen's compensation: ibid. 140–7.

of community care. In the provision of mental deficiency services, three factors mediated policy at the micro-political level: first, the structure and language of the forms, mental tests, and regulations used in everyday administration, and the ways in which this machinery was interpreted and manipulated; secondly, the moral, social, and medical attitudes of the social workers, local authority welfare workers, and psychologists who controlled the administration at the ground level and who defined mental deficiency in practice; and thirdly, the roles of the defectives and their families who may have actively requested aid, resisted provision, or encouraged the modification of policy to serve their needs. At this micro-political level, the weighting of power was heavily on the side of the authorities. Families accepted provision of services through a combination of deference to medical authority, confusion about the legislation, their own attitudes towards morality and social competence, and their own economic problems and difficulties in coping with and controlling defective dependants. This era of adjustment to democracy did not see the emergence of a consumer-driven mental welfare service—resort to state assistance generally remained a last resort—but there were signs of a shift away from the harsh moralism of the poor law towards a more sympathetic consideration of the interests and problems of defectives and their families.

8

Mental Deficiency and the Welfare State

INTRODUCTION

The creation of the Welfare State dominates the historiography of British social policy. It is seen as a fundamental crossroads, establishing a new relationship between the state and the people, and making a minimum of welfare a universal right. In T. H. Marshall's lecture of 1949, the Welfare State was portrayed as the culmination of a long process of democratization, which began with the granting of civil rights in the eighteenth century, progressed to the spread of political rights in the nineteenth century, but was not fully realized until the inauguration of an era of 'social citizenship' in 1945; only with social rights was it guaranteed that people would have the education, health, and social security to exercise their democratic freedom. In this chapter I shall reconsider this transition from the fresh perspective of mental deficiency policy. To what extent was there a transformation and democratization in this area of policy? And what were the implications of social citizenship for the definition of, and attitudes towards, mental deficiency?

Two main models have been put forward by historians to delineate the formation of the British Welfare State. The first model claims that it evolved out of a long series of *ad hoc* extensions of state services. For instance, Gilbert has argued that the extension of national insurance from 1911 onwards had created a virtual social security state by 1939,[1] while Fraser has described how a municipal hospital service was evolving out of the poor law in the interwar years, and Fox how the hospital system was undergoing a process of modernization towards the 'medical regional hierarchalism' of the NHS.[2] The second model emphasizes the discontinuity of the Welfare State with the past and the novelty of

[1] B. Gilbert, *The Evolution of National Insurance in Great Britain: The Origins of the Welfare State* (London, 1966), 308.

[2] D. Fraser, *The Evolution of the British Welfare State* (London, 1973); D. Fox, *Health Policies and Health Politics: The British and American Experience, 1911–1965* (Princeton, 1986), 50.

universalist, free, and state-sponsored provision.[3] Webster has launched a trenchant assault on those who have portrayed the NHS as a product of evolution and consensus, and paints a picture of party and ideological conflict in which the Labour victory of 1945 was crucial.[4] Others have argued that the Second World War was the springboard for radical changes in policy: for instance, Titmuss emphasized the importance of the practical challenges brought about by the war and the new sense of social solidarity which resulted; while Marwick has written of a war-induced 'revolution in values and expectations'.[5] I shall use the issue of mental deficiency to re-examine the strengths and weaknesses of each of these models in turn.

EVOLUTION TOWARDS THE WELFARE STATE?

The history of mental deficiency services lends some support to those who have argued that the Welfare State reforms emerged out of a gradual process of reform and modernization. In organization, many of the developments of mental deficiency policy mirrored the movement of general health care towards the nationally organized, rationalized, and medically dominated NHS. The passage of the 1913 Act was an early example of specialization of care, replacement of poor law by medical control, and of central government assuming an active role in regulation and funding of welfare. In the interwar years mental deficiency services underwent a process of rationalization, with the take-over and professionalization of voluntary work by the state, the amalgamation of mental deficiency with mental illness and public health work at the local level, and the redistribution of resources between different types of institutions. On the other hand, certain administrative features set mental deficiency apart from an imagined evolutionary trajectory towards the

[3] K. Morgan, *Labour in Power, 1945–51* (Oxford, 1984).

[4] C. Webster, 'Conflict and Consensus: Explaining the British Health Service', *Twentieth Century British History*, 1 (1990), 115–51; id., *Problems of Health Care: The National Health Service Before 1957* (London, 1988), 391–3.

[5] R. Titmuss, *Problems of Social Policy* (London, 1950); for a summary of Marwick/Titmuss debate see H. Smith, 'Introduction', in id. (ed.), *War and Social Change: British Society in the Second World War* (Manchester, 1986), p. ix. Most critics of Titmuss have concentrated on the conclusion of his book (p. 508). A more thorough reading shows a far more complex picture which is at times highly critical of government policy.

Welfare State. Psychiatrists involved in the care of mental defectives continued to form a backwater within the medical profession, their positions challenged by lay managers, and their institutions understaffed, overcrowded, and performing a housing rather than an effective curing or caring role. Central government regulation of care, promotion of research, and funding were all undermined by lack of resources. In reality, standards of care depended on local activity which varied widely, with rationalization of services in some areas of the country, but in others a confusing overlap between poor law, educational, and public health services, and a continuing dependence on voluntary activity. In fact, rather than moving towards the administrative structures of the NHS, what characterized the organization of interwar mental deficiency services was diversity and the sharing of responsibility between central state, local authorities, the voluntary sector, and the family in what has been termed a 'mixed economy of welfare'.[6]

In ideological terms, the Mental Deficiency Act was also set in a very different welfare paradigm to that of Marshall's social citizenship. The style of the Act was paternalistic, even authoritarian. Though consciously a step away from the pauperization of the poor law, care under the Act, with its language of 'ascertainment', 'supervision', and 'segregation', still carried a pervasive stigma. Welfare was provided for the protection of the defenceless and for the good of the community, nation, and race, rather than offered as a right of 'social citizenship'. This paternalistic/authoritarian paradigm continued—even after the expansion of citizenship of 1918 and the rise to power of a working-class party—for mental defectives, who were by definition inherently incapable, and continued to lie outside the boundaries of responsible citizenship. During the interwar period, the only movement in this relationship involved debates over where to place the borderline of mental deficiency. However, partly because of this, there was increasing interest in moving mental health services beyond the borderline of mental deficiency into the normal population, so as to construct good citizens through the regulation of mental hygiene; it is here, and not in mental deficiency policy, that one can begin to see a transition towards a paradigm of mental welfare as a right.

[6] J. Lewis, 'Family Provision of Health and Welfare in the Mixed Economy of Care in the late Nineteenth and Twentieth Centuries', *Social History of Medicine*, 8 (1995), 1–16.

WAR AND SOCIAL CHANGE

'Change as a Response to the Demands of War'

The demands of the Second World War did stimulate some changes in mental health care. At the start of the war there had been gloomy predictions that the strains of bombing campaigns, evacuation, and disruptions to civilian life, would cause widespread mental breakdown.[7] Even when these expectations were not realized, there was still a fear that the mental effects of the war might emerge in the postwar world. In response, the government withdrew C. P. Blacker from active service in the army to undertake a survey of civilian mental health problems and propose a blueprint for the services of the future.[8]

The disruption of war, and the diversion of state resources to warfare rather than welfare, reopened opportunities for voluntary groups and women welfare workers to utilize their flexibility in innovating new mental health services. A system of psychiatric social workers was established to deal with the mental problems which were exposed or caused by the evacuation of city children to the countryside, and hostels were set up to provide accommodation for 'problem children'.[9] Under the Emergency Medical Service, the Ministry of Health itself began to employ psychiatric social workers, and it pressurized local authorities, likewise, to employ more social workers to co-ordinate the process of evacuation and the complicated network of hostels.[10] The voluntary mental welfare groups, encouraged by the Ministry of Health to combine to provide a more rational and efficient system, amalgamated to form the National Association for Mental Health (NAMH). This had already been proposed by the Feversham Committee in the late 1930s, but the war had accelerated the process.[11] The voluntary groups developed new services to cope with wartime conditions: they set up a new training course away from the dangers of London, developed

[7] Titmuss, *Problems of Social Policy*, 337–41.

[8] C. P. Blacker, *Neurosis and the Mental Health Services* (London 1948); PRO MH 76/115, A. Lewis to F. R. Fraser, 4 Mar. 1942.

[9] J. Macnicol, 'The Effect of the Evacuation of School Children on Attitudes to State Intervention', in Smith (ed.), *War and Social Change*, 3–31; E. M. Hearnshaw, 'Some Psychological Difficulties of Evacuation', *Mental Health*, 1 (1940), 5–10; id., 'Observed Effects of Wartime Conditions on Children', *Mental Health*, 2 (1941), 93–8; C. A. Wollen, 'The Problem Child In the Hostel', *Mental Health*, 5 (1944), 28–31.

[10] Titmuss, *Problems of Social Policy*, 381–3.

[11] P. Norman, *In the Way of Understanding* (Godalming, 1983), 58–66.

schemes of agricultural hostels with the backing of the Ministry of Agriculture, and extended hostel care to the dull, the backward, and the maladjusted.[12]

Bearing in mind the reversals which had followed reconstruction after the First World War, the mental health lobby wanted to ensure that the psychiatric advances of this war would not suffer a similar fate. Mental hygienists tended to be slightly wary of pinning their hopes on the cautious medical profession and instead looked to lay participation for the enthusiasm and energy for the expansion of the movement. They wanted to extend their influence beyond the limited field of the mentally ill or handicapped to the mental health of the whole population, and therefore saw the social services as the key arena for their ideas and the social worker as the ideal vector.[13] The mental health movement had gained encouragement from developments in the army, where preventive psychiatry was used to avoid the shell-shock problems of the First World War, mental tests found an application in detecting both the dull and backward and potential officers, and group psychiatry was used to maintain morale and influenced some doctors who returned from the war to develop new methods of social psychiatry.[14] The metaphors of war were adopted in the battle for mental health. One of the boldest assertions was that voiced by the psychiatrist and mental hygienist J. R. Rees, author of *The Shaping of Psychiatry by War* and subsequently the first President of the World Federation for Mental Health, who advocated a 'fifth column' to infiltrate influential circles and disseminate psychiatric ideas. Under the slogan of 'efficiency and economy', Rees also appropriated the language of the wartime command economy, proposing to set up a 'Social Efficiency Board' in order to 'sell mental health', with the Minister of Labour, Ernest Bevin, as its chairman.[15]

Institutions were fairly successful in coping with wartime problems. The death rate in mental hospitals did not escalate as dramatically as in the previous war, and there is no sign that the Nazi policy of

[12] M. Rooff, *Voluntary Societies and Social Policy* (London, 1957), 151–3.

[13] C. H. Rogerson, 'Psychiatry in the National Health Service of the Future', *Mental Health*, 5 (1945), 45–9; D. W. Harding, 'Mental Health in Post-War Reconstruction', *Mental Health*, 2 (1941), 61–5; C. O. Stallybrass, 'Mental Health in Relation to the Family', *Public Health*, 61 (1948), 123–8.

[14] K. Soddy, 'Some Lessons in Wartime Psychiatry', *Mental Health*, 6 (1946), 32–3, 66–70; H. V. Dicks, *Fifty Years of the Tavistock Clinic* (London, 1970), 102–38.

[15] 'Strategic Planning in Mental Health', *Mental Health*, 1 (1940), 103–6. Obituary of Rees, *Munk's Roll: Lives of the Fellows of the Royal College of Physicians of London*, vi (Oxford, 1982), 387–90. J. R. Rees, *The Shaping of Psychiatry by War* (London, 1945).

exterminating patients was considered as a solution to the burden of care.[16] A fall in admissions and the increased use of licences enabled space to be given up for military casualties. At the Royal Eastern Institution licences were explicitly justified in terms of national economy, saving the institution £11,000 in a year. Wartime manpower shortages increased the chances of licensed or discharged defectives finding work in the outside community. Because of staff shortages and lack of resources, the defectives who remained were set to work in producing food, in the care of lower-grade defectives, and even in the civil defence of institutions.[17] In summary, the challenge of war had stimulated the development of mental health services and encouraged institutions to release high-grade defectives for work in the community.

In the long term, however, the response to the challenge of war was not as important as it might first appear. Much of the progress made by voluntary groups and psychiatric social workers was not maintained in the postwar world. Psychiatric social workers might have formed the backbone to a future mental health system; instead, the Ministry of Health ceased to employ them. Even Brock, who greatly admired the contribution of the mental health lobby, felt that they had overstated the case for their indispensability.[18] Expansion of the psychiatric social work profession was held back by the difficulty of training, the limited places on courses, a poor completion rate, and inadequate government aid.[19]

The worst-hit area was mental deficiency. Once mental welfare workers became trained in psychological and psychoanalytical theory they tended to lose interest in mental deficiency work, and moved to psychiatric clinics where their new skills found more use. At the same time, these clinics became more selective over the type of cases they accepted,

[16] C. P. Blacker, 'Eugenic Experiments Conducted by the Nazis on Human Subjects', *Eugenics Review*, 44 (1952), 9–19; in Britain the death-rate was held down to 6.8 per cent of the population per year (*BOC 1945*). In WWI, the rate in London reached 27 per cent in 1918: G. Ayers, *England's First State Hospitals and the Metropolitan Asylums Board, 1867–1930* (London 1971), app. III, table C.

[17] D. Turner, *Mental Health*, 1 (1940), 86; E. Benjacar, 'A Mental Deficiency Institution in Wartime', *Mental Health*, 1 (1940), 107–12; 'Agricultural Hostels For Defectives', *Mental Health*, 3 (1942), 49–51.

[18] PRO MH 77/72 Provisional National Council For Mental Health, proposals for the mental health services, 26 June 1944; Brock's reaction, 18 July 1944.

[19] N. Timms, *Psychiatric Social Work in Great Britain, 1939–1962* (London, 1964), 13–46; PRO MH 51/664.

concentrating on children who could be treated by psychoanalysis, rather than defectives, and backward or maladjusted children.[20] Reflecting this shift in priorities, psychiatric social work courses gradually abandoned much of their mental deficiency content, thereby perpetuating the low status of mental deficiency work.[21]

In the armed forces, progress made in the welfare of mental defectives was again limited. Until 1942 there was no systematic mental testing for defectives and the backward, with only 2 per cent of recruits rejected on psychological grounds, compared to 10 per cent in the United States. The increased use of testing after this date was provoked by concern that mentally weak soldiers were likely to become maladjusted and delinquent. However, dismissal was deterred by fear that it could cause resentment among the remaining troops, especially if the defectives returned home to well-paid work. There was also a feeling that it would be eugenically detrimental to keep the least able at home while sacrificing the fit in the war. Army discipline was seen as a preferable way to deal with these individuals, rather than returning them to place extra strain on the already disrupted civilian services.[22] Therefore, instead of being discharged, the dull and defective were often placed in the pioneer corps, which served as a dumping ground for the disaffected and less able.[23]

Military requirements also had a damaging impact on standards of care on the home front. In order to make beds available for an expected deluge of military casualties, mentally defective patients were discharged from institutions, crowded together, or placed in temporary annexes. As Titmuss pointed out, civilian groups of the chronic sick, such as mental defectives, paid the price for the progress of the Emergency Hospital Service.[24] The war worsened the staffing crisis in mental deficiency institutions as nurses left for better work or signed up in the armed forces. Eventually, a government order had to be issued to restrict the freedom to leave these posts.[25] Nevertheless, the labour of

[20] G. Keir, 'A History of Child Guidance', *Occasional Papers of the British Psychiatric Society*, 12 (1952), 5–29.

[21] M. Ashdown and S. C. Brown, *Social Service and Mental Health: An Essay on Psychiatric Social Workers* (London, 1953), 188–90; Timms, *Psychiatric Social Work*, 28–33.

[22] BL 18.3.39, C. P. Blacker, 'Rehabilitation of Neurotic Soldiers', memo to F. R. Frazer, 1942.

[23] R. H. Ahrenfeldt, 'Military Psychiatry', in A. S. MacNalty and W. Franklin Mellor (eds.), *The Medical Services in War* (London, 1968), 182–7; F. J. S. Esher, 'Military Service for Mental Defectives', *Mental Health*, 3 (1942), 14–18; Soddy, 'Lessons in Wartime Psychiatry', 32–3.

[24] Titmuss, *Problems of Social Policy*, 498.

[25] Mental Nurses (Employment and Offences) Order 1941, Board of Control, circular no. 899, Sept. 1941.

inmates became vital for the running of institutions, with higher-grade patients, in practice, acting as nurses. In some instances, patients were even recalled from licence in the community to compensate for staff shortages.[26]

The use of beds by the military caused a serious shortage of accommodation: 25,000 were evacuated from mental hospitals to provide 42,000 military casualty beds.[27] Initially, there had been plans to evacuate whole mental hospitals for military use. However, because of the expense and difficulty of evacuating seriously handicapped patients, casualties were often housed in vacated wings of otherwise occupied hospitals. The impact of this military occupation continued after the war, with 5,509 beds still not returned to mental patients by 1947.[28] This caused a crisis of accommodation, especially since 1946 saw the highest ever notification of mental defectives, at 35,585.[29] As this figure suggests, the war left an immediate legacy of increased demand for care, perhaps because it had only temporarily created an environment in which high-grade cases could cope in the community, but also because of the rundown of ascertainment services during the war. Even when mental deficiency beds were freed from military occupation, the demand for labour in other areas of postwar reconstruction meant that there was insufficient staff to make full use of them; in 1948, the bed shortage was estimated at 3,034, of which 2,689 was due to the shortage in nurses and ancillary staff. Overcrowding was estimated at 12.2 per cent and there was little prospect of improvement as building schemes were held back by postwar shortages of labour, materials, and finance.[30]

Lack of resources made it impossible to realize the dynamic conception of the colony as a training or rehabilitation centre. Critics began to argue that the training was of little value, and that the institutional regime had a negative effect, lessening the forethought and initiative of patients. The mental deficiency institution was attacked for being used as a special type of prison, from which there was no definite date of release, and where there was the added discomfort of being placed alongside people with serious mental disabilities.[31] Licensing and discharge continued to have only a limited success: between 1949 and 1952, of 1,934 licences issued, 854 had been recalled, 675 were still on

[26] PRO MH 51/457, letter 10 Dec. 1941; memo 5 Oct. 1942.
[27] Titmuss, *Social Policy*, 497; *BOC 1945*, p. 8.
[28] *BOC 1947*, p. 3. [29] *BOC 1946*, p. 1. [30] *BOC 1948*, p. 3.
[31] H. C. Günzburg, 'The Colony and the High-Grade Mental Defective', *Mental Health*, 9 (1950), 87–92.

licence, and just 405 had been discharged. Institutions continued to be sites for long-term care, as reflected by figures from 1953 showing that just 6 per cent of patients had been in residence for under one year, while about 50 per cent had been in institutional care for over ten years.[32] Therefore the war had not caused a marked shift towards community care. Although the cheap option of licensing increased during the war, use of guardianship and supervision rose more slowly than in the 1930s, the numbers under voluntary supervision actually fell, and occupation centres were cut back, falling from 191 in 1939 to 90 in 1945.[33] In fact, the rise in the institutionalized defective population, from 46,000 in 1939 to 60,000 in 1955, was greater than that of defectives in community care, which only rose from 70,000 to 79,000.[34] In summary, a problem of institutional stagnation and shortage of space, already serious before the war, had been further exacerbated by it.

'A Wartime Revolution in Values and Expectations'.

The most easily obtained evidence of changing wartime attitudes derives from groups directly involved in mental health care. Views expressed in mental welfare journals suggest that the war encouraged a shift in interest, from regulation of mental illness and deficiency, to promotion of mental health. British mental hygienists recognized that manipulation of the minds of the public through propaganda had been a powerful tool in the rise of fascism. In reaction, they hoped to play a part in defeating Nazism by using the new media of the radio and cinema to inculcate positive mental health among the British people.[35] Integrating the Christian socialist rhetoric of 'New Jerusalem' with their own psychiatric language, they called for a British spiritual and mental renewal to counter the 'German Psychological Machine'.[36]

The period also saw eugenic attitudes become less central among mental hygienists. No doubt this was partly in reaction to revelations

[32] *Roy. Com. 1954–7*, table 14, p. 323; table 11b, p. 321.

[33] *BOC 1945*, p. 21; Rooff, *Voluntary Societies*, 126.

[34] *Roy. Com. 1954–7*, p. 324.

[35] *Report of the Feversham Committee: The Voluntary Mental Health Services* (London, 1939), 181–9; by 1940 the mental hygienist film, *Make Friends with Fear*, was in production: *Mental Health*, 1 (1940), 51. Between 1947 and 1949 *Mental Health* reviewed popular films, both to comment on their depiction of psychiatric topics, and to vet them from a mental hygiene point of view: *Mental Health*, 7 (1947), 47–8.

[36] K. Walker, 'The Need for a Positive Philosophy of Life', *Mental Health*, 2 (1941), 102–8.

about Nazi eugenic policies.[37] However, the shift was also associated with the wartime valorization of the family unit and studies of the mental health of evacuated children, which indicated that maternal deprivation could be just as important a source of mental problems as inheritance.[38] The emphasis on social and psychological environment mirrored a growing interest in social medicine and was surely, at least in part, a response to a wartime heightened sense of social solidarity.[39]

It is more difficult to detect change in attitudes amongst the public at large. It seems likely that the war would have increased sympathy for the disabled.[40] Significantly, the Disabled Persons Employment Act was the first piece of reconstruction legislation to be passed. Groups were set up to represent the parents of handicapped children. Four new journals appeared which addressed the problems of parents, and correspondence in one of these journals, *Nursery World*, led to the foundation of the National Association for the Parents of Backward Children in 1947.[41] A spate of novels and films criticized conditions within mental hospitals and discrimination against the mentally ill.[42] Literature addressing the issue of mental defectives was rarer, usually written by those with personal experience of caring, such as parents and teachers, and aimed to break down discrimination and describe the difficulty of coping with a handicapped person.[43]

The emergence of a libertarian outcry over mental deficiency is probably the strongest evidence of a change in the attitudes of the general public.[44] Libertarianism itself was not new: it had been the basis for opponents such as Wedgwood and for much of the interwar public suspicion towards the Act. However, during the interwar years, libertarian feeling tended to be outweighed by the eugenic, moral, and

[37] Blacker, 'Eugenic Experiments', 9–19.

[38] J. Bowlby, *Maternal Care and Mental Health* (Geneva, 1951); I. Bowman, 'Maladjustment: The History of the Category', in W. Swann (ed.), *The Practice of Special Education* (London, 1981), 102–21; H. Jones, 'Group Sentiment and Delinquency', *Mental Health*, 8 (1948), 41–4.

[39] D. Porter, 'Changing Disciplines: John Ryle and the Making of Social Medicine in Twentieth-Century Britain', *History of Science*, 30 (1992), 119–47.

[40] K. Robinson, 'The Public and Mental Health', in R. Farndale & H. Freeman (eds.), *Trends in the Mental Health Services* (London, 1963), 16.

[41] *Mental Health*, 7 (1947), 13–14; 'The Mentally Defective Child At Home', *Mental Health*, 11 (1951), 3–4.

[42] Typical of this genre were: J. Vincent, *Inside the Asylum* (London, 1948); or the film *The Snakepit* reviewed in *Mental Health*, 8 (1949), 118–19.

[43] M. Halden, *First Year Out: The Diary of a Special School Teacher* (London, 1943); P. S. Buck, *The Child Who Never Grew* (London, 1951).

[44] G. Jones, *Social Hygiene in Twentieth-Century Britain* (London, 1986), 148–52.

social fears surrounding defectives. So the emergence of the postwar campaign may well have reflected an easing of such fears, as much as a revival of libertarianism. The campaign, orchestrated by the National Council for Civil Liberties (NCCL), began in 1947 with a letter in *Civil Liberty* which exposed exploitation and undue restriction of the liberty of patients within mental deficiency institutions.[45] Over the next decade the NCCL publicized the exploitation of patients in unpaid work, the practice of perpetually extending licences rather than giving defectives their freedom, the wrongful placement of the elderly in certified institutions, and the intransigent and unhelpful response of the Board of Control to complaints.[46] The Council catalogued cases of abuse and in 1950 exposed these in a book entitled *50,000 Outside the Law*. The campaign gathered pace as the press took up the issue, with banner headlines attacking 'Slave Labour', and in 1953 the government was forced to set up the Royal Commission on the Law Relating to Mental Illness and Mental Deficiency.[47] The NCCL gave highly critical evidence to the Commission, and was not fully satisfied by the reforms which resulted, voicing concern over the absence of a lay role in the process of certification and the loose use of the new medico-legal label 'psychopath'.[48] The history of the campaign provides a valuable insight into the nature of public opinion. It had proved possible to motivate a public outcry against abuses of liberty, but it was harder to attract the public to adopt a constructive position on the rights and conditions of those who still needed to have their liberty restricted in order to provide protective care. Moreover, the polemical nature of the libertarian campaign tended to ostracize potential supporters from within the psychiatric profession who might have established permanent structures to monitor rights. Therefore, after achieving the goal of reform—which ironically increased medical control and decreased legal safeguards—the campaign and public sympathy died away, arising again only fleetingly when provoked by periodic media revelations of continuing abuses.

The problem with much of the evidence on changing attitudes is that it is too often peripheral, superficial, or rhetorical, and tends to mask some far more fundamental continuities. For instance, although eugenic rhetoric had become less strident, defectives were still viewed with

45 'Forgotten Women', *Civil Liberty*, 7/11 (1947), 3.

46 'Margaret and the Magistrate', *Civil Liberty*, 9/8 (1949), 5.

47 *Civil Liberty*, 10/4 (1950), 5.

48 *NCCL Annual Report 1958–9*, p. 3.

mistrust as a eugenic, as well as a moral and social danger, and blamed for a whole range of social problems. The belief that mental defectives were the key to the spread of a 'social problem group' was translated into a new anxiety about 'problem families'.[49] Although the influence of the familial social and psychological environment, rather than genetic inheritance, was now emphasized, the concentration of social problems within specific families was still frequently explained by the low intelligence of the parents; even if this mental weakness was not genetically inherited by their children—which was not ruled out—these parents would still produce such miserable homes that their children would grow up backward and socially incompetent and would perpetuate the problem family as they, in turn, deprived their own children.[50] Eugenic anxieties also continued to influence the postwar debate over declining national intelligence. Although the mentally defective were no longer the centre of concern, the debate reflected the continuity of deep-seated moral, social, and eugenic fears that the mentally feeble, dull, and backward—and, more generally, the 'lower' orders—could not control their profligate breeding, and would thereby gradually dilute the mental fitness of the whole population.[51] Just as eugenic ideas were translated, rather than fully disappearing, many of the groups and individuals active in interwar social hygiene and mental deficiency now shifted into the new arena of mental health.[52] In the promotion of normative standards of mental health, they found a new tool with which to target the 'borderline' group which had always been their main object of concern.[53] The selection of Blacker, Secretary of the Eugenics Society, to produce a plan for a future mental health service is an indication of the strategic shift in eugenics, yet also of its persisting influence.

[49] C. P. Blacker, *Problem Families: Five Enquiries* (London, 1952), 12–14.

[50] Ibid. 26–7; and less pessimistically, the wartime surveys of the Pacifist Service Unit in T. Stephen (ed.), *Problem Families* (London, 1947).

[51] D. Thom, 'The 1944 Education Act: The Art of the Possible?', in Smith (ed.), *War and Social Change*, pp. 101–28; E. O. Lewis, 'Intelligence and Social Trends', *Mental Health*, 12 (1952), 18–19; C. Burt, *Intelligence and Fertility: The Effect of the Differential Birthrate on Inborn Mental Characteristics* (London, 1952).

[52] B. Schreiber, *The Men Behind Hitler* (London, *c*.1975), 64–9; Norman, *In the Way Of Understanding*, 151–216; Jones, *Social Hygiene*, 137–56.

[53] W. Langdon-Brown, 'The Task of Social Hygiene in War-Time and After', *Public Health*, 53 (1940), 253.

Central to the notion of a 'revolution in values and expectations' is the idea that a new welfare ideology of 'social citizenship' and 'universalism' emerged in the war and its aftermath.[54] However, there is little direct evidence of this in debates over mental deficiency policy. This may simply reflect the fact that mental deficiency was a low priority and therefore did not attract discussion. It also appears less remarkable in the light of recent historical work which has cast doubt on the extent to which the Welfare State rested on a consistent new philosophical foundation or a transformation in popular values.[55] For all the discussion about a new ideology of 'social citizenship', there was considerable continuity with that notion of the citizen's rights to welfare and responsibility to the state which had underpinned the liberal welfare reforms of 1906–14. Since this earlier period was also crucial in excluding the mental defective from the contract of citizenship, we should not be surprised by the persistence of this status under the Welfare State. In fact, an emphasis on social citizenship and universalism had the potential to place mental defectives in an even more disadvantageous position than previously. For instance, the idea of social citizenship lay at the heart of Beveridge's insurance principle, which already dominated British welfare provision during the interwar period and was now intended as the mainstay of the Welfare State. In this system, social security was to be a safety net, providing a minimum of social security until return to self-independence, and was to be financed by contributions from the working population. However, this vision would seem to exclude mental defectives: such individuals needed permanent, rather than temporary aid; they had special needs, rather than needs which would be detected or remedied by a safety net; and they were personally unable to contribute towards the funding of the system. Presumably, they were to be dealt with in a separate residual system for the socially and morally incompetent, through paternalism

[54] T. H. Marshall, *Citizenship and Social Class* (Cambridge, 1950); R. Titmuss, 'War and Social Policy', in id., *Essays on the Welfare State*, 2nd edn. (London, 1963; first pub. 1958), 75–87.

[55] J. Harris, 'Political Ideas and the Debate on State Welfare, 1900–45', in Smith (ed.), *War and Social Change*, 256–7; J. Harris, 'Did British Workers Want the Welfare State?: G. D. H. Cole's Survey of 1942', in J. Winter (ed.), *The Working Class in Modern British History: Essays in Honour of Henry Pelling* (Cambridge, 1983), 200–14; T. Mason and P. Thompson, '"Reflections on a Revolution"? The Political Mood in Wartime Britain', in N. Tiratsoo (ed.), *The Attlee Years* (London, 1991), 54–70; S. Fielding, '"Don't Know and Don't Care": Popular Political Attitudes in Labour's Britain', in Tiratsoo (ed.), *The Attlee Years*, 106–25.

or authoritarianism, rather than by right as citizens, and perhaps cared for by the voluntary groups who continued to have an important place in Beveridge's vision.[56] On the other hand, since Beveridge's plan was premised on the idea of full employment, it also brought the prospect of a substantial shift in the effective borderline of social efficiency, and therefore in an indirect sense it did have the potential to alleviate the problem of mental deficiency.

Perhaps the most radical idea associated with the Welfare State was that of universalism, though here again it is unclear whether this was beneficial for a group such as the defective. Universalism envisaged the replacement of systems of health care and education, open only to the poor and carrying a consequent social stigma, by systems open and equal for all. By bringing the middle classes into 'national' systems of care, universalism promised rising standards of services. However, because of the Welfare State's limited resources, mental defectives and other such groups, who had previously been a high priority of the welfare system, and who had special needs not catered for by universal services, were likely to lose out in the struggle for services and were least likely to benefit from rising standards.

One person who did directly address the implications of the new Welfare State for mental defectives was Secretary of the Eugenics Society, C. P. Blacker. He argued that the Welfare State could easily become overloaded because of excessive demands from members of the social problem group. However, he hoped that this could be prevented, since the administrative system of the Welfare State would also provide the most comprehensive system ever for surveillance and control of eugenically undesirable individuals.[57] Blacker's views are too extreme to be taken as representative, but they do reflect a feeling in some quarters that the Welfare State could be undermined by the burden of the irresponsible and least able members of the community, such as mental defectives, who placed constant demands on welfare services but made no contribution in return.

Finally, there are also reservations to seeing the war as marking a watershed for sympathetic and humanitarian attitudes towards mental deficiency. Evidence collected by the Royal Commission of 1954–7 suggests that popular attitudes towards those with serious mental

[56] J. Harris, *William Beveridge: A Biography* (Oxford, 1977); W. Beveridge, *Voluntary Action: A Report on Methods of Social Advance* (London, 1948).

[57] Blacker, *Neurosis and the Mental Health Services*, ch. 14; PRO MH 76/115, Blacker, 26 Apr. 1943.

problems continued to be dominated by prejudice and fear. Although there was growing sympathy towards those suffering from mental breakdown, a sharp dividing line was drawn between this group and the 'really mad'.[58] As this suggests, we need to be cautious in assessing the meaning of the NCCL campaign: the popular support for the campaign reflects libertarian concern over wrongful confinement of high-grade defectives, but there was far less interest in the seriously handicapped, who continued to be viewed as objects for paternal care and not as the possessors of rights.

In summary, the war and postwar years of welfare reform may have seen a slight shift in the boundary of mental deficiency, but attitudes towards those remaining below this line seem to have changed very little. The primary shift in attitudes, and one which does echo broader ideological tendencies in the formation of the Welfare State, was a growing interest in constructing the mental health of the whole population. However, this interest had negative implications for mental deficiency services, for it diverted limited resources to the broader population. Moreover, although a new emphasis on mental health education encouraged public interest in the area, this was usually lukewarm and superficial.[59] People were fascinated by the glamorized image of psychiatry which was such a frequent subject in the motion pictures of the era, and they were attracted to new cults, such as dianetics, which used some of the popular psychiatric language, but they remained fearful and ignorant of the certified mentally ill and defective.

THE LEGISLATION OF 1944–1948

Although mental health legislation was not directly reformed, between 1944 and 1948 a series of legislative changes had a significant impact on the care of mental defectives. The style of these reforms can be seen as progressive, in that they promised to make services more comprehensive yet voluntaristic, and they aimed to destigmatize mental health care and the position of the mentally defective in society. The Disabled

[58] *Roy. Com. 1954–7*, pp. 22–4; Robinson, 'Public and Mental Health', 12–20.

[59] C. Greenland, 'Psychiatry and the Community', *Mental Health*, 10 and 11 (1952), 14–19, 111–21. Attacks on popular psychiatry: *Mental Health*, 11 (1952): 'Editorial', 102; review of L. R. Hubbard's dianetic cult, 137.

Persons Employment Act of 1944 obliged local authorities to keep a register of physically and mentally disabled persons and to provide occupation centres for their rehabilitation and training. The Act also introduced a quota scheme which obliged large employers in certain trades to ensure that at least 3 per cent of their workforce were disabled persons.[60] The Education Act of 1944 extended special-school education up to the age of 16. Testing for mental deficiency was recommended from the age of 2 and notification of defectives on leaving school was made a duty rather than an option.[61] In the hope that teachers and parents could be made to view notification as less of a stigma than in the past, the Handicapped Pupils and Medical Services Regulations of 1945 replaced the nomenclature of 'mental defective' with that of 'educationally subnormal' and 'maladjusted' pupils and accepted that these children could be taught in normal schools if facilities were adequate.[62] The National Insurance Act also seemed to promise some less stigmatized alternatives to institutional care. The Act made it possible for parents of defectives to receive social security cash payments to help home care. In place of the abolished public assistance institutions, it proposed specialized residential care for various categories of dependants, including the elderly and maladjusted children. As such, it presented alternatives for many individuals previously placed in mental deficiency institutions.[63] Finally, the most significant piece of legislation for mental deficiency was the creation of the National Health Service in 1948. Under the NHS Act local authorities were replaced by regional health authorities as the managers of the hospital service (including mental institutions), the Ministry of Health replaced the Board of Control in charge of supervising hospitals, and central government's power to finance community care was extended to include voluntary supervision. At least superficially, therefore, there was an attempt to break down the stigma surrounding the mental health services by integrating them into the new welfare system.[64]

The primary problem with viewing the formation of the Welfare State as a turning-point for mental health care is that mental health

[60] Disabled Persons Employment Act, 1944 (7 & 8 Geo. VI, c. 10); H. Bolderson, 'The Origins of the Disabled Persons Employment Quota and its Symbolic Significance', *Journal of Social Policy*, 9 (1980), 169–86.

[61] Education Act, *1944* (7 & 8 Geo. VI, c. 31).

[62] *Handicapped Pupils and Medical Services Regulations, 1945* (London, 1945).

[63] National Assistance Act 1948 (11 & 12 Geo. VI, c. 81).

[64] National Health Services Act 1946 (9 & 10 Geo. VI, c. 81, ss. 49–52).

legislation itself was left unchanged—even though it had been marked out as a priority area for reform as early as 1938.[65] The omission of mental health reform was excused on the pretext that it was impossible without the recommendations of a Royal Commission.[66] In 1948, there was a minor review of the 1913 Act, to adjust for the administrative changes caused by the NHS, but the majority of the regulations, forms, and mental tests remained unchanged, perpetuating attitudes and standards set over thirty years earlier.[67] The Board of Control attempted to overcome some of these outdated practices via administrative circulars and orders. For instance, in 1952 it issued an order to allow mental defectives to enter institutions on a voluntary and temporary basis.[68] However, this type of executive action was always liable to be viewed as unconstitutional by libertarian critics and had to be used with great caution.[69]

Superficially, the NHS Act may have placed mental health under the same administrative structure as the rest of the health system, but in reality it remained apart. Although the Act placed responsibility for approval of building schemes and financing of mental hospitals under the Minister of Health, such duties were delegated to the Board of Control, which acted as an internal department of the Ministry.[70] This left the Board, with even less authority and staff than previously, to perform an enormous mass of inspectorial and regulatory duties.[71] If the Board did fail as a legal watchdog, this was not simply due to the failure of legalism, as the 1954–7 Royal Commission appeared to conclude, but was also the result of inadequate staffing and resources.

There is a case for arguing that the mental health services would have been better off had they been excluded from the National Health Service—a possibility which was seriously proposed as late as 1944. Leaving the services in municipal hands would have enabled a considerable saving of between £185m. and £130m.[72] And it was unclear

[65] PRO MH 80/82, Third Office Conference of Ministry of Health, 31 May 1938.

[66] PRO MH 77/25, Brock (consulted for the first time) 25 Aug. 1941; Brock to Maude, 4 Sept. 1941.

[67] *Roy. Com. 1954–7: Minutes of Evidence* (London, 1954), 832–4.

[68] Ministry of Health Circular 5/52 (1952); *Roy. Com. 1954–7*, p. 77; PRO MH 51/611, Barter to R. Rieu, 10 Mar. 1951.

[69] *Civil Liberty*, 5/3 (Oct. 1944), 1–2.

[70] PRO MH 51/678, 'The Future of the Board', Brock, 14 Aug. 1944.

[71] *Roy. Com. 1954–7*, p. 256.

[72] F. Honigsbaum, *Health, Happiness and Security: The Creation of the National Health Service* (London, 1989), 40, 195, 197.

whether the reorganization of the hospital system brought any benefit for the mental health services. As self-contained units, mental hospitals gained little from being grouped with general hospitals, and were in danger of having their resources diverted. Where control had been taken away from local authorities and placed in the hands of joint bodies during the interwar period, the results had not been promising.[73]

Among the leading critics of regionalization were those municipal authorities that lost power as a result. This included the LCC, which had been the largest hospital authority in the world, with 35,000 mental hospital beds costing £2 million per annum. Rural authorities may have benefited from combining together to provide adequate care, but for an area such as London it meant a dilution of standards, as administration was disrupted, services were shared with the outlying home counties, and bed provision fell from 26/1,000 to 17/1,000.[74] A second major critic was the Mental Hospitals Association (MHA), which represented the local authority committees responsible for monitoring mental deficiency and mental illness hospitals. The MHA argued that it was inappropriate and premature to include mental health services in the NHS before the more urgent issue of legislative reform had been tackled. It attacked the new administrative bifurcation between hospital care, run by the regions, and community care, still in the hands of the local authorities. It also condemned the proposed funding levels, which suggested that mental health would be more neglected than ever once it was placed alongside general public health.[75] The MHA found an ally in the Confederation of Health Service Employees (COHSE). Both organizations felt that medical persons were overrepresented on the administrative councils and committees of the regional hospital scheme. The MHA wanted lay representation from the old local authority

[73] PRO MH 77/25, Brock, 25 Aug. 1941; PRO MH 77/25, Brock to Maude, 4 Sept. 1941.

[74] PRO MH 77/25, F. W. Menzies, 'Memorandum on London Hospital Policy', Aug. 1941. LCC, *Report of the County Medical Officer of Health and School Medical Officer for 1948*, pp. 91–5; LCC *Report of the County Medical Officer and School Medical Officer for 1949*, pp. 79–81.

[75] PRO MH 77/71, MHA report on 'A National Health Service', Apr. 1946; J. Barclay, 'The Case for Advance Status', *Municipal Journal and Local Government Administrator* (5 Oct. 1945).

committees, and COHSE wanted Labour's Minister of Health, Aneurin Bevan, to live up to his rhetoric about 'industrial democracy' by placing members of the nursing and ancillary professions on to consultative committees.[76]

Although the Board of Control dismissed the MHA's criticism as self-interested, many of its concerns turned out to be justified.[77] The most serious problem was that the removal of hospital services from local authorities also separated them from community care services. This was particularly harmful to mental deficiency services, where a close interaction had developed between institutional and community care during the interwar period. The new hospital system enabled some co-operation to develop between mental illness hospitals and general hospital out-patient clinics, but there was little potential overlap in an area such as mental deficiency. Co-ordination with the outside community was much more important. However, with the regional hospitals attracting patients from numerous local authorities, each with their own community services, an integrated service was far less likely than under municipal control.[78] Finally, in making voluntary supervision the direct responsibility of local authorities, the NHS Act stimulated the demise of the voluntary sector, with the number of local associations for mental health falling from forty-two to twenty-six by 1948.[79] In an era of austerity these voluntary groups found it hard to rely on charitable income alone and were forced to close or to provide services on a profit-making basis. With responsibilities for community care taken away from these voluntary activists, there was perhaps more danger than before that any future neglect of this low priority area would go unnoticed.

The National Assistance Act, Disabled Persons Employment Act, and Education Act were also not as effective as promised. With limited space available in the residential homes under the National Assistance Act, the long-term mentally ill and defective had to remain in mental institutions.[80] Under the Disabled Persons Employment Act, mental defectives and the permanently disabled were a low priority compared

[76] PRO MH 77/71, joint memorandum MHA and COHSE, 11 June 1946; verbatim report of the interview between Chairman of the Board of Control and representatives of MHA and COHSE, 15 July 1946; proposals for a Joint Advisory Council of the Mental Health Services for England and Wales.

[77] PRO MH 77/71, P. Barter to Minister, 11 Feb. 1946.

[78] K. Jones, *Mental Health and Social Policy, 1845–1959* (London, 1960), 154–9.

[79] *Mental Health*, 8 (1948), 21.

[80] *Roy. Com. 1954–7*, pp. 214–16, 239–40.

to the rehabilitation of disabled servicemen, and the government was unwilling to upset employers or workers by enforcing the employment of mental defectives.[81] Finally, the move away from special schools to integration also had a negative side, since it became harder for parents to obtain specialized education for mentally defective children.[82] Likewise, although there was expansion of child guidance services for children with 'interesting' psychiatric problems and an increased interest in the 'maladjusted' child, parents of mentally defective children were often forced to struggle alone with little aid.[83]

The limitations of provision for mental health under the Welfare State are highlighted when contrasted to Blacker's proposals in *Neurosis and the Mental Health Services*. The government had originally commissioned this survey, withdrawing Blacker from the army and appointing him under the Board of Control. His initial remit had been to examine the organization of wartime psychiatric services, but the influence of the Beveridge Report changed the direction of the project to include consideration of postwar reorganization.[84] Blacker's plan can be seen as a mental hygienist design for a national mental health system. The mental deficiency and mental illness hospital was to be at the centre of a system which expanded out into the community via out-patient clinics, child-guidance clinics, and child-guidance centres. Psychiatric education was central to the scheme: integration of psychiatry into general medical training would encourage general practitioners to send patients to mental clinics; specialist courses would train the army of psychiatric social workers, psychologists, and psychiatrists needed to staff the clinics; and the high profile of psychiatry in teaching hospitals and universities would filter out into the medical profession and the educated classes as a whole. Finally, Blacker called for a substantial increase in facilities, recommending a national force of 1,000 psychiatric social workers, and 2,500 beds for defectives in each area of a million inhabitants.[85] The Ministry of Health admitted that the findings highlighted the unevenness of provision, but saw any large-scale

[81] Bolderson, 'Disabled Persons Employment Quota'.

[82] *Mental Health*, 7 (1947), 13–14.

[83] 'The Mentally Defective Child at Home', *Mental Health*, 11 (1951), 3–4. On interest in the 'maladjusted': E. Wellisch, 'A Colony for Maladjusted Children', *Mental Health*, 10 (1950), 6–8; Bowman, 'Maladjustment', 102–21.

[84] PRO MH 76/115; BL 4; BL 14.2–3.

[85] Blacker, *Neurosis and the Mental Health Services*, 94, 82.

extension of services as unfeasible. Solutions were to be limited to regionalization, rather than tackling real deficiencies.[86]

Blacker's was not the only voice calling for more fundamental reform. In 1943 the Medical Planning Committee of the RMPA proposed a series of significant expansions in the role of psychiatry.[87] It accepted the desirability of changing the title of the Board of Control, but wanted its role to continue, with extended powers and increased medical representation. At the local level, it called for the appointment of mental medical officers of health of equal status to the medical officers of health, regional health committees with co-opted psychiatric members to replace local authority control, and medical rather than lay control of institutions. Like Blacker, the RMPA wanted more psychiatry within general medical education. It also recommended that mentally disturbed children be placed under psychiatric, rather than educational control. Finally, it was in favour of an extension of community care, licensing, and hostels, but wanted these services under the control of psychiatrists as a complement to their institutional work.

To understand why the changes to the mental health services were so limited in comparison to the proposals of Blacker or the RMPA, one has to look at the dynamics of the policy-making process. Because of the decline in independence of the Board of Control since 1913, the low status of the mental health services, and the political fear that the issue, once raised, would expose abuses and prove controversial, reform of mental health legislation had become a low priority within government, and remained so, despite the interest promoted by the wartime challenges. Reflecting its marginal status within government, the Board of Control was evacuated to Lancashire during the war. Here, separated from the centre of government, its negotiating position was undermined. As the Board reported in 1946:

> [evacuation] gravely handicapped our work and militated against the effective handling of many urgent questions that arose, including some of the work on the proposed integration of the Mental Health Service in the National Health Service.[88]

The retirement of the Board's two most long-standing and respected officials, the Chairman, Brock, and Senior Commissioner, C. H. Bond, further weakened its position.

[86] PRO MH 76/115, Ministry of Health note on survey, 20 Mar. 1944.
[87] *JMS* (suppl.), 89 (Oct. 1943), 18–20.
[88] *BOC 1946*, p. 5.

Faced with a similar dilemma to that presented by the formation of the Ministry of Health in 1918, the Board again attempted to defend its independence, but could not resist integration with the NHS since this promised to help erode the stigma and isolation of the mental health services. A compromise was reached in which the Ministry of Health took on the Board's staff to compensate for its own lack of expertise.[89] The resulting loss of executive power, while remaining in day-to-day administrative control, placed the Board's administrators in a vulnerable position when competing for resources and power within the NHS. This was reflected in the limited number of psychiatric specialists appointed within the administrative structure of the NHS, with just four out of forty-one members on the proposed Central Advisory Council and a minimum of just two psychiatric specialists on Regional Hospital Boards.[90]

Similarly, the plans of the RMPA were undermined by the low status of psychiatry within the medical profession.[91] It was not directly consulted about the NHS Act, and its recommendations were dismissed by the Chairman of the Board as 'controversial' and 'disputable' 'pet proposals'.[92] Typically, any of its attempts to expand the reach of services were unsuccessful, as with the proposals for child psychiatry which were defeated by the educational lobby.[93] On the other hand, in alliance with the powerful medical profession, the RMPA did succeed in protecting and bolstering psychiatric control over the mental hospital system, with administration separated from the lay control of local authorities, and psychiatrists given representation on the new Regional Boards.[94] In many ways, this outcome was reminiscent of the victory of the doctors in negotiations over general health care; medical control over the psychiatric system was maintained at the expense of extension of mental health care into the mainstream of social welfare, just as general practitioners and hospital consultants benefited at the cost of health centres and public health services.

89 PRO MH 51/678, Brock, 'The Future of the Board', 14 Aug. 1944.

90 PRO MH 77/71, MHA/COHSE, 11 June 1946; C. Webster, 'Psychiatry and the Early National Health Service: The Role of the Mental Health Standing Advisory Committee', in G. Berrios and H. Freeman (eds.), *150 Years of Psychiatry, 1841–1991* (London, 1991), i. 103–16.

91 Webster, 'Psychiatry and the Early NHS'.

92 PRO MH 77/71, Brock, 13 Oct. 1943.

93 Ibid. Brock, 30 Jan. 1945; Bosworth Smith (Board of Education), 26 Apr. 1945.

94 *JMS* (suppl.), 89 (1943), Medical Planning Committee, 14–15; *JMS* (suppl.), 91 (July, 1945), 'Report of the Medical Planning Committee of November 1944', 19–26.

CONCLUSION

In this chapter, I began by considering the historiography on the development of the Welfare State. From the perspective of the history of mental deficiency, the models of both evolution and transformation are too simplistic, neither consistently withstanding scrutiny. The mental deficiency system was evolving, but towards a mixed economy of welfare rather than the centralized 'classic' Welfare State. Equally, legislative change, the impact of war, and changes in values and expectations all had some transformative effect, but just as remarkable were the continuities with the past. These included the persistence of the mental deficiency institutions, despite all their problems; the translation from the defective being seen as the source of the 'social problem group' to the source of the 'problem family'; and the continued stigmatization and moral, social, and eugenic fear surrounding the mentally defective.

Developments of mental health policy are better comprehended through an understanding of the changing dynamics of the policy-making process. For all its problems in the interwar period, mental health care was relatively well served, having its own central government department, active mental welfare pressure groups who lobbied government for more resources, and the potential for local government integration and innovation of community and institutional care. The war and the subsequent welfare settlement furthered a process of rationalization and centralization which eroded the independence of these groups. There was also a significant shift in priorities, with the result that welfare resources were redistributed from 'problem groups', such as the mentally defective, to the general population. The representatives of the interests of mental health care—in central government, the medical profession, and voluntary groups—all became far less influential in such circumstances. The most successful of these groups were the psychiatrists, who were able to protect and strengthen their control over institutional care. Those interested in developing integrated and preventive policies, and in extending mental health services to the whole population, were less successful, including mental hygienists and clinic-based psychiatrists, voluntary mental welfare groups, psychiatric social workers, and public health officers in local government. The policies which might have had a greater impact on the problem of mental deficiency, if fully realized, were those which dealt with the issue only indirectly, such as full employment, better housing, comprehensive social security, and family allowances; by raising standards of living, assisting families, and providing more opportunities for high-grade defectives to support themselves, these policies had the potential to raise many defectives over the vital barrier of social inefficiency, effectively redefining mental deficiency itself.

The history of mental deficiency in the formation of the Welfare State highlights the uneven nature of ideological change: the impact of 'democratization', 'universalism', and 'social citizenship' was mediated by status. The shift from a mixed economy of welfare towards a more centralized mental health service under the NHS did remove power from an unaccountable array of pressure groups and voluntary organizations, but it also concentrated power in the hands of central bureaucrats and the medical profession; as far as mental deficiency policy was concerned, the democratic implications were mixed. The universalism of the Welfare State raised particular problems for a group such as the mentally defective, who could not defend their own interests in the competition for welfare resources. Finally, since the theory of citizenship underlying the Welfare State was premised on notions of social responsibility, it maintained (perhaps even confirmed) mental defectives in the position of second-class citizens, and paid little attention to the special needs and rights of those who lay outside the boundaries of 'social citizenship'. From the perspective of mental deficiency, therefore, the advent of the welfare state was a mixed blessing, and it certainly did not mark an ideological watershed.

EPILOGUE

Although the main target of the NCCL campaign had been abuses within the mental deficiency system, the resulting Mental Health Act of 1959 addressed the whole of the mental health field; it was now acknowledged that the interwar attempt to clearly separate the problems of mental illness and deficiency was impossible.[95] The breakdown of what had always been a somewhat artificial barrier pointed the way to a new division, already tacitly emerging in the interwar period, between high-priority and well-resourced services for the curable, and a continuing decay and neglect of services for the incurable and chronic. The

[95] J. M. Crawford, 'Mental Deficiency: A Misnomer', *Mental Health*, 13 (1953), 9–12.

Act made voluntary treatment rather than legal certification the normal entry point into the mental health services. As such, the Act can be seen as a belated victory for the psychiatric profession, and one which brought the mental health services in line with the open access of the NHS.[96] Reflecting increased public confidence in medical expertise, psychiatrists now played the crucial role in deciding whether to treat and release patients. But as a consequence there were fewer legal safeguards to protect patients—an ironic twist of fate, considering the libertarian campaign which had led to reform. Moreover, with mental health care now simply a medical issue, the Board of Control's legal role was considered redundant and it was completely abolished, leaving mental patients without this governmental watchdog to monitor their rights.

There was a belief that problems could be solved by administrative reorganization, changing names to break down stigma, and placing even more responsibility into medical hands. Questions of resources and finance, vital to any effective caring policy, were either evaded or relied unrealistically upon reallocation of patients from hospital to community care, even though there were long waiting lists to quickly refill any emptied beds.[97] The Act replaced the terminology of 'mental defective', 'feeble-minded', imbecile', and 'idiot', with the equally problematic label of 'mental subnormality'. This linguistic sleight-of-hand was intended as a liberalizing exercise. However, the new legal category of 'psychopath' enabled medical control to reach that 'borderline' area, lying between mental illness and deficiency, which had always been of greatest moral, social, and eugenic concern to the mental deficiency and mental hygiene lobbies.[98]

The terminology of 'community care' replaced 'supervision', but again this linguistic change belied novelty, especially in the case of mental deficiency where there had been considerable development of care in the community during the previous forty years. By describing social security assistance as 'community care', an illusory picture of expansion was produced.[99] The attempt to shift people, previously catered for in mental deficiency institutions, into local authority residential homes under national assistance was akin to a return to the workhouse, especially as the accommodation was often located in old

[96] C. Unsworth, *The Politics of Mental Health Legislation* (Oxford, 1986), 230.

[97] *Roy. Com. 1954–7*, pp. 716–19.

[98] B. Wootton. *Social Science and Social Pathology* (London, 1959); Jones, *Social Hygiene*, 152–3.

[99] *Roy. Com. 1954–7*, pp. 49–65.

poor law buildings. This 'community care' was really residential care under a new name.[100]

Although community care seemed to offer a cheaper alternative to hospital care, there was little consideration of what either the cost or the meaning of effective community care would be for a group such as the mentally subnormal.[101] Unless there was a massive injection of effort, money, and manpower, community care was likely to result in abandonment by the state.[102] As Titmuss pointed out, the level of expenditure per head when inflation was taken into account had probably fallen over the 1950s, despite all the talk of community care innovation; shamefully, government expenditure in 1960 was less than that spent on compensation for fowl pest.[103] Community care on this scale offered little realistic prospect of providing an alternative to institutional care, and diverted attention from the serious need to improve institutional accommodation. In 1962 the Hospital Survey acknowledged that, although it might be possible to reduce drastically the number of beds for the mentally ill, increased bed space was in fact needed in institutions for the mentally subnormal, especially since the life span of residents was gradually increasing.[104]

Without a rationale for care based on rights—and as the desire to provide eugenic and moral control diminished, philanthropic zeal lessened, and libertarian concern was placated by legislative reform—there was a greater danger of neglect than ever. Mentally disabled people, lacking a voice of their own or special rights, were neither in a position to protect themselves against potentially coercive medical control, nor able to compete for the resources of the Welfare State.[105] The Mental Health Act of 1959 may therefore have

[100] Ibid., pp. 29–30.

[101] M. Bulmer, *The Social Basis of Community Care* (London, 1987), esp. 1–37.

[102] A. Scull, *Decarceration: Community Treatment and the Deviant—A Radical View*, 2nd edn. (Cambridge, 1984).

[103] R. M. Titmuss, 'Community Care: Fact or Fiction?', in Farndale and Freeman (eds.), *Trends in the Mental Health Services*, 229.

[104] M. Craft, 'A Modern Hospital Service for the Subnormal', in Farndale and Freeman (eds.), *Trends in the Mental Health Services*, 164. Falling mortality rates in institutions had produced an aged population. Those over 65 in mental hospitals had risen from 14.8 per cent of men and 19.7 per cent of women in 1938, to 19.1 per cent and 27.6 per cent respectively in 1948: *Annual Report of the Board of Control to the Lord Chancellor 1950*, p. 4.

[105] A critique of medical control and a call for greater attention to the rights of patients emerged in the reform movement of the 1970s: L. Gostin, *A Human Condition* (London, 1975). The result was a return to greater legalism in the mental health legislation of the 1980s: Unsworth, *Politics of Mental Health Legislation*, 315–51.

reformulated the problem of mental deficiency, but the longer-term problem of how to provide effective care for members of society with mental disabilities continued to be a considerable dilemma for the Welfare State.

Conclusion

It was no coincidence that mental deficiency became a 'problem' in the early twentieth century. The problem emerged in response to changing social conditions in the late nineteenth century which made it harder for mental defectives to cope in the community: in particular, changes in the labour market made it more difficult for defectives to contribute to their own support, and families found it increasingly difficult to provide care and control for mentally defective dependants at home. It also needs to be placed in the context of an ongoing political debate about the definition and construction of responsible citizenship, a debate which was heightened by the end of a property-based franchise, and by the extension of political citizenship to virtually all adults. Mental deficiency provided a new biological boundary-line for good citizenship, with social, moral, and intellectual incompetence all deemed to be manifestations of an underlying mental defect. The emergence of the problem was also intimately related to the dramatic expansion of institutional sites that revealed an apparently ineducable, unreformable, and unmanageable section of the population: the introduction of a universal and competitive system of elementary education was particularly important, but so too was the profusion and specialization of institutional care in the late nineteenth century, with mental defectives recognized in asylums, prisons, inebriate reformatories, rescue homes, industrial and reform schools, and poor law institutions. Though to a lesser degree than is sometimes claimed, the recognition of the problem was made possible by new technologies of detection—in particular, the complex apparatus of classification and mental testing. But perhaps most importantly, the recognition of the emergence of the problem of mental deficiency was rooted in the moral and ideological environment of the time. This was a far more complex ideological environment than that described in eugenics-centred accounts. Certainly, eugenics was an important factor in concerns about mental deficiency, but it was not always paramount. Equally important were humanitarian ideas about helping the incapable, concerns about the inherent social incompetence of these individuals, and moral anxieties about controlling those who lacked character and self-control. The most fundamental cause of concern about mental defectives was that they marked the boundary between responsible citizens and the naturally irresponsible. Like children, they deserved care, but also needed control. Unlike children, the care and control which they were to receive needed to be permanent.

The passage of the Mental Deficiency Act, by such a resounding majority, and with the support of conservatives and progressives alike, reflected the deep-rooted belief that there was a fundamental difference between the mentally defective and the normal. A policy of segregation, which was deemed to be unacceptable on libertarian grounds for virtually every other group in the population, including habitual criminals and inebriates, was nevertheless considered acceptable for mental defectives; by the very nature of their defect they were divorced from an otherwise universal liberal contract of individual rights. The provision of permanent care attracted conservatives who wanted to protect the community from a eugenic, social, and moral threat, but it also attracted progressives who regarded the specialization of care as a humane move and often looked favourably on eugenics as an example of the way scientific principles might forge a more modern social policy. Thus the ideological motivations behind the 1913 Act were complex, and its passage cannot be used as a simple litmus test for the influence of eugenics on British social policy; in fact, the Act's associations with the eugenics movement generally hampered its chances of success in a political environment which was highly suspicious of acting on the recommendation of the 'expert'.

Interpretations of the sterilization debate of the 1930s have also been found wanting in a number of respects. It is right to emphasize the increasing importance of suspicions of eugenics within the labour movement. However, Labour opposition in the parliamentary vote of 1931 was far from being clear cut, and over the decade as a whole the staunch opposition from Catholics was perhaps equally important; both would have been far less significant if the governments of the period had not been so politically cautious. But more crucial still was the fact that sterilization did not provide a simple solution to the problem of mental deficiency: support for segregation and community care meant that sterilization would only ever be a supplement, and not an alternative. A second problem with interpretations of the sterilization campaign is that they tend to assume that its failure can be used as a simple gauge for the limited appeal of eugenics in Britain. This ignores the fact that the extensive policies of segregation and community care which remained in place were themselves deeply imbued with eugenic principles. Moreover, eugenicists saw the sterilization campaign as only one part of a broader tactical reorientation—a reorientatation which did have a degree of success—and their interest in the issue waned as the nature and aims of the eugenics movement shifted in the 1930s. Ultimately, sterilization was not defeated by lack of support for eugenics; it was defeated because it infringed fundamental ideas about individual rights and social responsibilities within the British democracy. There was a deep-rooted caution and suspicion among most politicians about accepting policies that depended on 'expert' knowledge, especially any which was as scientifically and ethically controversial as eugenics, or which placed undue powers of intervention and control into the hands of state officials. Equally important, however, was the recognition that, although it was reasonable to remove all liberty from mental defectives in order to provide them with permanent care, there was no justification for wounding them physically through sterilization or for casting them out of institutions with no protection or care; mental defectives may have lacked the right to freedom, but it was society's duty to provide care and control in what was deemed to be their own best interests.

Once the 1913 Act and the sterilization debate are reframed in such a way—with the issues of rights, duties, and responsibility highlighted, and that of eugenics given a less prominent and more realistic significance than is often the case—a series of confusions and apparent contradictions can be explained. We begin to understand why segregation was possible for mental defectives, but not for the host of other groups for whom it was proposed in the early twentieth century. We are no longer forced to account for the anomaly of eugenics apparently having such a considerable influence on policy in 1913, but so little in the 1930s. And it becomes clearer why so many progressives could support the policy of segregation.

Perhaps the chief problem with current histories is that they produce snapshots of the policy debates of 1908–13 and the early 1930s, and leave us with little definite understanding of the periods before, between, or after. Lack of thorough research has led to a series of confusions in the historiography. When we go beyond the rhetoric of debate, it is clear that the popular myth that the Mental Deficiency Act was simply a tool to control young women with illegitimate children is a gross caricature: such women—if they could also be proved to be intellectually defective—were a critical concern, but in total, women did not outnumber men disproportionately. Men were just as prone to be certified because they transgressed accepted gender role as were women: in the male case, this usually involved unemployability or habitual criminal behaviour. The idea that plans for segregation largely failed, and that this was why calls for sterilization emerged in the 1930s, is again misleading. For obvious reasons, segregation was slow to develop at first, hindered as it was by both the war and the priorities of the postwar reconstruction. However, by the start of the 1930s, over 30,000 mental defectives had been segregated, there were 25,000 under forms of community care and control, and, in total, nearly 86,000 had been ascertained. The calls for sterilization were directly related to the problems of these two areas of care, but in the final analysis the feasibility of sterilization was undermined by the relative success of the segregative and community care solutions—both areas of considerable importance in the history of interwar British social policy, but areas which have received virtually no historical attention.

What the snapshot approach also ignores is that policy emerges out of a complex and ongoing political process. There was often a gap between the intention, and the effect of policy; between legislation and the reality of policy at the point of delivery. Because of the problems posed by adjusting to democracy—adjusting to the increased business and more complex legislation demanded by the democracy, and to the new political restraints on legislation brought about by increased power for the working class and for women—the legislature avoided the issue of mental deficiency, and effectively delegated responsibility and influence to lower levels of the policy-making process. Civil servants were given considerable influence in interpreting ambiguous legislation, in distributing finance, in regulating standards, and in advising local authorities through administrative orders. Further down the scale, the professionals and social workers who actually implemented policy at the ground level had a large say in deciding who the Act was to be directed at, and what form this care would take. We can descend even further than this, for in certain respects the political interaction and negotiation between defectives, their families, local voluntary groups, and local authority officials also shaped policy in the process of implementation.

This analysis of policy-making as a multilevel process has a number of implications for understanding the development of welfare during the period. It has shown how political and social interpretations can usefully be integrated. Policy, here, has not simply been represented as the outcome of legislative or civil service 'high politics'; the 'low politics' or social history of family care played an integral role in the policy process. Equally, to understand the 'high politics' of mental deficiency, we need to look at the social history of the participants, for it was often their professional backgrounds, religion, or gender, as much as party, which accounted for their political positions on 'non-party' issues.

The multilevel analysis reveals problems with the rather simplistic notion of the state which is sometimes adopted in 'state-centred' accounts of welfare development. Here, the state has been revealed to be a complex and at times rather fluid entity. The analysis of central government exposes the divisions which could run between and within government departments, and it demonstrates that a body such as the Board of Control could occupy an uneasy territory, neither wholly within nor outside government, and could act as a conduit for pressure groups and professions to influence and even enter government. The notion that one can confine an analysis of the state during this period to the central level of government has also been questioned. In an area such as mental deficiency policy, local government remained crucially important in both the implementation and formation of policy. And at this local level, the boundaries between the official and non-official, or the statutory and voluntary, remained even more fluid than at the central level. In fact, it would be impossible to understand the construction of mental deficiency policy if we confined ourselves to the state, strictly defined, and ignored the important roles played by voluntary, professional, and pressure groups.

Though the influence of voluntary and professional groups was increasingly important in the construction of policy, some serious limitations have been exposed in the idea that social policy was falling under the control of 'professional society'. Voluntary organizations remained important throughout the period, and professionalization was often held back because it increased demands on resources. Sterilization, segregation, and community care were all shaped by a common set of 'professional' desires for efficient, modern, and scientific solutions to social problems, but these ideals were highly malleable and could be adopted by all sides of the policy debate; simply because the rhetoric of eugenics and national efficiency were present in the policy debate does not mean that they were either dominant or even coherent sets of ideas. Ultimately, professional self-interests, as often as any 'professional ideal', guided the policy position of professional groups. Professional society remained divided by such self-interests and by the political fault lines of gender, class, and religion which cut across all political strata of the era. The professional ideal was increasingly important in social policy, but it was matched by concerns over liberal rights, social duties, and morality. Moreover, the professionals themselves were not immune from such concerns: their public support for the expansion of expert-driven social policy was tempered by a far more traditional desire for respectability and 'gentlemanly status', which turned them away from overt political lobbying on morally controversial and interventionist policies, such as sterilization. In a certain sense, the most important contribution of the professional ideal was much more indirect: in advocating a new meritocratic social order, it helped to position intellectual achievement at the very centre of contemporary social concern, and it thereby helped to provide the environment in which intellectual incompetence would be pathologized, and in which mental deficiency would be constructed into a major social problem.

In contrast to the centralization and rationalization of administration under the Welfare State, interwar mental deficiency policy was characterized by a mixed economy of welfare. It was also characterized by being a fairly high-profile issue and by being relatively well resourced during the interwar period. The existence of a special government department, a national network of voluntary organizations, and specialized departments of local government ensured that mental deficiency would be protected from neglect. With their disappearance in the postwar period, this was far less likely. In the early twentieth century it had been those groups regarded as helpless or dangerous who were most likely to attract the resources of the state. Since the mentally defective qualified on both counts, funding of their care and control was relatively generous. However, with the state's responsibility extending to the provision of universal services under the Welfare State, the special needs of the mentally defective were an increasingly low priority; so much so, that essential reform of mental deficiency legislation was virtually ignored in the legislative programme of 1944–8. In some respects, the fluid nature of the interwar administration benefited mental deficiency policy: it enabled the activities of voluntary groups to be engaged, it facilitated experimentation, and it provided a network of care which stretched between institution and community, knitting together the educational, poor law, prison, and mental health services into a web of provision for the mentally defective. Where local conditions were favourable, such as in London, this could result in impressive services; however, where conditions were less favourable, provision could be threadbare. In reaction to this uneven geography of provision and to the often chaotic structure of local administration, and because of broader processes of bureaucratic and professional modernization, the interwar period itself saw a significant evolution towards the more rationalized structures of the postwar world. However, the inflexibility of these structures—epitomized by the administrative separation between hospital and community care under the Welfare State—could prove to be a disadvantage when dealing with a group such as the mentally defective. Thus, in a number of ways, the transitions which took place under the Welfare State were not necessarily progressive steps for the mentally defective.

Comparing the period after 1913 to that which preceded it, there was a greater degree of continuity than one might at first expect. The passage of the 1913 Act, and the creation of a new central government department, and of new local government authorities, institutions, and expert groups, can all be seen as part of a broader process of specialization of welfare provision, and as a reflection of the growing authority of medical knowledge within social policy. However, this process of specialization reached back to the last third of the nineteenth century, when the feeble-minded first rose to prominence as a major social problem, the first voluntary groups were set up, local authority special education began to emerge, and the vogue for specialized institutionalization reached a climax. Even the Board of Control was less a model of governmental specialization for the future, than it was a continuation of the nineteenth-century inspectorial model which had already been developed by the Commissioners in Lunacy. Moreover, beneath the new rhetoric, many of the ideas which shaped interwar policy were less original than they might at first appear: mental tests recodified ideas about good character and social competence; psychiatric social workers adopted a new psychological language, but continued to base their actions and judgements on ideas about individual responsibility and morality; the Board of Control continued to place issues of individual liberty above medical interests; psychiatrists gained control of care in colonies, but they adopted commonly held ideas about the value of fresh air, physical activity, work, and discipline, rather than any new medical therapies; the policy debate continued to embrace the themes of rights, duties, and responsible citizenship, which had dominated political discourse in late nineteenth-century Britain; even the biological debate about the inheritance of mental deficiency was rooted in a late nineteenth-century evolutionary and degenerationist discourse. Therefore, in terms of the ideas surrounding the problem of mental deficiency, and in terms of the policy solutions which resulted, there were strong continuities running from about 1870 through to the formation of the Welfare State, and in some ways beyond into the 1950s.

Much of the historical confusion which has surrounded evaluations of mental deficiency policy has arisen because of the way it uneasily straddles our preconceptions about the 'traditional' and the 'modern', and the 'conservative' and the 'progressive'. In this book I have attempted to steer away from thinking in terms of such dichotomies. Instead, I have used the problem of mental deficiency to show that 'traditional' ideas were often integral to processes of modernization. I have attempted to show why 'progressives' could support policies, such as segregation and even sterilization, which appear highly conservative to us today. Rather than imposing our own judgements to portray mental deficiency policy as a sign of undemocratic tendencies in early twentieth-century social policy, I have suggested that we should view it in its own context, and explore the many ways in which it was related to the contemporary process of adjusting to democracy. This political context was fundamentally important in shaping the problem of mental deficiency: the very category of the mental defective was intimately related to the contemporary desire to construct democratic subjects; the political debate about mental deficiency was conducted within the parameters of a broader debate about defining the rights and duties of citizens within this democracy; and the construction and administration of policy can only be understood if we take into account the emergence of new democratic constraints from both above and below. However, adjusting to democracy also led to a series of anxieties and reactions which now appear to be far less progressive, and which were equally important in shaping the problem of mental deficiency: it led to the search for new ways to exclude the socially irresponsible and unfit from participation as political citizens; it induced fears of an inexorable rise in social expenditure in response to mass demand from the unfit; it provoked serious consideration of policies, such as eugenic segregation, sterilization, and community supervision, which might remedy these problems by limiting the growth of the residuum; and it led to political constraints which forced government to delegate responsibility to outside groups and professionals who were not under direct democratic control. Thus what we now see as a contradictory mix of the conservative and progressive in the history of mental deficiency was in fact two sides of the same coin, both integral to the process of making and implementing social policy during the adjustment to democracy in late nineteenth- and early twentieth-century Britain.

APPENDIX: NUMBER OF MENTAL DEFECTIVES UNDER VARIOUS FORMS OF PROVISION

Year[a]	State institutions (dangerous and criminal)	Certified institutions (local authority)	Approved (Sect. 37 Poor Law) Institutions	Certified houses (private)	Approved houses (private)	Guardian-ship	Subtotal	Statutory super-vision	Total ascertained[b]
1920	158	7,011	2,100	312	258	290	10,129		
1921	323	8,513	4,052	218	286	416	13,810		
1922	331	10,017	4,391	275	313	459	15,786		
1923	386	11,044	5,064	288	322	538	17,642		
1924	446	12,115	5,530	281	363	641	19,376		
1925	514	12,766	6,372	264	381	785	21,082		
1926	582	13,342	7,035	271	432	1,077	22,739	16,143	
1927	690	14,099	7,288	279	456	1,357	24,169	17,894	61,522
1928	785	14,840	7,844	273	465	1,602	25,809	19,206	66,458
1929	884	15,664	8,374	277	498	1,817	27,514	22,706	71,439
1930	1,043	17,200	9,219	238	536	2,120	30,354	24,710	85,963
1931	1,122	19,964	9,027	211	560	2,472	33,356	27,970	97,310
1932	1,196	22,493	8,880	226	564	2,799	36,158	29,753	102,345
1933	1,231	24,494	9,262	215	592	3,049	38,343	31,921	106,439
1934	1,268	26,782	9,128	219	590	3,327	41,314	33,377	108,382
1935	1,319	28,933	9,162	206	636	3,645	43,901	34,840	112,906
1936	1,376	30,798	9,408	202	657	3,978	46,419	36,307	117,147
1937	1,445	32,653	9,301	201	663	4,430	48,693	37,817	121,977
1938	1,475	34,290	9,445	173	671	4,841	50,895	39,009	125,859
1939									
1940									
1941									
1942									
1943									
1944									
1945	1,577	40,794	9,724	166	527	5,016	57,804	41,804	132,247
1946	1,596	41,548	9,614	159	444	5,172	58,333	43,272	133,967

[a] The Board of Control Reports were not produced during the two World Wars.

[b] The number of ascertained defectives is substantially higher than the total of those under institutional care and statutory supervision; a substantial part of the difference was made up by defectives who were under informal 'voluntary' supervision.

Source: *BOC Annual Reports*.

SELECT BIBLIOGRAPHY

Primary Sources

Archives

Addison Papers, Bodleian Library, Oxford.

Arthur Greenwood Papers, Bodleian Library, Oxford.

Blacker Archive, Wellcome Unit for the History of Medicine, Oxford.

Cambridge County Record Office.

Commonwealth Fund Archives, Rockefeller Archive Centre, Tarrytown, New York.

Conservative Party Archive, Bodleian Library, Oxford.

Eugenics Society Archive, Contemporary Medical Archives Centre, Wellcome Institute for the History of Medicine, London.

Family Welfare Association Collection, Goldsmith's Library, London University.

Greater London Record Office: London County Council Records, Metropolitan Asylums Board Records, Charity Organization Society Archive, Letitia Fairfield Archive, W. H. Dickinson Papers, Caterham Hospital Records, Leavesden Hospital Records.

History of the London School of Economics Papers (BLPES).

John Johnson Collection, Bodleian Library, Oxford.

Karl Pearson Papers, University College London.

Labour Party Archive (microfilm), Bodleian Library, Oxford.

Lionel Penrose Papers, University College London.

Local Government Records, Wellcome Unit for the History of Medicine, Oxford.

National Association for Mental Health (MIND).

National Institute for Industrial Psychology Papers, British Library of Political and Economic Science (BLPES).

New Survey of London Papers (BLPES).

Offprints Collection, Fawcett Library, City of London Polytechnic.

Oxford City Council Record Office.

Political pamphlet collection, Nuffield College Library, Oxford.

Public Record Office, Kew: Ministry of Health, Board of Education, Home Office, Treasury, and Cabinet Papers.

Royal College of Psychiatrists.

Social Hygiene Council Archive, Wellcome Unit for the History of Medicine, Oxford.

Society of Medical Officers of Health records, Wellcome Unit for the History of Medicine, Oxford.
Titmuss papers, Wellcome Unit for the History of Medicine, Oxford.

Official Publications

Central Government

Reports of Commissions (in chronological order)

Report of the Royal Commission on the Blind, the Deaf, and the Dumb (C. 5781, 1889)
Report of the Departmental Committee on Defective and Epilectic Children (C. 8746, 1898).
Report of the Interdepartmental Committee on Physical Deterioration (Cd. 2175, 1904).
Royal Commission on the Care and Control of the Feeble-minded (Cd. 4215–4221, i–vii; *Report*, Cd. 4202, viii, 1908).
Report of the Royal Commission on the Poor Laws and Relief of Distress (Cd. 4499, 1909).
Report of the Commission on the Machinery of Government (*Haldane Commission*) (Cd. 9230, 1918).
Mental Deficiency Act 1913: Suggestions and Plans Relating to the Arrangements of Institutions for Defectives (Board of Control, London, 1919)
Report of the Committee on the Administration of Public Mental Hospitals (Cobb Report), (Cmd. 1730. viii. 185, 1922).
Lunacy Reform: Report of the Proceedings of the Conference Convened by Sir F. Willis (Board of Control, London, 1922).
The Nursing Services in Mental Hospitals: Report of the Proceedings of the Conference Convened by Sir F. Willis (Board of Control, London, 1925).
The Provision of Mental Hospital Accommodation: Report of the Proceedings of the Conference Convened by Sir F. Willis (Board of Control, London, 1925).
Report of the Royal Commission on Lunacy and Mental Disorder (Cmd. 2700, xiii. 373, 1926).
Royal Commission on Lunacy and Mental Disorder: Minutes of Evidence (London, 1926).
Report of the Interdepartmental Committee on Mental Deficiency, 1925–1929 (Wood Report), 3 vols., 4 pts. (London, 1929).
Report of the Departmental Committee on Colonies for Mental Defectives (Hedley Report) (London, 1931).
Report from the Select Committee on Estimates, 1932–33 (v.197, 1932–3).
Committee on Local Expenditure in England and Wales (Ray Committee) (Cmd. 4200, xiv, 1932–3).

Report of the Departmental Committee on Sterilisation (Brock Report) (Cmnd. 4485, xv, 1934).
Interdepartmental Committee on Nursing Services Interim Report (Athlone Report) (London, 1939).
Royal Commission on the Law Relating to Mental Illness and Mental Deficiency, 1954–7: Minutes of Evidence (London, 1954).
Report of the Royal Commission on the Law Relating to Mental Illness and Mental Deficiency, 1954–7 (Cmnd. 169, 1957).
Community Care with Special Reference to Adult Mentally Ill and Mentally Handicapped Persons (Second Report of the Social Services Committee) (London, 1985).

Periodical and Annual Publications

Annual Reports of the Board of Control to the Lord Chancellor, 1947–59.
Annual Reports of the Chief Medical Officer of the Board of Education, 1909–20.
Annual Reports of the Chief Medical Officer of the Ministry of Health, 1921–39.
Board of Control Annual Reports, 1914–46.
Census of England and Wales, General Report, 1931.
The Health of the School Child, Annual Reports of the Chief Medical Officer of the Board of Education, 1921–38.
Ministry of Health Local Taxation Returns: Local Government Financial Statistics, 1919–39.
Parliamentary Debates (Commons), 5th series.
Parliamentary Debates (Lords), 5th series.
Registrar General's Statistical Review of England and Wales, 1950–1 (London, 1955).
Reports of the Industrial Fatigue Board.
Reports of the Medical Research Council.

Legislation (in chronological order)

Lunatics Act, 1845 (8 & 9 Vic., c. 100).
Idiots Act, 1886 (49 & 50 Vic., c. 25).
Lunacy Act, 1891 (54 & 55 Vic., c. 65).
Elementary Education (Defective and Epileptic Children) Act, 1899 (62 & 63 Vic., c. 32).
Children Act, 1908 (8 Edw. VII, c. 67).
Asylum Officers Superannuation Act, 1909 (9, Edw. VII, c. 48)
Mental Deficiency Act, 1913 (3 & 4 Geo. V, c. 28).
Elementary Education (Defective and Epileptic Children) Act, 1914 (4 & 5 Geo. V, c. 45).
Mental Deficiency (Amendment) Act, 1927 (17 & 18 Geo. V, c. 33).
Local Government Act, 1929 (19 Geo. V, c. 17).
Mental Treatment Act, 1930 (20 & 21 Geo. V, c. 23).

Disabled Persons Employment Act, 1944 (7 & 8 Geo. VI, c. 10).
Education Act, 1944 (7 & 8 Geo. VI, c. 31).
National Health Service Act, 1946 (9 & 10 Geo. V, c. 81).
National Assistance Act, 1948 (11 & 12 Geo. VI, c. 81).
Mental Health Act, 1959 (7 & 8 II Eliz., c. 72).

Local Government

Cambridgeshire County Council Medical Officer's Annual Reports.
City of Oxford Council Reports.
City of Oxford Education Committee Reports.
Lancashire Mental Hospitals Board, County Mental Hospitals Reports 1934–8.
London County Council Annual Reports 1900–59.
Metropolitan Asylums Board Annual Reports 1900–30.
Oxfordshire County Council Medical Officer's Annual Reports.

Organizational Reports and Publications

Annual Conference Reports of the Central Association for Mental Welfare.
Annual Reports of the Central Association for Mental Welfare.
Annual Reports of the London Association for Mental Welfare.
Annual Reports of the National Council for Civil Liberties.
Annual Reports of the National Association for Promoting the Welfare of the Feeble-Minded.
Annual Reports of the National Society for the Prevention of Cruelty to Children.
Child Guidance Council Annual Reports.
Committee for Legalizing Voluntary Sterilization publications:
Report of the Voluntary Eugenic Sterilization Conference held on 23 May 1932 (London, 1932).
Eugenic Sterilization (London, 1930).
Better Unborn (London, 1931).
Binney, C., *The Law as to Eugenic Sterilization*, 2nd edn. (London, 1931).
Rüdin, E., *Psychiatric Indications for Sterilization* (London, *c.*1931).
Handbook for Speakers (London, 1934).
Labour Party publications:
Labour and the New Social Order (London, 1918).
The Hospital Problem (London, 1924).
The Labour Movement and Preventive and Curative Medical Services (London, 1922).
The Labour Movement and the Hospital Crisis: A Statement of Policy (London, 1922).
Labour Party Conference Reports.
TUC Conference Reports.
Reports of the National Conference of Labour Women.

Report of the Feversham Committee: The Voluntary Mental Health Services (London, 1939).
World Federation for Mental Health, *Mental Health and World Citizenship (International Congress on Mental Health)* (London, 1948).

Primary Literature

Journals and Newspapers

British Medical Journal.
British Journal of Medical Psychology.
British Journal of Psychology.
Catholic Times.
Charity Organisation Quarterly.
Charity Organisation Review.
Civil Liberty.
Civil Service Gazette.
Contemporary Review.
County and Municipal Record.
Daily Chronicle.
Daily Herald.
Eugenics Review.
Health and Empire.
Howard Journal.
John Bull.
Journal of Mental Hygiene.
Journal of Mental Science.
Journal of the Royal Sanitary Institute.
Journal of State Medicine.
The Lancet.
Local Government Journal.
Manchester Guardian.
Medical Officer.
Medical Press and Circular.
Medico-Legal and Criminological Review.
Mental Health.
Mental Hospital and Institution Workers' Journal.
Mental Welfare.
Morning Post.
Municipal Journal.
The Nation.
National Asylum Workers' Union Magazine.
Nature.

New Statesman and Nation.
Nineteenth Century.
Public Administration.
Public Health.
Social Service Review.
Studies in Mental Inefficiency.
The Times.

Books and Articles

ADDISON, C., *Four and a Half Years*, i (London, 1934).

AIKMAN, A. K., 'The Treatment of Mental Defect', *Medical Press and Circular*, 193 (1936), 270.

ALLEN, C. K., *Bureaucracy Triumphant* (London, 1931).

ANDERSON, A., 'The Home Teaching of Defectives', *Mental Welfare*, 17 (1936), 33–9.

ARMSTRONG, C. W., *Survival of the Unfittest* (London, 1927).

ASHER, C., 'The Relationship of Birth Weight to Mental Development', *Archives of Disease in Childhood*, 23 (1948), 142.

ATTLEE, C., *The Social Worker* (London, 1920).

AUDEN, G., 'The Difficult Child in Industry', *Mental Welfare*, 12 (1931), 51–60.

BALDWIN, S., *This Torch of Freedom* (London, 1935).

BARR, M. W., *Mental Defectives: Their History, Treatment and Training* (Philadelphia, 1904).

BEACH, L., 'The CAMW Guardianship Scheme', *Mental Welfare*, 13 (1932), 29–38.

—— 'Beyond Hospital Gates', *Mental Welfare*, 16 (1935), 49–59.

BENJACAR, E., 'A Mental Deficiency Institution in Wartime', *Mental Health*, 1 (1940), 107–12.

BERRY, R. J. A., *The Mental Defective* (London, 1931).

—— (ed.), *Stoke Park Monographs on Mental Deficiency* (London, 1933).

—— 'What To Do With the Mental Defective in Private Practice', *Medical Press and Circular*, 193 (1936), 441–3.

BEVERIDGE, W., *Voluntary Action: A Report on Methods of Social Advance* (London, 1948).

BINNEY, C., *The Law as to Eugenic Sterilization*, 2nd edn. (London, 1931).

BLACKER, C. P., *Voluntary Sterilization* (London, 1934).

—— (ed.), *A Social Problem Group?* (Oxford, 1937).

—— *Eugenics in Retrospect and Prospect* (London, 1945).

—— 'Social Problem Families in the Limelight', *Eugenics Review*, 19 (1946), 117–27.

—— *Neurosis and the Mental Health Services* (London, 1948).

—— 'Eugenic Experiments Conducted by the Nazis on Human Subjects', *Eugenics Review*, 44 (1952), 9–19.

—— *Eugenics: Galton and After* (London, 1952).

—— *Problem Families: Five Enquiries* (London, 1952).
—— 'Voluntary Sterilization: The Last Sixty Years', *Eugenics Review*, 54 (1962), 9–23.
BOND, C. H., 'The Need for Schools of Psychiatry', *JMS* 66 (1920), 14–22.
BOND, C. J., *Essays and Addresses By a Surgeon* (London, 1930).
BOSANQUET, H., *Social Work in London, 1869–1912: A History of the Charity Organisation Society* (London, 1914).
BOURDILLON, A. F. C., *Voluntary Social Services: Their Place in the Modern State* (London, 1945).
BOWES, G. K., 'Mental Defect and Mental Degeneracy in a Rural Area', *The Lancet* (16 August 1924), 313–18.
BOWLBY, J., *Maternal Care and Mental Health* (Geneva, 1951).
BRACKENBURY, H., 'Maternity in its Sociological Aspects', *Social Service Review*, 18 (1937), 37–47.
BROCK, L., 'Mental Health Workers: An Appreciation', *Mental Welfare*, 12 (1931), 65–7.
—— 'Towards a Unified Mental Health Service', *Mental Welfare*, 20 (1939), 93–7.
—— 'Psychiatry and the Public Health Service', *JMS* 92 (1946), 287–304.
BROWN, S. C., 'A Visit to Gheel Colony Belgium', Mental Welfare, 14 (1933), 3–7.
—— 'Family Case Work and Mental Health', *Charity Organisation Quarterly*, NS 13 (1939), 45–50.
—— 'Methods of Social Case Workers', in F. C. Bartlett, M. Ginsburg, et al. (eds.), *The Study of Society* (London, 1939), 379–401.
BUCK, M., 'Co-operation in Mental Health Administration', *Mental Welfare*, 17 (1936), 104–8.
BUCK, P. S., *The Child Who Never Grew* (London, 1951).
BURDETT, H. C., *Hospitals and Asylums of the World*, 2 vols. (London, 1891 and 1893).
BURNS, C. L. C., 'Birmingham Child Guidance Clinic: A Review of Two Years' Work', *Mental Welfare*, 16 (1935), 3–6.
BURT, C., 'Psychology and Social Work', *Charity Organisation Review*, NS 43 (1918), 5–9, 51–60.
—— 'Delinquency and Mental Defect', *British Journal of Medical Psychology*, 3 (1923), 168–78.
—— 'The Causes of Sex Delinquency in Girls', *Health and Empire*, 1 (1926), 251–71.
—— 'The Contribution of Psychology to Social Hygiene', *Health and Empire*, 1 (1926), 25–33.
—— *The Young Delinquent*, 4th edn. (London, 1945; first pub. 1925).
—— *The Backward Child*, 2nd edn. (London, 1946; first pub. 1937).

BURT, C., *Intelligence and Fertility: The Effect of the Differential Birthrate on Inborn Mental Characteristics* (London, 1952).

—— *The Subnormal Mind*, 3rd edn. (London, 1955; first pub. 1935).

CARHILL, H. F., 'Administrative Habits of Mind', *Public Administration*, 3 (1930), 119–30.

CARLILE, W., and CARLILE, V. W., *The Continental Outcast: Land Colonies and Poor Law Relief* (London, 1907).

CARR, H. SAUNDERS, A. M., 'Eugenics in the Light of Population Trends', *Eugenics Review*, 37 (1935), 11–14.

—— and WILSON, P., *The Professions* (London, 1933).

CATTELL, R. B., 'Is National Intelligence Declining?', *Eugenics Review*, 28 (1936), 181–202.

—— *The Fight For Our National Intelligence* (London, 1937).

CHAPPLE, W., *The Fertility of the Unfit* (London, 1905).

CHARITY ORGANISATION SOCIETY, *The Feeble-Minded Child and Adult* (London, 1893).

CHARLES, E., *The Twilight of Parenthood* (London, 1934).

CHESTERTON, G. K., *Eugenics and Other Evils* (London, 1922).

CLOKIE, H. M., and ROBINSON, J. W., *Royal Commissions of Inquiry: The Significance of Investigations in British Politics* (London, 1937).

CLOUSTON, T. S., 'The Diploma in Psychiatry', *JMS* 57 (1911), 207–18.

COLE, G. D. H., *Fabian Socialism* (London, 1943).

CONQUEST, J., *The Naked Truth . . . About the Slums* (London, 1935).

COSENS, M., 'Psychology and Social Case-Work', *Charity Organisation Quarterly*, 7 (1933), 63–73.

CRAWFORD, J. M., 'Mental Deficiency: A Misnomer', *Mental Health*, 13 (1953), 9–12.

CRICHTON-MILLER, H., 'Mental Hygiene and Preventive Medicine', *Mental Hygiene*, 2 (1936), 160.

DALE, H. E., *The Higher Civil Service of Great Britain* (Oxford, 1941).

DARWIN, L., *The Need for Eugenic Reform* (London, 1926).

DAVIES, S. P., *Social Control of the Mentally Deficient* (London, 1930).

DAVIS, K., 'Mental Hygiene and the Class Structure', *Psychiatry*, 1 (1938), 55–66.

DENDY, M., 'On the Training and Management of Feeble-minded Children', appendix to C. P. Lapage, *Feeble-mindedness in Children of School Age* (Manchester, 1911), 243–96.

DICEY, A. V., *Law and Opinion in England*, 2nd edn. (London, 1914).

DICKINSON, R. L., *Sterilization without Unsexing* (New York, 1929).

DOLL, E. A., 'Community Control of Mental Deficiency in the USA', *Mental Welfare*, 14 (1933), 65–6.

—— 'Impressions of Mental Deficiency Work in England', *Mental Welfare*, 15 (1934), 17–20.

EICHOLZ, A., 'The Treatment of the Feeble-Minded', *British Medical Journal*, 2 (1902), 683–7.

ELLIS, H., 'The Sterilisation of the Unfit', *Eugenics Review*, 1 (1909), 204–6.

—— *BMJ* 1 (1923), 835.

ESHER, F. J. S., 'Military Service for Mental Defectives', *Mental Health*, 3 (1942), 14–18.

FAIRFIELD, L., *The Case Against Sterilisation* (London, 1934).

FARIS, R. E. L., and DUNHAM, H. W., *Mental Disorders in Urban Areas* (Chicago, 1939).

FINER, H., *English Local Government* (London, 1945).

FISHER, R. A., 'The Elimination of Mental Defect', *Eugenics Review*, 16 (1926), 114–15.

—— 'Eugenics Academic and Practical', *Eugenics Review*, 27 (1935), 105–7.

FOX, E., 'National and Municipal Demands for Local and Voluntary Work', *Charity Organisation Review*, 36 (1914), 62.

—— *The Work of the Local Associations for the Care of the Mentally Defective* (London, 1920).

—— 'The Mental Deficiency Act', *Charity Organisation Quarterly*, 8 (1924), 134–7.

—— 'The Mental Deficiency Act', *Eugenics Review*, 10 (1918), 1–17.

—— 'Modern Developments in Mental Welfare Work', *Eugenics Review*, 30 (1938), 165–78.

FREMANTLE, F. E., *The Health of the Nation* (London, 1927).

GIBBON, I. G., and BELL, W. H., *A History of the London County Council, 1889–1939* (London, 1939).

GIBSON, G., 'The Boarding Out System in Scotland', *JMS* 71 (1925), 253–63.

GIBSON, W., 'The Hostel Method for Feeble-minded Young Men and Women', *Mental Welfare*, 11 (1930), 75–7.

GILLESPIE, R. D., 'Mental Hygiene as a National Problem', *Mental Hygiene*, 1/4 (1931), 1–9.

GODDARD, H., *Feeble-mindedness: Its Causes and Consequences* (New York, 1914).

GORDON, R. G. (ed.), *A Survey of Child Psychiatry* (London, 1939).

GOSNEY, E. S. and POPENOE, P., *Sterilisation for Human Betterment* (New York, 1929).

GREENLAND, C., 'Psychiatry and the Community', *Mental Health*, 10 and 11 (1952), 14–19, 111–21.

GÜNZBURG, H. C., 'The Colony and the High Grade Mental Defective', *Mental Health*, 9 (1950), 87–91.

HALDANE, J. B. S., *Heredity and Politics* (London, 1938).

HALDEN, M., *First Year Out: The Diary of a Special School Teacher* (London, 1943).

HALL, S., *Adolescence—Its Psychology* (New York, 1904).

HALLIDAY, L., 'Psychoneuroses as a Cause of Incapacity among Insured Persons', *BMJ* (9 March 1935).
HARDING, D. W., 'Mental Health in Post War Reconstruction', *Mental Health*, 2 (1941), 61–5.
HARE, E. H., 'The Ecology of Mental Disease', *JMS* (1952), 579–94.
HARGROVE, A., 'Work for the Mentally Defective', *Womens' Employment* (26 February 1925), 5–6.
HAW, G., *From Workhouse to Westminster: The Life Story of Will Crooks, MP* (London, 1907).
HEARNSHAW, E. M., 'Some Psychological Difficulties of Evacuation', *Mental Health*, 1 (1940), 5–10.
—— 'Observed Effects of Wartime Conditions on Children', *Mental Health*, 2 (1941), 44–8.
HERD, H., *The Diagnosis of Mental Deficiency* (London, 1930).
HEWART, LORD, OF BURY, *The New Despotism* (London, 1929).
HODSON, C., *Human Sterilisation Today: A Survey of the Present Position* (London, 1933).
HOGBEN, L., 'The Interaction of Heredity and Environment', *JMS* 79 (1933), 590–601.
—— *Dangerous Thoughts* (London, 1939).
HYSLOP, T. B., 'The Mental Deficiency Bill 1912', *JMS* 58 (1912), 548–97.
IRELAND, W. W., *Mental Affections of Children*, 2nd edn. (London, 1900).
JEFFRIES, M., 'Mental Defectives Under the Poor Law', *Studies in Mental Inefficiency*, 4 (1923), 35–8.
JONES, D. C. (ed.), *The Social Survey of Merseyside* (London, 1934).
JONES, H., 'Group Sentiment and Delinquency', *Mental Health*, 8 (1948), 41–4.
Justice in England, by a Barrister (London: Left Book Club, 1938).
KIMBER, W. J. T., 'Social Values in Mental Hospital Practice', *JMS* 85 (1939), 29–44.
KIRBY, H., 'The Feeble-Minded and Voluntary Effort', *Eugenics Review* (1909), 87.
LAIRD, I., 'Some Psychological Problems in Institutionalising Defectives', *Mental Welfare*, 19 (1938), 1–7.
LANDTMAN, B., 'On the Relationship Between Maternal Conditions During Pregnancy and Congenital Malformations', *Archives of Disease in Childhood*, 23 (1948), 237–46.
LAPAGE, C. P., *Feeble-mindedness in Children of School Age* (Manchester, 1911).
LEWIS, A., 'Ageing and Senility: A Major Problem of Psychiatry', *JMS* (1946), 150–2.
LEWIS, E. O., 'Mental Deficiency and Social Medicine', *JMS* 94 (1948), 258.
—— 'Intelligence and Social Trends', *Mental Health*, 12 (1952), 18–19.
LIDBETTER, E. J., *Heredity and the Social Problem Group* (London, 1933).
LINDSAY, T., 'Mental Deficiency Practice at Caterham', *JMS* 80 (1934), 397–408.

LOCH, C. S., 'Charity Organisation and the Feeble-minded', *Charity Organisation Review*, 33 (1913), 6–8.

LOMAX, M., *Experiences of an Asylum Doctor* (London, 1921).

LONDON COUNTY COUNCIL, *Handbook on Mental Health Social Work* (London, 1947).

LORD, J. R., *Social Workers and the Insane* (London, 1923).

—— *Mental Hospitals and the Public* (London, 1927).

—— '"After-care" and Other Aspects of Social Services as an Adjunct to Mental Treatment', *JMS* 76 (1930), 622–31.

—— 'American Psychiatry', *JMS* 76 (1930), 456–95.

MACADAM, E., *The New Philanthropy: A Study in the Relations between the Statutory and Voluntary Social Services* (London, 1934).

MCBRIDE, E. W., 'Cultivation of the Unfit', *Nature*, 11 (January 1936).

MCCOULL, G., 'Miner's Nystagmus in Neuronic Inefficiency', in G. Berry (ed.) *Stoke Park Monographs on Mental Deficiency* (London, 1933), 223–30.

MCCUTCHEON, A. M., 'Institutional Treatment of Mental Defectives, with Special Reference to Occupation', *JMS* 71 (1925), 694–703.

MCDOUGALL, M., *Introduction to Social Psychology* (London, 1908).

MCLACHLAN, H., *Records of a Family* (Manchester, 1935).

M'GONIGLE, G. C. M., and KIRBY, J., *Poverty and Public Health* (London, 1936).

MAIER, H., 'Practical Experience of Sterilization in Switzerland', *Eugenics Review*, 26 (1934), 19–25.

MAPOTHER, E., 'Safeguards in Eugenic Sterilization', *Eugenics Review*, 26 (1934), 15–17.

MARKHAM, V., 'Robert Morant: Some Personal Reminiscences', *Public Administration*, 28 (1950), 249–62.

MARTINDALE, H., *Women Servants of the State*, 1870–1938 (London, 1938).

MAUDSLEY, H., *The Physiology and Pathology of Mind* (London, 1868).

—— *Responsibility in Mental Disease* (London, 1874).

Mental Health and World Citizenship (London,1948).

MESS, H., *Voluntary Social Services Since 1918* (London, 1948).

MILLER, B. S., 'Mental Deficiency and the Local Government Bill', *Mental Welfare*, 10 (1929), 29–40.

MOSLEY, O., *The Greater Britain* (London, 1934).

—— *Fascism: 100 Questions Asked and Answered* (London, 1936).

MOTT, F. W., 'Heredity and Insanity', *Eugenics Review*, 2 (1910), 257.

MPA, *Handbook for Attendants on the Insane*, 6th edn. (London, 1912).

MUNRO, T. A., 'Consanguinity and Mental Disorder', *JMS* 84 (1938), 708–14.

NATIONAL COUNCIL FOR CIVIL LIBERTIES, *50,000 Outside the Law* (London, 1950).

NEUSTATTER, W. L., 'Social and Economic Factors in Psychiatry', *Eugenics Review*, 34 (1942), 55–61.

NEVILLE-ROLFE, S., *Social Biology and Social Welfare* (London, 1949).

NORMAN, P., *In the Way of Understanding* (Godalming, 1982).

PACIFIST SERVICE UNIT, *Problem Families*, ed. T. Stephen (London, 1945).

PADDLE, K. C. L., 'Lumbar Puncture and the Cerebro-Spinal Fluid in 2,000 Cases of Mental Deficiency', *JMS* 80 (1934), 674–80.

—— 'The Control of Dysentery by Prophylactic Innoculation', *JMS* 84 (1938), 134–9.

PEARL, R., 'Sterilization of Degenerates and Criminals', *Eugenics Review*, 11 (1919), 1–6.

PEARSE, I. H., and CROCKER, L. H., *The Peckham Experiment* (London, 1943).

PELHAM, H. S., *The Training of a Working Boy* (London, 1914).

PENROSE, L., 'Mental Deficiency: The Subcultural Group', *Eugenics Review*, 24 (1932), 289.

—— 'The Complex Determinants of Amentia', *Eugenics Review*, 26 (1934), 121.

—— *A Clinical and Genetic Study of 1280 Cases of Mental Defect* (London, 1938).

—— 'Some General Problems in Mental Deficiency', *JMS* 84 (1938), 693–707.

—— 'Eugenic Progress with Respect to Mental Deficiency', *Eugenics Review*, 31 (1939), 35.

—— *The Biology of Mental Defect* (London, 1949).

PIERCE, B., 'The Status of British Psychiatry and of Medical Officers', *JMS* 59 (1913), 699–701.

PINSENT, E., 'On the Permanent Care of the Feeble-minded', *The Lancet*, 1 (1903), 513–15.

—— 'Social Responsibility and Heredity', paper to Church Congress (September 1910).

—— *The Mental Health Services in Oxford City, Oxfordshire, and Berkshire*, Oxford Survey of Social Services (Oxford, 1937).

POLITICAL AND ECONOMIC PLANNING, *Report on the British Health Services* (London, 1937).

—— *Report on the British Social Services* (London, 1937).

POWELL, A., *The Metropolitan Asylums Board and its Work, 1867–1930* (London, 1930).

RAVEN, A., 'The Training for Occupational Therapy', *Mental Welfare*, 11 (1930), 34–5.

REES, J. R., 'Strategic Planning in Mental Health', *Mental Health*, 1 (1940), 103–6.

—— *The Shaping of Psychiatry by War* (London, 1945).

ROBINSON, F., 'Local Administration of the Mental Deficiency Act 1913', *Public Health*, 29 (1916), 78–93.

ROBSON, W., *Development of Local Government* (London, 1930).

—— *The British Civil Servant* (London, 1937).

—— *The Government and Misgovernment of London* (London, 1939).

ROGERSON, C. H., 'Psychiatry in the National Health Service of the Future', *Mental Health*, 5 (1945), 45–9.

ROSS, N., 'Occupational Therapy', *Mental Welfare*, 17 (1936), 12–15.

ROWNTREE, B. S., *Poverty—A Study of Town Life* (London, 1941).

RYLE, J. A., 'Medicine and Eugenics', *Eugenics Review*, 30 (1938), 9–19.

SAVAGE, S. W., 'Intelligence and Infant Mortality in Problem Families', *BMJ* 1 (1946), 86–7.

SAVAGE, SIR W., 'Our Future: Presidential Address to the Society of Medical Officers of Health', *Public Health*, 49 (1935), 42–7.

SÉGUIN, E., *Idiocy and Its Treatment by the Physiological Method* (New York, 1866).

SELBY-BIGGE, L. A., *The Board of Education*, 2nd edn. (London, 1934).

SHERLOCK, E. B., *The Feeble-minded* (London, 1911).

—— *Minds in Arrear* (London, 1932).

SHRUBSALL, F., and WILLIAMS, A. C., *Mental Deficiency Practice* (London, 1932).

SHUTTLEWORTH, E., *Mentally Deficient Children: Their Treatment and Training* (London, 1900).

—— and POTTS, D. C., *The Mentally Defective Child*, 3rd edn. (London, 1910).

SIMON, E. D., *A City Council From Within* (London, 1926).

SLATER, E., 'The Inheritance of Manic Depressive Insanity and its Relation to Mental Defect', *JMS* 82 (1936), 693–707.

SLAUGHTER, J. W., *The Adolescent* (London, 1911).

SMITH, H. L. (ed.), *The New Survey of London Life and Labour* (London 1934).

SODDY, K., 'Some Lessons in Wartime Psychiatry', *Mental Health*, 6 (1946), 32–3, 66–70.

SPENCE, J. B., 'Assistant Medical Officers in Asylums', *JMS* 59 (1913), 263–73.

SPRANKLING, REVD C. J., 'The Work of the Metropolitan Asylums Board for the Mentally Defective', in *CAMW Conference Report* (London, 1920), 83–93.

STALLYBRASS, C. O., 'Mental Health in Relation to the Family', *Public Health*, 54 (1948), 123–8.

STANSFIELD, T. E. K., 'Heredity and Insanity', *JMS* 57 (1911), 55–63.

—— 'The Villa or Colony System for the Care and Treatment of Mental Diseases', *JMS* 60 (1914), 30–7.

STEPHENS, H. F., 'Cheaper Institutional Care', *Mental Welfare*, 13 (1933), 8–15.

STRÖM-OLSEN, R., 'The Therapeutic Value in Mental Illness of Physical Fitness through Exercise', *Mental Welfare*, 18 (1937), 1–10.

THOMSON, J. B., 'The Hereditary Nature of Crime', *Journal of Mental Science*, 15 (1869), 487–98.

The Times, *The Nation's Health* (London, 1937).

TITMUSS, R., *Poverty and Population* (London, 1938).

TOWNSEND, P. H., 'The Care of the Feeble-minded', *Charity Organisation Society Review*, 44 (1900), 151–8.
TREDGOLD, A. F., 'A Note on Sterilisation of Defectives', *Mental Welfare*, 11 (1930), 9–14.
—— *Mental Deficiency (Amentia)*, 6th edn. (London, 1937: first pub. 1908).
TURNER, F. D., 'Notes about Institutions for Defectives', *Studies in Mental Inefficiency*, 1 (1920), 32–8.
—— 'Aims of a Mental Deficiency Institution', *JMS* 74 (1928), 465–73.
—— 'Forty Years in an Institution', *Mental Health*, 7 (1946), 1–5.
VALENTINE, C. W., 'Present-day Trends and Dangers in the Field of Psychology', *British Journal of Educational Psychology*, 13 (1948), 134–47.
VAUGHAN, K., 'The Shape of the Pelvic Brim as the Determining Factor in Childbirth', *BMJ* 2 (1931), 939–41.
VINCENT, J., *Inside the Asylum* (London, 1948).
WALKER, K., 'The Need For a Positive Philosophy of Life', *Mental Health*, 2 (1941), 102–8.
WEDGWOOD, J., *Memoirs of a Fighting Life* (Plymouth, 1940).
WELLISCH, E., 'A Colony for Maladjusted Children', *Mental Health*, 10 (1950), 6–8.
WILDENSKÖV, H. O., 'Denmark's Care of Mental Defectives', *Mental Welfare*, 18 (1937), 37–44.
WINTERTON, P., *Mending Minds: The Truth About Our Mental Hospitals* (London, 1938).
WOLLEN, C. A., 'The Problem Child In the Hostel', *Mental Health*, 5 (1944), 28–31.
WOODHEAD, SIR G., and VARRIER-JONES, P. C., *Industrial Colonies and Settlements for the Consumptive* (Cambridge, 1920).
WOMEN'S GROUP ON PUBLIC WELFARE, *Out Towns: A Close Up* (London, 1943).
WORMWALD, J., and WORMWALD, S., *A Guide to the Mental Deficiency Act 1913* (London, 1913).
YELLOWLEES, H., 'Modern Psychiatry and Mental Hospitals', *JMS* 81 (1935), 250–5.

SECONDARY LITERATURE

Published Books and Articles

ABEL-SMITH, B., *A History of the Nursing Profession* (London, 1960).
—— *The Hospitals, 1800–1948* (London, 1964).
ABRAMS, P., 'The Failure of Social Reform', *Past and Present*, 24 (1963), 53–64.
ADAMS, F., 'From Association to Union: Professional Organisation of Asylum Attendants, 1869–1919', *British Journal of Sociology*, 20 (1969), 11–26.

ADAMS, M. (ed.), *The Wellborn Science: Eugenics in Germany, France, Brazil, and Russia* (Oxford, 1990).

ADDISON, P., *Churchill on the Home Front, 1900–1955* (London, 1992).

AHRENFELDT, R. H., 'Military Psychiatry', in A. S. MacNalty and W. Franklin Mellor (eds.), *The Medical Services in War* (London, 1968), 182–7.

ALBROW, M., *Bureaucracy* (London, 1970).

ANDERSON, M., *Family Structure in Nineteenth-Century Lancashire* (Cambridge, 1971).

ANDREWS, J., 'Identifying and Providing for the Mentally Disabled in Early Modern England', in D. Wright and A. Digby (eds.), *From Idiocy to Mental Deficiency: Historical Perspectives on People with Learning Disabilities* (London, 1996), 65–92.

ANNAN, LORD, 'The Intellectual Aristocracy', in J. H. Plumb (ed.), *A Tribute to Trevelyan* (London, 1955), 241–87.

ARMSTRONG, D., *The Political Anatomy of the Body* (Cambridge, 1983).

ASHDOWN, M., and BROWN, S. C., *Social Service and Mental Health: An Essay on Psychiatric Social Workers* (London, 1953).

ASHFORD, D., *The Emergence of the Welfare States* (Oxford, 1986).

AUSTOKER, J., 'Walter Morley Fletcher and the Origins of a Basic Biomedical Research Policy', in J. Austoker and L. Bryder (eds.), *Historical Perspectives on the MRC* (Oxford, 1989), 23–33.

—— and BRYDER, L. (eds.), *Historical Perspectives on the MRC* (Oxford, 1989).

AYERS, G., *England's First State Hospitals and the Metropolitan Asylums Board, 1867–1930* (London, 1971).

BAILEY, V., *Delinquency and Citizenship: Reclaiming the Young Offender, 1914–1948* (Oxford, 1987).

BALDWIN, P., *The Politics of Social Solidarity: Class Bases of the European Welfare State, 1875–1975* (Cambridge, 1990).

BALDWIN, S., 'From Community to Neighborhood', in S. Baldwin and J. Hattersley (eds.), *Mental Handicap: Social Science Perspectives* (London, 1991), 145–66.

BARKER, D., 'How to Curb the Fertility of the Unfit: The Feeble-minded in Edwardian Britain', *Oxford Review of Education*, 9 (1983), 197–211.

—— 'The Biology of Stupidity: Genetics, Eugenics and Mental Deficiency in the Inter-War Years', *BJHS* 22 (1989), 347–75.

BARKER, R., *Education and Politics: A Study of the Labour Party* (Oxford, 1972).

BARNES, J., *Ahead of his Age: Bishop Barnes of Birmingham* (London, 1979).

BARUCH, G., and TREACHER, A., *Psychiatry Observed* (London, 1978).

BAYLEY, M., *Mental Handicap and Community Care* (London, 1973).

BEETHAM, D. *Max Weber and the Theory of Modern Politics* (Oxford, 1974).

BEHLMER, G. K., *Child Abuse and Moral Reform in England, 1870–1908* (Stanford, 1982).

BELLAMY, C., *Administering Central–Local Relations, 1891–1919* (Manchester, 1988).

BELLAMY, J. M., and SAVILLE, J., *Dictionary of National Labour Biography*, 9 vols. (London, 1972–93).

BELOFF, M., 'The Whitehall Factor: The Role of the Higher Civil Service', in G. Peele and C. Cook (eds.), *The Politics of Reappraisal* (London, 1975), 209–31.

BENSON, J., *The Working Class in Britain, 1850–1939* (London, 1989).

BENTLEY, M., and STEVENSON, J. (eds.), *High and Low Politics in Modern Britain* (Oxford, 1983).

BERRIOS, G., 'Mental Retardation', in G. Berrios and R. Porter (eds.), *A History of Clinical Psychiatry* 225–38.

BERRIOS, G., and FREEMAN, H. (eds.), *150 Years of British Psychiatry, 1841–1991*, 2 vols. (London, 1991 and 1996).

—— and PORTER, R., *A History of Clinical Psychiatry: The Origin and History of Psychiatric Disorders* (London, 1995).

BIAGINI, E., *Liberty, Retrenchment and Reform: Popular Liberalism in the Age of Gladstone, 1860–1880* (Cambridge, 1992).

BLAND, L., '"Guardians of the Race" or "Vampires upon the Nation's Health"? Female Sexuality and its Regulation in Early Twentieth-Century Britain', in E. Whitelegg et al. (eds.), *The Changing Experience of Women* (Oxford, 1982), 373–89.

BOCK, G., 'Antinatalism, Maternity and Paternity in National Socialist Racism', in G. Bock and P. Thane (eds.), *Maternity and Gender Politics: Women and the Rise of the European Welfare States, 1880–1950* (London, 1991), 233–55.

BOLDERSON, H., 'The Origins of the Disabled Persons Employment Quota and its Symbolic Significance', *Journal of Social Policy*, 9 (1980), 169–86.

BOND, C. H., 'After-Care in Cases of Mental Disorder, and the Desirability of its More Extended Scope', *JMS* 59 (1913), 274–83.

BOWMAN, I., 'Maladjustment: The History of the Category', in W. Swann (ed.), *The Practice of Special Education* (London, 1981), 102–21.

BRASNETT, M., *Voluntary Social Action: A History of the NCSS, 1919–1969* (London, 1969).

BROWN, C., 'Urbanization and Living Conditions', in R. Pope (ed.), *Atlas of British Social and Economic History* (London, 1989), 176–9.

BROWN, I., 'Who were the Eugenicists?', *History of Education*, 17 (1988), 295–307.

BROWN, J., 'Charles Booth and the Labour Colonies', *Economic History Review*, 2nd ser., 21 (1968), 49–70.

BROWN, K. D. (ed.), *The First Labour Party, 1906–1914* (Beckenham, 1985).

BRYDER, L., 'The First World War: Healthy or Hungry?', *History Workshop Journal*, 24 (1987), 141–57.

—— *Below the Magic Mountain* (Oxford, 1989).

—— 'Public Health Research and the MRC', in J. Austoker and L. Bryder, *Historical Perspectives on the MRC* (Oxford, 1989), 59–81.

BULMER, M., 'The Development of Sociology and Empirical Research in Great Britain', in M. Bulmer (ed.), *Essays in the History of British Sociological Research* (Cambridge, 1985).

—— *The Social Basis of Community Care* (London, 1987).

—— BALES, K., and KISH SKLAR, K. (eds.), *The Social Survey in Historical Perspective, 1880–1940* (Cambridge, 1991).

BURLEIGH, M., '"Euthanasia" in the Third Reich: Some Recent Literature', *Social History of Medicine*, 4 (1991), 317–28.

—— 'Psychiatry, German Society, and the Nazi "Euthanasia" Programme', *Social History of Medicine* 7 (1994), 213–28.

BURNHAM, J. C., 'The Struggle Between Physicians and Paramedical Personnel in American Psychiatry 1917–41', *Journal of the History of Medicine*, 29 (1974), 93–106.

BUSFIELD, J., *Managing Madness: Changing Ideas and Practice* (London, 1986).

—— 'The Female Malady? Men, Women and Madness in Nineteenth Century Britain', *Sociology*, 28 (1994), 259–77.

BUTLER, D., and BUTLER, G., *British Political Facts* (London, 1986).

BYNUM, W. F., PORTER, R., and SHEPHERD, M. (eds.), *The Anatomy of Madness*, 3 vols. (London, 1985–8).

CARPENTER, M., 'Asylum Nursing Before 1914', in C. Davies (ed.), *Rewriting Nursing History* (London, 1980).

—— *Working for Health: The History of COHSE* (London, 1988).

CATERALL, P., 'Morality and Politics: The Free Churches and the Labour Party between the Wars', *Historical Journal*, 36 (1993), 67–85.

CLARK, M., 'The Rejection of Psychological Approaches to Mental Disorder in Late Nineteenth-Century British Psychiatry', in A. Scull (ed.), *Madhouses, Mad-Doctors and Madmen* (London, 1981), 271–312.

CLARKE, A. D. B., and CLARK, A. M. (eds.), *Readings From Mental Deficiency: The Changing Outlook* (London, 1978).

COCHRANE, D., '"Humane, Economical and Medically Wise": The LCC as Administrators of Victorian Lunacy Policy', in W. F. Bynum et al. (eds.), *The Anatomy of Madness*, iii (London, 1988), 247–71.

COHEN, S., and SCULL, A., *Social Control and the State: Historical and Comparative Essays* (Oxford, 1983).

COLLINI, S., 'The Idea of "Character" in Victorian Political Thought', *Transactions of the Royal Historical Association*, 35 (1985), 29–50.

COOK, C., 'Liberals Labour and Local Elections', in G. Peele and C. Cook (eds.), *The Politics of Reappraisal, 1918–1939* (London, 1975), 166–88.

—— *The Age of Alignment: Electoral Politics in Britain 1922–1929* (London, 1975).

—— and STEVENSON, J., *Longman Handbook of Modern British History, 1714–1987* (London, 1988).

COX, P., 'Girls, Deficiency, and Deliquency', in D. Wright and A. Digby (eds.), *From Idiocy to Mental Deficiency* (London, 1996), 184–206.

CRAFT, M., 'A Modern Hospital Service for the Subnormal', in R. Farndale and H. Freeman (eds.), *Trends in the Mental Health Services* (London, 1963), 164–74.

CRAIG, F. W. S. (ed.), *British General Election Manifestos, 1918–1966* (Chichester, 1970).

CRONIN, A. J., 'The British State and the Structure of Political Opportunity', *Journal of British Studies*, 27 (1988), 199–231.

CROWTHER, A., *The Workhouse System, 1834–1929: The History of an English Social Institution* (London, 1981).

—— 'Family Responsibility and State Responsibility in Britain Before the Welfare State', *Historical Journal*, 25 (1982), 131–45.

CRUIKSHANK, M., 'Mary Dendy 1855–1923', *Journal of Educational and Administrative History*, 8 (1976), 26–9.

DAVIDSON, R., and LOWE, R., 'Bureaucracy and Innovation in British Welfare Policy, 1870–1945', in W. J. Mommsen (ed.), *The Emergence of the Welfare State in Britain and Germany* (London, 1981), 263–95.

DAVIES, C., 'The Health Visitor as Mothers' Friend: A Woman's Place in Public Health, 1900–1914', *Social History of Medicine*, 1 (1988), 39–60.

DAVIS, J., *Reforming London: The London Government Problem, 1885–1900* (Oxford, 1988).

—— 'The Progressive Council', in A. Saint (ed.), *Politics and the People of London* (London, 1989), 27–48.

DAY, K., and JANCAR, J., 'Mental Handicap and the Royal Medico-Psychological Association: A Historical Association, 1841–1991', in G. Berrios and H. Freeman (eds.), *150 Years of British Psychiatry, 1841–1991* (London, 1991).

DE SWANN, A., *In Care of the State: Health Care, Education, and Welfare in Europe and the USA in the Modern Era* (Oxford, 1988).

DICKS, H. V., *Fifty Years of the Tavistock Clinic* (London, 1970).

DIGBY, A., and BOSANQUET, N., 'Doctors and Patients in an Era of National Health Insurance and Private Practice', *Economic History Review*, 2nd ser., 41 (1988), 74–94.

DINGWALL, R., 'Collectivism, Regionalism, and Feminism: Health Visiting and British Social Policy', *Journal of Social Policy*, 6 (1977), 291–315.

—— WEBSTER, C., and RAFFERTY, A. M., *An Introduction to the Social History of Nursing* (London, 1988).

DOLL, E., 'Historical Review of Mental Retardation, 1800–1965', *American Journal of Mental Deficiency*, 72 (1967), 165–89.

DONAJGRODZKI, A. P. (ed.), *Social Control in Nineteenth-Century Britain* (London, 1977).

DRIVER, F., 'The Historical Geography of the Workhouse System', *Journal of Historical Geography*, 15 (1989), 269–89.

—— 'Bodies in Space: Foucault's Account of Disciplinary Power', in C. Jones and R. Porter (eds.), *Reassessing Foucault: Power, Medicine and the Body* (London, 1994), 113–31.

DWYER, J., 'The Catholic Press, 1850–1950', in G. A. Beck (ed.), *The English Catholics* (London, 1930), 473–514.

DYER, C., 'Sterilisation of Mentally Handicapped Women', *British Medical Journal*, 2 (1987), 825.

ECKSTEIN, H., *Pressure Group Politics: The Case of the British Medical Association* (London, 1960).

EVANS, P., RUESCHEMEYER, D., and SKOCPOL, T. (eds.), *Bringing the State Back In* (Cambridge, 1985).

EVANS, B., and WAITES, B., *IQ and Mental Testing* (London, 1981).

FAIR, J. D., and HUTCHESON, J. A., 'British Conservatism in the Twentieth Century: An Emerging Ideological Tradition', *Albion*, 19 (1987), 549–78.

FARNDALE, R., and FREEMAN, H. (eds.), *Trends in the Mental Health Services* (London, 1963).

FELDMAN, D., and JONES, G. S., *Metropolis London: Histories and Representations since 1800* (London, 1989).

FIELDING, S., '"Don't Know and Don't Care": Popular Political Attitudes in Labour Britain', in N. Tiratsoo (ed.), *The Attlee Years* (London, 1991), 106–25.

FINLAYSON, G., *Citizen, State, and Social Welfare in Britain, 1830–1990* (Oxford, 1990).

—— 'A Moving Frontier: Voluntarism and the State in British Social Welfare, 1911–1949', *Twentieth Century British History*, 1 (1990), 183–206.

FINNANE, M., 'Asylums, Families, and the State', *History Workshop Journal*, 20 (1988), 134–48.

FISSELL, M., 'The Sick and Drooping Poor in Eighteenth-Century Bristol and its Region', *Social History of Medicine*, 2 (1989), 35–58.

FORTY, A., 'The Modern Hospital in England and France: The Social and Medical Uses of Architecture', in A. King (ed.), *Buildings and Society: Essays on the Social Development of the Built Environment* (London, 1980).

FOX, D., *Health Policies and Health Politics: The British and American Experience, 1911–1965* (Princeton, 1986).

FRASER, D., *The Evolution of the British Welfare State*, 2nd edn. (London, 1973).

FREEDEN, M., *The New Liberalism: An Ideology of Social Reform* (Oxford, 1978).

—— 'Eugenics and Progressive Thought: A Study in Ideological Affinity', *Historical Journal*, 22 (1979), 645–71.

—— *Liberalism Divided: A Study in British Political Thought, 1914–1939* (Oxford, 1986).

FREEDEN, M., 'The Stranger at the Feast: Ideology and Public Policy in Twentieth-Century Britain', *Twentieth Century British History*, 1 (1990), 9–34.

FREIDSON, E., *Professional Dominance: The Social Structure of Medical Care* (Chicago, 1970).

—— 'The Theory of Professions: The State of the Art', in R. Dingwall (ed.), *The Sociology of the Professions* (London, 1983), 19–37.

FRY, G. K., *Statesmen in Disguise: The Changing Role of the Administrative Class of the British Home Civil Service, 1853–1966* (London, 1969).

GARLAND, D., *Punishment and Welfare: A History of Penal Strategies* (Aldershot, 1985).

—— *Punishment and Modern Society* (Oxford, 1990).

GARSIDE, P., 'Unhealthy Areas: Town Planning, Eugenics, and the Slums, 1890–1945', *Planning Perspectives*, 3 (1988), 24–46.

GARSIDE, W. R., 'Juvenile Unemployment and Public Policy Between the Wars', *Economic History Review*, 2nd. ser., 30 (1977), 322–39.

GASKELL, M. (ed.), *The Slums* (Leicester, 1990).

GILBERT, B. B., *The Evolution of National Insurance in Great Britain: The Origins of the Welfare State* (London, 1966).

GILLIS, J., *Youth and History: Traditions and Change in European Age Relations, 1770–Present* (London, 1974).

GITTINS, D., *Fair Sex: Family Size and Structure, 1900–1939* (London, 1982).

GORDON, L., *Heroes of Their Own Lives: The Politics and History of Family Violence* (London, 1989).

GOSTIN, L., *A Human Condition* (London, 1975).

GOULD, S. J., *The Mismeasure of Man* (London, 1981).

GRAHAM, L., 'Science and Values: The Eugenics Movement in Russia and Germany in the 1920s', *American Historical Review*, 52 (1977), 1133–64.

GREEN, E. H. H., 'The Strange Death of Tory England', *Twentieth Century British History*, 2 (1991), 67–88.

GREENAWAY, J. R., 'Warren Fisher and the Transformation of the British Treasury, 1919–39', *Journal of British Studies*, 23 (1984), 125–42.

GREENLEAF, W. H., *The British Political Tradition*, 3 vols. (London, 1983–8).

GRIFFITHS, R., *Fellow Travellers of the Right* (Oxford, 1983).

GRIGGS, C., 'Labour and Education', in K. Brown (ed.), *The First Labour Party 1906–1914* (Beckenham, 1985), 158–82.

GROB, G., *Mental Illness and American Society, 1875–1940* (Princeton, 1983).

—— *From Asylum to Community: Mental Health Policy in Modern America* (Princeton, 1991).

—— 'Marxian Analysis and Mental Illness', *History of Psychiatry*, 1 (1990), 223–32.

GROSSMAN, T., 'Feminist Debates About Women and National Socialism', *Gender and History*, 3 (1991), 350–8.

HALSEY, A. H. (ed.), *Traditions of Social Policy* (Oxford, 1976).

HARRIS, H., 'Lionel Penrose', *Biographical Memoirs of the Fellows of the Royal Society*, 19 (1973), 521–61.

HARRIS, J., *Unemployment and Politics: A Study in English Social Policy, 1886–1914* (Oxford, 1972).

—— *William Beveridge: A Biography* (Oxford, 1977).

—— 'Did British Workers Want the Welfare State?: G. D. H. Cole's Survey of 1942', in J. Winter (ed.), *The Working Class in Modern British History: Essays in Honour of Henry Pelling* (Cambridge, 1983), 200–14.

—— 'The Transition to High Politics', in M. Bentley and J. Stevenson, *High and Low Politics in Modern Britain* (Oxford, 1983), 58–79.

—— 'Political Ideas and the Debate on State Welfare 1940–45', in H. Smith (ed.), *War and Social Change* (Manchester, 1986), 233–63.

—— 'Enterprise and Welfare States: A Comparative Perspective', *Transactions of the Royal Historical Society*, 5th ser., 40 (1991), 175–95.

—— 'Political Thought and the Welfare State, 1870–1940: An Intellectual Framework for British Social Policy', *Past and Present*, 135 (1992).

—— *Private Lives, Public Spirit: A Social History of Britain, 1870–1914* (Oxford, 1993).

HARRISON, B., *Peaceable Kingdom: Stability and Change in Modern Britain* (Oxford, 1982), 309–77.

—— 'Women in a Men's House: The Women MPs, 1919–1945', *Historical Journal*, 29 (1986), 623–54.

—— *Prudent Revolutionaries: Portraits of British Feminists Between the Wars* (Oxford, 1987).

HARRISON, E., 'The Changing Meaning of Social Work', in A. H. Halsey (ed.), *Traditions of Social Policy* (Oxford, 1976), 81–99.

HARVEY, A. and ABRAMS, S., *For the Welfare of Mankind: The Commonwealth Fund and American Medicine* (Baltimore, 1986).

HAYES, P., 'The Contribution of British Intellectuals to Fascism', in K. Lunn and R. Thurlow (eds.), *British Fascism: Essays on the Radical Right in Inter-War Britain* (London, 1980), 168–86.

HEALY, D., 'Irish Psychiatry', in G. Berrios and H. Freeman (eds.) *150 Years of British Psychiatry* (London, 1991), 314–20.

HEARNSHAW, L. S., *Cyril Burt, Psychologist* (London, 1979).

HENDRICK, H., *Images of Youth: Age, Class, and the Male Youth Problem, 1880–1920* (Oxford, 1990).

HENNOCK, E. P., 'Poverty and Social Theory in England: The Experience of the 1880s', *Social History*, 1 (1976), 67–91.

HERVEY, N., 'A Slavish Bowing Down: The Lunacy Commission and the Psychiatric Profession, 1845–60', in W. F. Bynum et al. (eds.), *The Anatomy of Madness*, ii (London, 1985), 98–131.

—— 'Advocacy or Folly: The Alleged Lunatics' Friend Society, 1845–63', *Medical History*, 30 (1986), 245–75.

HIMMELFARB, G., *The Idea of Poverty: England in the Early Industrial Age* (London, 1985).

—— *Poverty and Compassion: The Moral Imagination of the Late Victorians* (New York, 1991).

HIRST, J. D., 'A Failure "Without Parallel": The School Medical Service and the London County Council, 1907–12', *Medical History*, 25 (1981), 281–300.

—— 'The Growth of Treatment Through the School Medical Service, 1908–18', *Medical History*, 33 (1989), 318–42.

HOLLIS, P., *Ladies Elect: Women in English Local Government, 1865–1914* (Oxford, 1987).

HONIGSBAUM, F., *The Division in British Medicine: A History of the Separation of General Practice from Hospital Care, 1911–1968* (London, 1979).

—— *Health, Happiness and Security: The Creation of the National Health Service* (London, 1989).

HOWARD, C., 'Expectations Born to Death: Local Labour Party Expansion in the 1920s', in J. Winter (ed.), *The Working Class in Modern British History* (Cambridge, 1983), 65–81.

HUMPHRIES, S., *A Secret World of Sex: Forbidden Fruit: The British Experience, 1900–1950* (London, 1988).

IGNATIEFF, M., 'Total Institutions and the Working Classes', *History Workshop Journal*, 15 (1983), 167–73.

INGLEBY, D. (ed.), *Critical Psychiatry: The Politics of Mental Health* (London, 1981).

—— 'Mental Health and Social Order', in S. Cohen and A. Scull (eds.), *Social Control and the State* (Oxford, 1983), 141–88.

JACKSON, M., 'Images of Deviance: Visual Representations of Mental Defectives in Early Twentieth-Century Medical Texts', *British Journal for the History of Science*, 28 (1995), 319–37.

—— 'Institutional Provision for the Feeble-Minded in Edwardian England: Sandlebridge and the Scientific Morality of Permanent Care', in D. Wright and A. Digby (eds.), *From Idiocy to Mental Deficiency* (London, 1996).

JEFFRIES, S., *The Spinster and Her Enemies: Feminism and Sexuality, 1880–1930* (London, 1985).

JOHNSON, P., 'Class Law in Victorian England', *Past and Present*, 141 (1993), 147–69.

JONES, G., *Social Darwinism and English Thought* (Brighton, 1980).

—— 'Eugenics and Social Policy Between the Wars', *Historical Journal*, 25 (1982), 717–28.

—— *Social Hygiene in Twentieth-Century Britain* (London, 1986).

JONES, G. S., *Outcast London: A Study in the Relationship Between Classes in Victorian Society* (London, 1971).

—— 'Working-Class Culture and Working-Class Politics in London, 1870–1900:

Notes on the Remaking of the Working Class', *Journal of Social History*, 7 (1974).

JONES, G. W., *Borough Politics: A Study of the Wolverhampton Town Council, 1888–1964* (London, 1966).

JONES, H., 'Conservatives and the LCC after 1934', in A. Saint (ed.), *Politics and the People of London* (London, 1989), 237–50.

JONES, K., *Mental Health and Social Policy, 1845–1959* (London, 1960).

—— *A History of the Mental Health Services* (London, 1972).

—— *Experience in Mental Health: Community Care and Social Policy* (London, 1988).

JONES, S. G., 'Labour, Society and the Drink Question in Britain, 1918–1939', *Historical Journal*, 30 (1987), 105–22.

JORDANOVA, L., 'The Social Construction of Medical Knowledge', *Social History of Medicine*, 8 (1995), 361–81.

JOYCE, P., *Democratic Subjects: The Self and the Social in Nineteenth-Century England* (Cambridge, 1994).

—— 'The end of social history?', *Social History*, 20 (1995), 73–91.

JOYNSON, R. B., *The Burt Affair* (London, 1989).

KANNER, L., 'Medicine in the History of Mental Retardation, 1800–1965' *American Journal of Mental Deficiency*, 72 (1967), 170–5.

KEIR, G., 'A History of Child Guidance', *British Journal of Educational Psychology*, 12 (1952), 5–29.

KEOWN, J., *Abortion, Doctors, and the Law: Some Aspects of the Legal Regulation of Abortion in England from 1803 to 1982* (Cambridge, 1988).

KEVLES, D., *In the Name of Eugenics: Genetics and the Uses of Human Heredity* (London, 1985).

KEYNES, M., *Leonard Darwin, 1850–1943* (Cambridge, 1943).

KINGDOM, E., 'Consent, Coercion, and Consortium: The Sexual Politics of Sterilisation', *Journal of Law and Society*, 12 (1985), 19–29.

KIRK, N., 'Traditional Working Class Culture and the Rise of Labour', *Social History*, 16 (1991), 203–16.

KOVEN, S., and MICHEL, S., 'Womanly Duties: Maternalist Politics and the Origins of Welfare States in France, Germany, Great Britain, and the United States, 1880–1920', *American Historical Review*, 95/4 (1990), 1076–108.

—— —— (eds.), *Mothers of the New World: Maternalist Politics and the Origins of Welfare States* (London, 1993).

KRAFT, I., 'Edouard Séguin and Nineteenth-Century Moral Treatment of Idiots', *Bulletin of the History of Medicine*, 35 (1961), 393–416.

LANDSBOROUGH THOMSON, A., *Half a Century of Medical Research*, 2 vols., (London, 1973 and 1975).

LANGAN, M., and SCHWARTZ, B., *Crises in the British State, 1880–1930* (London, 1985).

LARSON, E. J., 'The Rhetoric of Eugenics: Expert Authority and the Mental Deficiency Bill', *BJHS* 24 (1991), 45–60.

LEE, R., 'Uneven Zenith: Towards a Geography of the High Period of Municipal Medicine in England and Wales', *Journal of Historical Geography*, 14 (1988), 260–80.

LEES, R., 'Social Work, 1925–50: The Case for a Reappraisal', *British Journal of Social Work*, 1 (1971), 371–9.

LEVITT, I., *Poverty and Welfare in Scotland, 1890–1948* (Edinburgh, 1988).

LEWIS, D. S., *Illusions of Grandeur: Mosley, Fascism and British Society, 1931–81* (Manchester, 1987).

LEWIS, J., *The Politics of Motherhood: Child and Maternal Welfare in England, 1900–1939* (London, 1980).

—— (ed.), *Labour and Love: Women's Experience of Home and Family, 1880–1940* (Oxford, 1986).

—— *What Price Community Medicine?* (Brighton, 1986).

—— 'The Working Class Wife and Mother and State Intervention, 1870–1918', in J. Lewis (ed.), *Labour and Love* (Oxford, 1986), pp. 109–15.

—— 'Myrdal, Klein, *Women's Two Roles* and Post War Feminism, 1945–60', in H. Smith (ed.), *British Feminism in the Twentieth Century* (London, 1990).

—— 'The Public's Health: Philosophy and Practice in Britain in the Twentieth Century', in E. Fee and R. Acheson (eds.), *A History of Education and Public Health: Health That Mocks the Doctors' Rules* (Oxford, 1991), pp. 202–16.

—— 'Gender, the Family, and Women's Agency in the Building of "Welfare States": The British Case', *Social History*, 19 (1994), 37–55.

—— 'Family Provision of Health and Welfare in the Mixed Economy of Care in the late Nineteenth and Twentieth Centuries', *Social History of Medicine*, 8 (1995), 1–16.

LILLY, M., *The National Council for Civil Liberties: The First 50 Years* (London, 1984).

LOUDON, I., 'Deaths in Childbirth from the Eighteenth Century to 1935', *Medical History*, 30 (1986), 1–41.

—— 'On Maternal and Infant Mortality, 1900–60', *Social History of Medicine*, 4 (1991), 29–74.

LOUGHLIN, M., GELFAND, M., and YOUNG, K., *Half a Century of Municipal Decline, 1935–1985* (London, 1985).

LOWE, N., 'The Legal Status of Fatherhood Past and Present', in M. O'Brien and L. McKee (eds.), *The Father Figure* (London, 1982), 26–42.

LOWE, R., 'The Erosion of State Intervention in Britain, 1917–24', *Economic History Review*, 2nd ser., 31 (1978), 270–86.

—— 'Bureaucracy Triumphant or Denied? The Expansion of the British Civil Service, 1919–39', *Public Administration*, 62 (1984), 291–310.

—— *Adjusting to Democracy: The Role of the Ministry of Labour in British Politics, 1916–1939* (Oxford, 1986).

LOWE, R. A., 'Eugenicists, Doctors and the Quest for National Efficiency: An Educational Crusade 1900–1939', *History of Education*, 8 (1979), 293–306.

LOWMAN, J., MENZIES, R., and PALYS, T. S., *Transcarceration: Essays in the Sociology of Social Control* (Cambridge, 1987).

LUCAS, B. K., and RICHARDS, P. G., *A History of Local Government in the Twentieth Century* (London, 1978).

LUCKIN, B., 'Towards a History of Institutionalization', *Social History*, 8 (1983), 87–94.

LUMMIS, T., 'The Historical Dimensions of Fatherhood: A Case Study, 1890–1914', in M. O'Brien and L. Mckee (eds.), *The Father Figure* (Tavistock, 1982).

MCBRIAR, A., *Fabian Socialism and English Politics, 1884–1918* (New York, 1962), 120.

MCCANDLESS, P., 'Build! Build! The Controversy over the Care of the Chronically Insane in England, 1850–70', *Bulletin of the History of Medicine*, 53 (1979), 553–74.

MCCORD, N., 'Ratepayers and Social Policy', in P. Thane, *The Origins of British Social Policy* (London, 1978), 21–35.

MACDONAGH, O., 'The Nineteenth-Century Revolution in Government: A Reappraisal', *Historical Journal*, 1 (1958), 52–67.

—— *A Pattern of Government Growth, 1800–1860: The Passenger Acts and their Enforcement* (London, 1961).

MCDONALD, A., 'The Geddes Committee and the Formulation of Public Expenditure Policy, 1921–1922', *Historical Journal*, 32 (1989), 643–74.

MACKENZIE, C., 'Women and Psychiatric Professionalization, 1780–1914', in London Feminist History Collectives (eds.), *The Sexual Dynamics of History* (London, 1983).

MACKENZIE, D., *Statistics in Britain, 1865–1930: The Social Construction of Scientific Knowledge* (Edinburgh, 1980).

MCKIBBIN, R., 'Class and Conventional Wisdom: The Conservative Party and the "Public" in Interwar Britain', in id., *The Ideologies of Class* (Oxford, 1990), 259–93.

MACLEOD, R. M., 'Statesmen Undisguised', *American Historical Review*, 78 (1973), 1386–405.

—— (ed.), *Government and Expertise: Specialists, Administrators and Professionals, 1860–1919* (Cambridge, 1988).

MACNICOL, J., *The Movement for Family Allowances, 1918–45: A Study in Social Policy Development (London,* 1981).

—— 'In Pursuit of the Underclass', *Journal of Social Policy*, 16 (1987), 293–318.

—— 'The Effect of the Evacuation of School Children on Attitudes to State Intervention', in H. Smith (ed.), *War and Social Change* (Manchester, 1988), 3–31.

MACNICOL, J., 'Eugenics and the Campaign for Voluntary Sterilization in Britain Between the Wars', *Social History of Medicine*, 2 (1989), 147–70.

MAIER, C. (ed.), *Changing Boundaries of the Political: Essays on the Evolving Balance Between the State and Society, Public and Private in Europe* (Cambridge, 1987).

MANDLER, P. (ed.), *The Uses of Charity: The Poor on Relief in the Nineteenth-Century Metropolis* (Philadelphia, 1990).

MARK-LAWSON, J., SAVAGE, M., and WARDE, A., 'Gender and Local Politics: Struggles Over Welfare Policies, 1918–39', in L. Murgatroyd et al. (eds.), *Localities, Class, and Gender* (London, 1985), 195–231.

MARKHAM, V., 'Robert Morant—Some Personal Reminiscences', *Public Administration*, 28 (1950).

MARSHALL, T. H., *Citizenship and Social Class* (Cambridge, 1950).

MARWICK, A., 'Middle Opinion in the Thirties: Planning, Progress, and Political Agreement', *English Historical Review*, 79 (1964), 285–98.

—— 'The Labour Party and the Welfare State in Britain 1900–1948', *American Historical Review*, 73 (1967), 380–403.

MASON, T., and THOMPSON P., '"Reflections on a Revolution"? The Political Mood in Wartime Britain', in N. Tiratsoo (ed.), *The Attlee Years* (London, 1991), 54–70.

MATTHEW, C., KAY, J., and MCKIBBIN, R., 'The Franchise Factor in the Rise of the Labour Party', *English Historical Review*, 91 (1976) 723–52.

MAZUMDAR, P., *Eugenics, Human Genetics, and Human Failings* (London, 1992).

The Medical Directory (London, 1910–40).

MELLETT, D. J., 'Bureaucracy and Mental Illness: The Commissioners in Lunacy, 1845–90', *Medical History*, 25 (1981), 221–50.

—— *The Prerogative of Asylumdom* (London, 1982).

MERSKEY, H., 'Shell-Shock', in G. Berrios and H. Freeman (eds.), *150 Years of British Psychiatry*, (London, 1991), 245–67.

MILLER, E., 'Mental Retardation', in G. Berrios and R. Porter (eds.), *A History of Clinical Psychiatry*, 212–24.

MOMMSEN, W., '"Toward the Iron Cage of Future Serfdom"? On the Methodological Status of Max Weber's Ideal-Typical Concept of Bureaucratization', *Transactions of the Royal Historical Society*, 5th ser., 30 (1980).

—— (ed.), *The Emergence of the Welfare State in Britain and Germany, 1850–1950* (London, 1981).

MOORE, M. J., 'Social Work and Social Welfare: The Organisation of Philanthropic Resources in Britain, 1900–1914', *Journal of British Studies*, 16 (1977), 85–104.

MORGAN, K., *Consensus and Disunity: The Lloyd George Coalition Government, 1918–1922* (Oxford, 1979).

—— *Labour in Power, 1945–51* (Oxford, 1984).

—— 'The High and Low Politics of Labour: Keir Hardie to Michael Foot', in

M. Bentley and J. Stevenson (eds.), *High and Low Politics* (Oxford, 1983), 285–312.

MORT, F., 'Purity, Feminism, and the State: Sexuality and Moral Politics, 1880–1914', in M. Langan and B. Schwarz, *Crises in the British State, 1880–1930* (London, 1985), 209–25.

—— *Dangerous Sexualities: Medico-Moral Politics in England Since 1830* (London, 1987).

MOWAT, C. L., *The Charity Organisation Society, 1869–1913: Its Ideas and Work* (London, 1961).

Munk's Roll: Lives of the Fellows of the Royal College of Physicians of London, iv–vii (London, 1955–84).

OPPENHEIM, J., *'Shattered Nerves': Doctors, Patients, and Depression in Victorian England* (New York, 1991).

PARFITT, D., 'A Mental Hospital in 1929', in G. Berrios and H. Freeman (eds.), *150 Years of British Psychiatry*, ii, *The Aftermath* (London, 1996), 465–76.

PARRY, N., and PARRY J., 'Social Work, Professionalism, and the State', in N. Parry, M. Rustin, and C. Satyamurti, *Social Work, Welfare, and the State* (London, 1979), 26–40.

PARRY-JONES, W., 'The Model of Gheel Lunatic Asylum and its Influence on the Nineteenth-Century Asylum System in Britain', in A. Scull (ed.), *Madhouses, Mad-Doctors and Madmen* (London, 1981), 201–17.

PAUL, D., 'Eugenics and the Left', *Journal of the History of Ideas*, 45 (1984), 561–90.

PEACOCK, A. T., and WISEMAN, J., *The Growth of Public Expenditure in the United Kingdom* (London, 1961).

PEDEN, G. C., 'The Treasury as the Central Department of Government, 1919–1939', *Public Administration*, 61 (1983), 371–85.

PELLEW, J., *The Home Office, 1848–1914: From Clerks to Bureaucrats* (London, 1982).

PELLING, H., 'The Working Class and the Welfare State', in id., *Popular Politics and Society in Late-Victorian Britain* (London, 1968), 1–18.

PELLING, M., and SMITH, R. (eds.), *Life, Death, and the Elderly: Historical Perspectives* (London, 1991).

PENNYBACKER, S., 'Reconsidering London Progressivism, 1889–1907', in D. Feldman and G. Stedman Jones (eds.), *Metropolis*, 129–62.

PERETZ, E., 'The Professionalization of Childcare', *Oral History Journal*, 12 (1989), 22–7.

PERETZ, L., 'Maternity Services for England and Wales: Local Authority Maternity Care in the Inter-War period in Oxfordshire and Tottenham', in J. Garcia, R. Kilpatrick, and M. Richards (eds.), *The Politics of Maternity Care: Services For Childbearing Women in Twentieth-Century Britain* (Oxford, 1990), 30–46.

PERKIN, H., *The Rise of Professional Society: England Since 1880* (London, 1989).

PICK, D., *Faces of Degeneration: A European Disorder, c.1848–1918* (Cambridge, 1989).

PICKSTONE, J., 'Psychiatry in District General Hospitals: History, Contingency and Local Innovation in the Early Years of the National Health Service', in J. Pickstone (ed.), *Medical Innovations in Historical Perspective* (London, 1992).

PIMLOTT, B., *Labour and the Left in the 1930s* (Cambridge, 1977).

PINDER, J. (ed.), *Fifty Years of Political and Economic Planning* (London, 1981).

PLANT, R., and VINCENT, A., *Philosophy, Politics, and Citizenship: The Life and Thought of the British Idealists* (Oxford, 1984).

POPE, R. (ed.), *Atlas of British Social and Economic History Since c.1700* (London, 1989).

PORTER, D., 'Enemies of the Race: Biologism, Environmentalism, and Public Health in Edwardian Britain', *Victorian Studies*, 34 (1991), 159–79.

—— 'Stratification and its Discontents: Professionalisation and Conflict in the British Public Health Service', in E. Fee and R. Acheson, *Education in Public Health* (Oxford, 1991), 83–113.

—— and FEE, E., 'Public Health, Preventive Medicine and Professionalisation' in E. Fee and R. Acheson, *Education in Public Health* (Oxford, 1991), 38–40.

—— 'Changing Disciplines: John Ryle and the Making of Social Medicine in Twentieth-Century Britain', *History of Science*, 30 (1992), 119–47.

—— and PORTER, R., 'What was Social Medicine? An Historiographical Essay', *Journal Of Historical Sociology*, 1 (1988), 90–106.

PORTER, R., *Mind-Forg'd Manacles: A History of Madness in England from the Restoration to the Regency* (London, 1990), 222–8.

—— 'History of Psychiatry in Britain', *History of Psychiatry*, 2 (1991), 271–9.

POTTS, M., and FIDO, R., 'Its Not True What Was Written Down', *Oral History*, 17 (1989), 31–4.

POWELL, M., 'The Geography of English Hospital Provision in the 1930s', *Journal of Historical Geography*, 18 (1992), 307–16.

PRITCHARD, D. G., *Education and the Handicapped, 1760–1960* (London, 1963).

PROCHASKA, F., *Women and Philanthropy in Nineteenth Century Britain* (Oxford, 1980).

—— *The Voluntary Impulse in Modern Britain: Continuity and Change, 1880–1987* (London, 1988).

—— 'Philanthropy', in F. M. L. Thompson (ed.), *The Cambridge Social History of Britain, 1750–1950*, iii (Cambridge, 1991), 357–93.

PUGH, M., 'Popular Conservatism in Britain: Continuity and Change, 1880–1987', *Journal of British Studies*, 27 (1988), 254–84.

RADZINOWICZ, L., and HOOD, R., *The Emergence of Penal Policy in Victorian and Edwardian England* (Oxford, 1990).

RAMON, S., *Psychiatry in Britain: Meaning and Policy* (London, 1985).

RAMSDEN, J., *The Age of Balfour and Baldwin, 1902–40* (London, 1978).
—— *The Making of Conservative Party Policy: The Conservative Party Research Department Since 1929* (London, 1980).
RAPP, D., 'The Early Discovery of Freud by the British General Public, 1912–1919, *Social History of Medicine*, 3 (1990), 217–43.
RAY, L. J., 'Models of Madness in Victorian Asylum Practice', *European Journal of Sociology*, 22 (1981), 229–64.
—— 'Eugenics, Mental Deficiency, and Fabian Socialism', *Oxford Review of Social History*, 9 (1983).
REID, A., *Social Classes and Social Relations in Britain, 1850–1914* (London, 1992).
REILLY, P. B., *The Surgical Solution: A History of Involuntary Sterilization in the United States* (Baltimore, 1991).
RIVETT, G., *The Development of the London Hospital System* (London, 1986).
ROBERTS, E., 'Women's Strategies, 1890–1940', in J. Lewis (ed.), *Labour and Love* (Oxford, 1986), 223–48.
ROBINSON, K., 'The Public and Mental Health', in R. Farndale and H. Freeman (eds.), *Trends in The Mental Health Services* (London, 1963), 12–20.
RÖLL-HANSEN, N., 'The Progress of Eugenics', *BJHS* 26 (1988), 295–331.
ROOFF, M., *Voluntary Societies and Social Policy* (London, 1957).
ROSE, N., *The Psychological Complex: Psychology, Politics and Society in England, 1869–1939* (London, 1985).
ROSS, E., 'Survival Networks: Women's Neighborhood Sharing in London Before World War II', *History Workshop Journal*, 15 (1983), 4–27.
ROUTH, G., *Occupational Pay in Great Britain, 1906–60* (Cambridge, 1965).
ROWAN, C., 'Women in the Labour Party, 1906–20', *Feminist Review*, 12 (1982), 74–91.
RUSHTON, P., 'Lunatics and Idiots: Mental Disorder, the Community, and the Poor Law in North East England 1600–1800', *Medical History*, 32 (1988), 34–50.
SAINT, A. (ed.), *Politics and the People of London: The London County Council, 1889–1965* (London, 1989).
SAUNDERS, J., 'Quarantining the Weak-Minded: Psychiatric Definitions of Degeneracy and the Late-Victorian Asylum', in W. F. Bynum et al., *The Anatomy of Madness*, iii (London, 1988), 273–96.
SCHREIBER, B., *The Men Behind Hitler* (London, *c.*1975).
SCULL, A., *Museums of Madness: The Social Organisation of Insanity in Nineteenth-Century England* (London, 1979).
—— 'A Convenient Place To Get Rid Of Inconvenient People', in A. King (ed.), *Buildings and Society: Essays on the Social Development of the Built Environment* (London, 1980), 39–60.
SCULL, A., *Madhouses, Mad-Doctors and Madmen: The Social History of Psychiatry in the Victorian Era* (London, 1981).

SCULL, A., *Decarceration: Community Treatment and the Deviant—A Radical View*, 2nd edn. (Cambridge, 1984).

—— *Social Order/Mental Disorder: Anglo-American Psychiatry in Historical Perspective* (London, 1989).

—— (ed.), *The Asylum as Utopia: W. A. F. Browne and the Mid-Nineteenth Century Consolidation of Psychiatry* (London, 1991).

—— 'Psychiatry and its Historians', *History of Psychiatry*, 2 (1991), 239–50.

—— *The Most Solitary of Afflictions: Madness and Society in Britain, 1700–1900* (New Haven, 1993).

—— '*Museums of Madness* Revisited', *Social History of Medicine*, 6 (1993), 3–23.

—— and COHEN, S. (eds.), *Social Control and the State* (Oxford, 1983).

SEARLE, G. R., *The Quest for National Efficiency* (Oxford, 1971).

—— 'Eugenics and Politics in the 1930s', *Annals of Science*, 36 (1979), 159–69.

—— 'Eugenics and Class', in C. Webster (ed.), *Biology, Medicine and Society, 1840–1940* (Cambridge, 1991), 217–42.

SEMMEL, B., *Imperialism and Social Reform: English Social-Imperial Thought, 1895–1914* (London, 1960).

SHARP, S., and SUTHERLAND, G., 'The Fust Official Psychologist in the Wuurld', *History of Science*, 18 (1980), 181–208.

SHELDRAKE, J., 'The LCC Hospital Service', in A. Saint (ed.), *Politics and the People of London: The London County Council, 1889–1965* (London, 1981), 187–98.

SHORTER, E., *The Making of the Modern Family* (London, 1976).

SHOWALTER, E., *The Female Malady: Women, Madness, and English Culture, 1830–1980* (London, 1987).

SIM, J., *Medical Power in Prisons: The Prison Medical Service in England, 1774–1984* (Milton Keynes, 1990).

SIMMONS, H. G., 'Explaining Social Policy: The English Mental Deficiency Act of 1913', *Journal of Social History*, 11 (1978), 387–403.

SIMON, B., *Education and the Labour Movement, 1870–1929* (London, 1965).

SKIDELSKY, R., *Politicians and the Slump: The Labour Government of 1929–1931* (London, 1970).

—— *Oswald Mosley* (London, 1975).

SMITH, H. (ed.), *War and Social Change: British Society in the Second World War* (Manchester, 1986).

—— (ed.), *British Feminism in the Twentieth Century* (Aldershot, 1990).

SMITH, L. D., 'Behind Closed Doors: Lunatic Asylum Keepers, 1800–1860, *Social History of Medicine*, 1 (1988), 301–27.

SOLOWAY, R., 'Counting the Degenerates: The Statistics of Race Deterioration in Edwardian England', *Journal of Contemporary History*, 17 (1982), 137–64.

—— *Birth Control and the Population Question in England, 1877–1930* (London, 1986).

—— *Demography and Degeneration: Eugenics and the Declining Birthrate in Twentieth-Century Britain* (London, 1990).

SRETZER, S., 'Healthy Government? Britain 1850–1950', *Historical Journal*, 34 (1991), 401–503.

STARR, P., and IMMERGUT, E., 'Health Care and the Boundaries of the Politics', in C. Maier (ed.), *Changing Boundaries of the Political* (Cambridge, 1987).

STEEDMAN, C., *Childhood Culture and Class in Britain: Margaret McMillan, 1860–1931* (London, 1990).

STENTON, M., and LEES, S., *Who's Who of British Members of Parliament*, ii–iv (Brighton, 1976–81).

STEVENSON, J., *British Society, 1914–45* (London, 1984).

STONE, M., 'Shellshock and the Psychiatrists', in W. Bynum, R. Porter, and M. Shepherd (eds.), *The Anatomy of Madness* (London, 1985–8) ii. 242–71.

SUTHERLAND, G. (ed.), *Studies in the Growth of Nineteenth-Century Government* (London, 1972).

—— *Ability, Merit, and Measurement: Mental Testing and English Education, 1880–1940* (Oxford, 1984).

SWIFT, R., and GILLEY, S., *The Irish in the Victorian City, 1815–1939* (London, 1985).

TANNER, D., *Political Change and the Labour Party, 1900–1918* (Cambridge, 1990).

TAYLOR, J., *The Hospital and Asylum Architecture in England, 1840–1914* (London, 1991).

TAYLOR, T., 'An Early Arrival of the Fascist Mentality: Robert Morant's Rise to Power', *Journal of Educational Administration and History*, 17 (1985), 48–62.

THANE, P. (ed.), *The Origins of British Social Policy* (London, 1978).

—— *Foundations of the Welfare State* (London, 1982).

—— 'The Working Class and State Welfare', *Historical Journal*, 27 (1984), 877–900.

—— 'The Labour Party and State Welfare', in K. D. Brown (ed.), *The First Labour Party* (Beckenham, 1985), 183–216.

—— 'The Women of the British Labour Party and Feminism, 1906–1945', in H. Smith (ed.), *British Feminism in the Twentieth Century* (Aldershot, 1990), 136–8.

—— 'Labour and Local Politics: Radicalism, Democracy, and Social Reform, 1880–1914', in E. Biagini and A. Reid (eds.), *Currents of Radicalism: Popular Radicalism, Organised Labour, and Party Politics in Britain, 1850–1914* (Cambridge, 1991), 244–70.

THANE, P. 'Visions of Gender in the Making of the British Welfare State: The Case of Women in the British Labour Party and Social Policy, 1906–1945', in G. Bock and P. Thane (eds.), *Maternity and Gender Politics: Women and the Rise of the European Welfare States, 1880–1950* (London, 1991), 93–118.

THOM, D., 'The 1944 Education Act: The Art of the Possible?', in H. Smith (ed.), *War and Social Change* (Manchester, 1986), 101–28.

—— 'Wishes, Anxieties, Play, and Gestures: Child Guidance in Inter-War England', in R. Cooter (ed.), *In the Name of the Child* (London, 1992), 200–19.

THOMPSON, F. M. L., 'Social Control in Nineteenth-Century Britain', *Economic History Review*, 2nd ser., 34 (1981), 189–208.

—— *The Rise of Respectable Society: A Social History of Victorian Britain, 1830–1900* (London, 1988).

—— (ed.), *The Cambridge Social History of Britain, 1750–1950*, 3 vols. (Cambridge, 1991).

THOMSON, D., 'The Decline of Social Welfare: Falling Support for the Elderly since Early Victorian Times', *Ageing and Society*, 4 (1984), 451–82.

THOMSON, G. O., 'Legislation and Provision for the Mentally Handicapped in Scotland since 1906', *Oxford Review of Education*, 9 (1983), 233–40.

THOMSON, M., 'Sterilisation, Segregation, and Community Care: Ideology and the Problem of Mental Deficiency in Inter-War Britain', *History of Psychiatry*, 3 (1992), 473–98.

—— 'Social Policy and the Management of the Problem of Mental Deficiency in Interwar London', *London Journal*, 18 (1993), 129–42.

—— and WEINDLING, P., 'Sterilisationspolitik in Grosbritannien und Deutschland' (The Politics of Sterilisation in Britain and Germany), in F. W. Kersting, K. Teppe, and B. Walter (eds.), *Nach Hadamar* (Paderborn, Germany, 1993).

—— 'Family, Community, and State: The Micro-Politics of Mental Deficiency', in A. Digby and D. Wright (eds.), *From Idiocy to Mental Deficiency: Historical Perspectives on People with Learning Disablities* (London, 1996), 207–30.

—— '"Though Ever the Subject of Psychological Medicine": Psychiatrists and the Colony Solution for Mental Defectives', in G. Berrios and H. Freeman (eds.), *150 Years of British Psychiatry*, ii, *The Aftermath* (London, 1996).

—— 'Community Care and Control of Mental Defectives in Interwar Britain', in P. Horden and R. Smith (eds.), *The Locus of Care* (forthcoming).

THORNHILL, W., *Central Government and the Local Authorities: A Study of Methods of Control* (Sheffield, 1954).

TIMMS, N., *Psychiatric Social Work in Great Britain, 1939–62* (London, 1964).

TIRATSOO, N. (ed.), *The Attlee Years* (London, 1991).

TITMUSS, R., *Problems of Social Policy* (London, 1950).

—— 'Community Care—Fact or Fiction ?', in R. Farndale and H. Freeman (eds.) *Trends in the Mental Health Services* (London, 1963), 221–5.

—— *Essays on the Welfare State*, 2nd edn. (London, 1963).

TIZARD, J., and GRAD, J. C., *The Mentally Handicapped and Their Families* (London, 1961).

TOMLINSON, S., 'The Expansion of Special Education', *Oxford Review of Education*, 11 (1985), 157–65.

TRENT, J., JR., *Inventing the Feeble Mind: A History of Mental Retardation in the United States* (London, 1994).

TROMBLEY, S., *The Right to Reproduce: A History of Coercive Sterilisation* (London, 1988).

TURNER, T., 'Henry Maudsley—Psychiatrist, Philosopher and Entrepreneur', *Psychological Medicine*, 18 (1988), 551–74.

UNSWORTH, C., *The Politics of Mental Health Legislation* (Oxford,1986).

WALK, A., 'The Royal College of Psychiatrists', *St Barthomew's Hospital Journal*, 77 (May 1973), 135–9.

WALMSLEY, T., 'Psychiatry in Scotland', in G. Berrios and H. Freeman (eds.), *150 Years of British Psychiatry* (London, 1991), 294–305.

WALTON, J., 'Lunacy in the Industrial Revolution: A Study of Asylum Admissions in Lancashire, 1848–50', *Social History*, 13 (1979), 1–22.

WALTON, J., *The English Seaside Resort: A Social History, 1750–1914* (Leicester, 1983).

—— 'Casting Out and Bringing Back in Victorian England: Pauper Lunatics, 1840–70', in W. Bynum et al. (eds.), *Anatomy of Madness* (London, 1985–8), ii. 132–46.

WALTON, R. G., *Women in Social Work* (London, 1975).

WARD, S. V., *The Geography of Interwar Britain: The State and Uneven Development* (London, 1988).

WATSON, S., 'Applying Foucault: Some Problems Encountered in the Application of Foucault's Methods to the History of Medicine in Prisons', in C. Jones and R. Porter (eds.), *Reassessing Foucault: Power, Medicine, and the Body* (London, 1994).

WEBSTER, C. (ed.), *Biology, Medicine and Society, 1840–1940* (Cambridge, 1981).

—— 'Healthy or Hungry Thirties', *History Workshop Journal*, 13 (1982), 110–29.

—— 'The Health of the School Child During the Depression', in N. Parry and D. Mcnair (eds.) *The Fitness of the Nation* (Leicester, 1983).

—— 'Health, Welfare and Unemployment during the Depression', *Past and Present*, 46 (1985), 204–29.

—— 'Labour and the Origins of the National Health Service', in N. A. Rupke (ed.), *Science, Politics, and the Public Good: Essays in Honor of Margaret Gowing* (London, 1988), 184–202.

WEBSTER, C., *Problems of Health Care: The National Health Service Before 1957* (London, 1988).

—— 'Conflict and Consensus: Explaining the British Health Service', *Twentieth Century British History*, 1 (1990), 115–51.

—— 'Psychiatry and the Early National Health Service: The Role of the

Mental Health Standing Advisory Committee', in G. Berrios and H. Freeman (eds.), *150 Years of British Psychiatry, 1841–1991* (London, 1991), 103–16.

WEDGWOOD, C. V., *The Last of the Radicals: Josiah Wedgwood MP* (London, 1951).

WEINDLING, P., 'Compulsory Sterilization in National Socialist Germany', *German History*, 5 (1987), 10–24.

—— *Health, Race and German Politics between National Unification and Nazism, 1870–1945* (Cambridge, 1989).

WEIR, M., ORLOFF, A. S., and SKOCPOL, T., *The Politics of Social Policy in the United States* (Princeton, 1988).

WERSKEY, G., *The Visible College* (London, 1978).

Whitaker's Almanack (London, 1910–40).

WHITE, J., *Campbell Bunk: The Worst Street in North London* (London, 1986).

Who Was Who, vols. i–iii (London, 1920–41).

WIENER, M., *English Culture and the Decline of the Industrial Spirit, 1850–1980* (Cambridge, 1981).

—— *Reconstructing the Criminal: Culture, Law, and Policy in England, 1830–1914* (Cambridge, 1990).

WILLIAMS, G., 'Compulsory Sterilisation of Welsh Miners, 1936', *Llafur*, 3 (1981), 67–73.

WILLIAMS, R., *The Country and the City* (London, 1975).

WILSON, C. H. (ed.), *Essays on Local Government* (Oxford, 1948).

WINTER, J. M., 'Unemployment, Nutrition, and Infant Mortality in Britain, 1920–50', in id. (ed.), *The Working Class and Modern British History* (Cambridge, 1983), pp. 232–56.

WOODCOCK, J., '75 Years of Voluntary Endeavour: A Review of the Work of the Cambridge Mental Welfare Association, 1908–1983' (unpublished paper held at Cambridge County Record Office).

WOODHOUSE, J., 'Eugenics and the Feeble-minded', *History of Education*, 2 (1982), 127–37.

WOODROOFE, K., *From Charity to Social Work in England and the United States* (London, 1962).

WOOLDRIDGE, A., *Measuring the Mind: Education and Psychology in England, c.1860–1990* (Cambridge, 1994).

WOOTTON, B., *Social Science and Social Pathology* (London, 1959).

WOOTTON, G., *Pressure Groups in Britain, 1720–1970* (London, 1975).

WRIGHT, D., '"Childlike in his Innocence": Lay Attitudes to "Idiots" and "Imbeciles" in Victorian England', in D. Wright and A. Digby (eds.), *From Idiocy to Mental Deficiency* (London, 1996), 118–33.

—— and DIGBY, A. (eds.), *From Idiocy to Mental Deficiency: Historical Perspectives on People with Learning Disabilities* (London, 1996).

WRIGHT, P., and TREACHER, A., *The Problem of Medical Knowledge: Examining the Social Construction of Medicine* (Edinburgh, 1982).

WYNN, M., and WYNN, A., *Prevention of Handicap and the Health of Women* (London, 1979).
YEO, E. and YEO, S., 'On the Uses of "Community": From Owenism to the Present', in S. Yeo (ed.), *New Views of Cooperation* (London, 1988), 229–58.
ZEDNER, L., *Women, Crime and Custody in Victorian England* (Oxford, 1991).
ZIMMECK, M., 'Strategies and Strategems for the Employment of Women in the British Civil Service, 1919–1939', *Historical Journal*, 4 (1984), 901–24.
—— 'The "New Woman" in the Machinery of Government: A Spanner in the Works?', in R. Macleod (ed.), *Government and Expertise* (Cambridge, 1988), 185–202.
—— 'Marry in Haste, Repent in Leisure: Women, Bureaucracy, and the Post Office, 1870–1920', in M. Savage and A. Witz (eds.), *Gender and Bureaucracy* (Oxford, 1992), 65–93.

Unpublished Theses

AUSTOKER, J., 'Biological Education and Social Reform: The British Social Hygiene Council, 1925–42', MA thesis (London, 1981).
BARRETT, M. A., 'From Education to Segregation: An Inquiry Into the Changing Character of Special Provision for the Retarded in England, 1846–1910', Ph.D. thesis (Lancaster, 1986).
BRAGAR, M., 'The Feeble-Minded Female: An Historical Analysis of Mental Retardation as a Social Definition, 1890–1920', Ph.D. thesis (Syracuse University, 1977).
BRYDER, L., 'The Problem of Tuberculosis in England and Wales, 1906–50', D.Phil. thesis (Oxford, 1985).
FISHER, D., 'The Impact of American Foundations on the Development of British University Education, 1900–1939', Ph.D. thesis (University of California, 1977).
GELBAND, S., 'Mental Retardation and Institutional Treatment in Nineteenth-Century England, 1845–1866', Ph.D. thesis (University of Maryland, 1979).
GRANT, M. L., 'Perceptions of Propaganda in Interwar Britain: The Development of Government Publicity in the Domestic Sphere', D.Phil. thesis (Oxford, 1988).
MARKS, L., 'Irish and Jewish Women's Experiences of Childbirth and Infant Care in East London, 1870–1939: The Responses of Host Society and Immigrant Communities to Medical Welfare Needs', D.Phil. thesis (Oxford, 1990).
NEUGEBAUER, R., 'Mental Illness and Government Policy in Sixteenth and Seventeenth Century England', Ph.D. thesis (Columbia University 1976).
NOLAN, P., 'Psychiatric Nursing Past and Present: The Nurses' Viewpoint', Ph.D. thesis (University of Bath, 1989).

ROWETT, J. S., 'The Labour Party and Local Government: Theory and Practice in the Interwar Years', D.Phil. thesis (Oxford, 1979).
SANDLER, E. O., 'Working Class Adolescents in Birmingham: A Study in Social Reform, 1900–1914', D.Phil. thesis (Oxford, 1987).
STACEY, S. R., 'The Ministry of Health, 1919–29: Ideas and Practice in a Government Department', D.Phil. thesis (Oxford, 1984).
STONE, M. J., 'The Military and Industrial Roots of Clinical Psychology in Britain, 1900–1945', Ph.D. thesis (London, 1985).
THOMSON, M., 'The Problem of Mental Deficiency in England and Wales *c*.1913–1946', D.Phil. thesis (Oxford, 1992)
WATKINS, D., 'The English Revolution in Social Medicine', Ph.D. thesis (London, 1984).
WATSON, S., 'The Moral Imbecile: A Study of the Relations Between Penal Practice and Psychiatric Knowledge of the Habitual Offender', Ph.D. thesis (University of Lancaster, 1988).
WELSHMAN, A. J., 'The School Medical Services in England and Wales, 1907–39', D.Phil. thesis (Oxford, 1988).
WOOLDRIDGE, A., 'Child Study and Educational Psychology in England, *c*.1880–1950', D.Phil. thesis (Oxford, 1985).
WRIGHT, D., 'The National Asylum for Idiots, Earlswood, 1847–1886', D.Phil. thesis (Oxford, 1993).
YOUNG, T., 'The Eugenics Movement and the Eugenic Idea in Britain, 1900–1914: A Historical Study', Ph.D. thesis (London, 1980).

INDEX